DARK POOL OF LIGHT

VOLUME ONE

THE NEUROSCIENCE, EVOLUTION, AND ONTOLOGY OF CONSCIOUSNESS

Other books in the series

Dark Pool of Light

VOLUME TWO

CONSCIOUSNESS IN PSYCHOSPIRITUAL AND PSYCHIC RANGES

Dark Pool of Light

VOLUME THREE

THE CRISIS AND FUTURE OF CONSCIOUSNESS

DARK POOL OF LIGHT

VOLUME ONE

THE NEUROSCIENCE, EVOLUTION, AND ONTOLOGY OF CONSCIOUSNESS

The Convergence of Physical, Philosophical, Psychological, Psychospiritual, and Psychic Views

Richard Grossinger

Forewords by Nick Herbert and Jeffrey J. Kripal

North Atlantic Books
Berkeley, California

Published by
North Atlantic Books
Berkeley, California

Cover photo by Richard Grossinger, "Beaver (Invisible) Swimming Through Reflection of Moon, Manset, Maine, 2011"
Cover and book design by Susan Quasha
Printed in the United States of America

Dark Pool of Light Volume One: The Neuroscience, Evolution, and Ontology of Consciousness is sponsored and published by the Society for the Study of Native Arts and Sciences (dba North Atlantic Books), an educational nonprofit based in Berkeley, California, that collaborates with partners to develop cross-cultural perspectives, nurture holistic views of art, science, the humanities, and healing, and seed personal and global transformation by publishing work on the relationship of body, spirit, and nature.

North Atlantic Books' publications are available through most bookstores. For further information, visit our website at www.northatlanticbooks.com or call 800-733-3000.

Library of Congress Cataloging-in-Publication Data

Dark pool of light.
    p. cm. — (Reality and consciousness)
  Summary: "Explores and compares neuroscientific and philosophical views of reality and human consciousness"—Provided by publisher.
  Includes bibliographical references and index.
  ISBN 978-1-58394-434-9 — ISBN 1-58394-434-6
1. Consciousness. I. Grossinger, Richard, 1944–
  B808.9.D37 2012
  128'.2—dc23

                                2012011839

2 3 4 5 6 7 8 9 SHERIDAN 20 19 18 17 16
Printed on recycled paper

For John Friedlander

*For Kimmie Hills, Erin Honeycutt, Monique Drinkwater,
Jake Wartell, Jesse Graham, Zef Maguire, Jed Bickman,
Gabe Weiss, and their tribes*

Acknowledgments

Thanks to Hisae Matsuda for her objective and clear-minded editorial oversight and project management; to Kathy Glass for her uncannily observant editing; to Paula Morrison for her production supervision; to Susan Quasha for her design and framing; to Victoria Baker for her hands-on index; to Ron Milestone, Fredrick Ware, Nick Herbert, Dusty Dowse, Neil Stillings, and Terrence Deacon for content feedback.

"I regard consciousness as fundamental. I regard matter as derivative from consciousness. We cannot get behind consciousness. Everything that we talk about, everything that we regard as existing, postulates consciousness."

—MAX PLANCK

"Does anyone have the foggiest idea of how a bunch of firing neurons in any kind of network produces consciousness?"

—BARCLAY MARTIN

"The body of a worm and the face of a man alike have to be taken as chemical responses."

—SIR CHARLES SHERRINGTON

"I'm an apeman, I'm an ape, apeman, oh I'm an apeman./I'm a King Kong man, I'm a voodoo man, oh I'm an apeman."

—RAYMOND DOUGLAS DAVIES, THE KINKS

"You, your joys and your sorrows, your memories and your ambitions, your sense of personal identity and free will are, in fact, no more than the behaviour of a vast assembly of nerve cells and their associated molecules."

—FRANCIS CRICK

"Reality must take precedence over public relations, for nature cannot be fooled."

—RICHARD FEYNMAN

"What the anvil? What dread grasp?"

—WILLIAM BLAKE, "THE TYGER"

"Unerringly impelling this dead, impregnable, uninjurable wall, and the most buoyant thing within; there swims behind it all a mass of tremendous life, only to be adequately estimated as piled wood is—by the cord; and all obedient to one volition, as the smallest insect."

—HERMAN MELVILLE

"The only thing in this universe that suggests the reality of consciousness is consciousness itself."

—SAM HARRIS

"There is no ghost in the organic machine and no inner intender serving as witness to a Cartesian theater. The locus of self-perspective is a circular dynamic, where ends and means, observing and observed, are incessantly transformed from one to another.... Intelligence is about making adaptively relevant responses to complex environmental contingencies, whether conscious or unconscious."

—TERRENCE DEACON

"Nowhere in the laws of physics or in the laws of the derivative sciences chemistry and biology is there any reference to consciousness or mind. This is not to affirm that consciousness does not emerge in the evolutionary process, but merely to state that its emergence is not reconcilable with the natural laws as at present understood."

—JOHN ECCLES

"Brains and neurons obviously have everything to do with consciousness, but how such mere objects can give rise to the eerily different phenomenon of subjective experience seems utterly incomprehensible. Despite this, I can't go so far as to conclude that mind poses some insurmountable barrier to materialism...."

—H. ALLLEN ORR

"I think we know quite well what consciousness is; what I maintain is that we don't understand how consciousness can arise from merely electrical and chemical properties of the brain...."

—COLIN McGINN

"Nothing happens. Nothing happens forever."

—JUSTIN TORRES

Contents

Foreword

by Nick Herbert

AUTHOR OF *QUANTUM REALITY, PHYSICS ON ALL FOURS,* AND *ELEMENTAL MIND*

"Wake up. Wake up. You are living in a miracle."

In his numerous books and journals written during a long and creative life, Richard Grossinger has consistently proclaimed the same message: "Wake up!" In previous books he's expressed this theme through the lenses of anthropology, bio-regionalism, esoteric philosophies, psychoanalysis, personal biography, cosmology, embryology, dreams, and various uncategorizables. In this book he tackles the number-one unsolved problem of science, religion, philosophy, and day-to-day living. What is consciousness?

"Wake up! Wake up!" Grossinger shouts. "You are living in a miracle." He's right. And to this miracle, we, his readers, are (mostly) anesthetized. Grossinger does not underestimate the importance of his topic. In the spectrum of mysteries, here's where he places the problem of mind:

> Subjective existence is the Holy Ghost; it cannot be made
> any more metaphysical or impart anything that is stranger
> than its very existence, so telekinesis and metempsychosis are
> well down on the overall strangeness list. Consciousness, by
> lounging about and conveying information, is already para-
> psychological and telekinetic. A universe with consciousness
> is more different from a universe without consciousness than
> a universe with remote viewing and telepathy is from the uni-
> verse we have—and by a long shot. Consciousness is the lu-
> minosity that illuminates everything (dreams included)—even
> the physical world, even what it is not supposed to illuminate,

and that is a big one, the biggest one of all. It reduces every other coup or experiment to sleight of hand (or sleight of mind if you prefer). Turning the light on is ~~the Big~~ Kahuna.

To the problem of consciousness, Grossinger does not provide a solution. Nobody could. "What is consciousness?" is the most profound question beings at our stage of development can ask. Confronted with this question we are all children: Nobel laureates, skateboarders, acid heads, musicians, terrorists, robber barons, Burning Man-ers, multiple mothers, police officers, Catholic priests, or PhDs—all babbling babies in the face of the giant mystery of mind. *Dark Pool of Light* is Grossinger's latest attempt to break through your waking hypnosis and to rub your noses in Reality.

Read it and go back to sleep.

Or, "Wake up! Wake up!

You are living in a miracle."

Nick Herbert
Boulder Creek, California
March 7, 2012

Foreword

by Jeffrey J. Kripal

J. Newton Rayzor Chair in Philosophy and Religious Thought and Chair of the Department of Religious Studies at Rice University

The Next Big Thing

I am an academic. I live, breathe, read, write, and imagine in an academic world, in my case a School of Humanities in a private research university. I love this world. It is a utopia of sorts. Not really, of course. But pretty darn close.

This academic world is very good at what I like to call, alliteratively and with capitals, Culture and Cognition. In the present volumes, Richard Grossinger gives many examples of both skill sets, particularly in his unblinking accounts of his PhD training within an anthropology department at the University of Michigan in the 1960s. Both these categories of Culture and Cognition are radically relativizing, destabilizing, and, well, just a bit scary, since, when deployed with sufficient nuance and depth, they have a very strong tendency of pulling the proverbial rug out from people who assume, quite falsely, that the particular worldview and way of thinking in which they were born just happens to be the singularly correct one, the one the rest of us should accept as universally true. Not so. No way.

In regards to Culture, we are very good at demonstrating that any human experience, including every religious experience, can be understood and interpreted by locating it in a very particular place and time, in a historical context, as we like to say. Such experiences are not absolute. They are historical, constructed events, like every other event. In regards to Cognition, we are very good at showing how something like a mythology, a religion, a piece of literature, or a language—really any human creation—is organized along particular cognitive grids and follows certain implicit rules, usually for some mundane social, economic, or political purpose or "function," as we like to say.

What we are very bad at is the third C—Consciousness. We haven't a clue what it is, and what we usually do is try our best to assume that it isn't. We thus talk and write endlessly about the "death of the subject" in the postmodern world; about the utter inaccessibility of "the Other"; about the vices of the "metaphysics of presence," that is, any talk or assumption of a being that exists in its own right and on its own terms and cannot be reduced to some local network of language, sign, or relationship; and about the impossibility of "universals," that is, talking and thinking about things that exist across disparate cultures and times. The latter is, or so we claim, very bad, since human beings are really, at base, expressions of local cultures and particular times and should never imagine that they can transcend these local cultures and times. "Stay down," we shout in so many words. "Be small!" "Be really, really small." Basically, we have talked, thought, and written human nature into a corner, and an increasingly shrinking one at that.

Oh, some things, really many things, have changed in academia from the 1960s. There are some very serious, and very fine, books published on paranormal themes by university presses now, for example. You would be surprised, I think. Perhaps this is because many of the young, hip, emerging scholars have grown up in a furiously plural and fluid world where phrases and words like "spiritual but not religious" and "paranormal" are in the water, if not in the baby bottle (their grandparents, after all, are baby boomers). Or perhaps this is because many of the most senior and accomplished professors in a field like the comparative study of religion found their callings deep within the American counterculture, which still shows up in their writing in all sorts of subtle, and not so subtle, ways. Some scholarship can be downright psychedelic.

As two simple signs of these changes, I might note in passing that two new journals have appeared these most recent days entitled *Paranthropology* and *Praeternature*, which feature, respectively, edgy, creative, brave essays that bring together the languages and methods of parapsychology and anthropology and rich historical studies of all things occult, magical, and liminal. A recent piece in *The Chronicle of Higher Education* (10 May 2012), a kind of standard-bearer of the academic world, celebrated the latter journal's appearance with a lovely little essay entitled "Unicorns and Sex? Out of this World." For what it's worth (maybe not much), my own mystical life began in a Benedictine seminary with a dream encounter with a unicorn and the mysteries of sex, an imaginal initiation which I wrote about over a decade ago now in a university press book entitled *Roads of Excess, Palaces of Wisdom*.

So, no, we're not all talking heads in ivory towers. Some of us are closet erotic mystics, unicorns in hiding.

●

I write all of this not to sound like a resume, but simply to own my perspective as I read—really fall into—a text like *Dark Pool of Light*. I hesitate to call this trilogy a book, or even three books. I mean, what *is* it? A double treatise on the limitations of scientific materialism and the renunciatory logic of Buddhism? A celebration of the possibilities of a shameless, liberated, lid-off theosophical imagination? A spiritual autobiography? An ethnographic report on some "psychic studies" in Berkeley? The answer, of course, is that it is all of these things, and more. I will not attempt to summarize all of the trilogy's themes. That is not what a foreword is for. Instead, allow me to isolate and celebrate just two themes that weave themselves throughout all three volumes, which I consider particularly important and refreshing.

Let us call the first theme the "mysticism of science," by which I do not mean the science of mysticism (although it includes and embraces that as well). This first theme begins with the conviction that no present or future worldview can afford not to engage, seriously and deeply, the methods and findings of modern science, *whatever* these findings are. The simple truth is that we know immeasurably more about the physical world than we ever have, and that this knowledge base conflicts really and seriously with most all (okay, all) traditional religious worldviews. This does not mean, as Richard shows, that the religious traditions have nothing to say back. Quite the contrary. But it does mean that we moderns, or postmoderns, cannot possibly live in some past mythical landscape that we can clearly see now does not, and cannot, map onto the physical universe as we know it and map it with the mathematics. For heaven's sake, let us just admit that our premodern worldviews were, and are, wrong in their most basic understanding of the natural world and move on. Let us let go.

This is only the beginning of the mysticism of science, though. This first theme, after all, is not simply a negative one. It's not just, "Let go." It is also, "Loop back." For fields like cosmology, evolutionary biology, and quantum mechanics have not simply resulted in an abandonment of traditional cosmologies and beliefs. In the twentieth- and now twenty-first centuries, they have also—and this is really what I am saying—made possible other spiritual understandings, other imaginal encounters, other mystical teachings, perhaps even future, and more effective, meditative and contemplative practices.

And why not? Would not our mystical practices be more effective if they actually engaged nature, including and especially human nature, as it is? And is there not a new microcosm-macrocosm correspondence to claim as our own here? We can think about psi in the context of quantum entanglement, and even note that the countercultural "hippie physicists" who kept alive the latter (including our own Nick Herbert) were deeply interested in the former.* Parapsychology and quantum physics have long been entangled in more ways than one. We can also think about precognition and clairvoyance via the models of space-time and multiple dimensions. We can think about anomalous human abilities as evolutionary buds, as early mutations signaling the future of the body. And we can imagine, indeed predict, the presence of countless other species and worlds strewn throughout a physical universe that make anything in our premodern mythologies look downright tame and banal. We may, of course, be finally incorrect about some, or all, of this new modeling, but so what? This is how new forms of knowledge and new worldviews are born. And besides, we may not be incorrect. We may also be onto something big here. Really, really big.

---

*See David Kaiser, *How the Hippies Saved Physics: Science, Counterculture, and the Quantum Revival* (New York: W. W. Norton, 2011); and my *Esalen: America and the Religion of No Religion* (Chicago: University of Chicago Press, 2007).

Note here that it is not simply a matter of big brother science confirming (or dismissing) little brother mysticism. That would be the science of mysticism. It also goes the other way, with little brother mysticism turning out to be onto something after all (and first to the punch), even if that something needs to be separated, intellectually at least, from the mythical, cultural, and historical frames in which it has been traditionally expressed (and that's where the whole enterprise needs, desperately, the social sciences and humanities).

Fritjof Capra's intuitions were dead-on. As were those of Niels Bohr, Wolfgang Pauli, and Carl Jung. There is a weird correspondence or parallelism between mysticism and science at work here, even if, yes, this cannot be framed as a simple identity. Mysticism and science are fusing fast, probably because they are the two most exquisite expressions of Mind beaming through the left and right brains of a single skull. And *that* is exciting. We are back to saying something positive, something universal, something grounded in both deepest spiritual and material natures. Even if nothing is quite clear, we can at least see now how we might be whole in the future. We can also see a way to be religiously honest. No more duplicity. No more bullshit masquerading as eternal certainty.

The second theme I see glowing and giggling throughout these three books is a particular, but very consistent, celebration of existence, of *this* world and *this* body, against the world-rejections and renunciations of traditional religion, particularly those of Buddhism and, to a lesser extent, Hinduism. Richard expresses this celebration of a divine universe most clearly in volume two, with these lines:

> If enlightenment is the unquestioned goal of all existence,
> what about the problem of timeless nondual being, with all
> of eternity to be nondual, yet somehow and for some purpose
> choosing to manifest intricately and perniciously dual forms?
> Why? Why has energy chosen to generate the human plat-
> form, to place ego identities inside subjective containers for
> experience ...? How to explain the source of a relatively low-
> end predatory situation on worlds and through incarnations
> like ours? How did this inferior state of being make it through

Buddhic customs and get sanctioned? Is it possible that this world's dualistic density and set of cosmic filters is essential to a different sacred agenda, perhaps to the creation of novelty (as Whitehead and crew would have it). Not exactly counter-enlightenment but rich and profound in a whole other way.

Richard's sources here are theosophical (in his terms) and esoteric, but they resonate deeply with all sorts of other movements and figures in the American religious countercultural scene over the last half century, including the human potential movement, about whose similar world-affirming spiritualities and mysticisms of science I have written.

Matter and mind, transcendence and immanence, soul and body, spirit and sex—these are all dualisms that are useful and perhaps even necessary at certain points in the human journey, but they are also finally false. The uni-verse did not spend fourteen frick'n billion years putting us together so that we could simply just opt out and leave the stage. The show wants to show. Clearly, something else is afoot here. Something big. *Dark Pool of Light*, for this reader anyway, is finally not about conclusions, certainties, or yet more unbelievable beliefs. It is about intuiting, imagining, guessing that next big thing.

JEFFREY J. KRIPAL
Houston, Texas
May 15, 2012

Introduction

D*ark Pool of Light* began inadvertently in March 2010 as a journal of my psychic studies. In May I noticed a book on my shelf, *The Nature of Consciousness: Philosophical Debates,* which I had received as a gift from the publisher in 2005 but hadn't paid attention to. I proceeded to read most of the anthology, in fact twice: all 800-plus pages' worth, as it addressed a riddle and mystery that had eluded me for decades while I was absorbed in related matters. I gradually conceived a bodacious alternative to "The Nature of Consciousness" that would blend my psychic journals with summaries and deconstructions of academic arguments *against* consciousness.

It did not start as a three-volume work, but consciousness is a large and omnivorous topic and it will swallow pretty much anything allowed near it. By the time I decided to rein in this baby at around a thousand pages, my Word file had become an unmanageable spool. I realized then that one binding wouldn't work, so I took an extra four months to make each of its movements into a full book in its own right.

•

What—when you really think about it—is consciousness? I had long understood that this is the mega-question of both science and philosophy, but I had not appreciated either the tautology of epicycles that neuroscience had projected onto relations between the brain and the mind or the full boggle of philosophy's forays in search of subjective being. Our own awareness is the thing in the universe most difficult to understand or vindicate. As one chemist friend emailed me upon learning that I was writing this book, "I personally have no idea what the mind is. I know what a brain is, but I have no idea how the two are related."[1] This is a plaint of the sort that I have heard so often that I tend now to hear its undercurrent too:

"Can you believe that the relationship got so weird we can't fit mind into the universe anymore or, for that matter, even find it?"

Most working scientists share this perplexity; yet they cannot entertain the notion that consciousness might have a source beyond entropy and its work-arounds. It is a lynchpin of modernity that we cannot be *intrinsically and independently* conscious. Of course even the folks who propose that we cannot *cannot*. The only way they can adduce or uphold their own mind-edness is by overdetermining or misplacing concreteness in one form or another and then assigning it a basis in their own brain. Failing (of course) to uncover such a basis, they concoct one.

Physicists and biologists will no more admit the contraversions in their theories than, say, Israeli settlers will cop to the fact that they are building villages on other folks' land or Chinese officials will confess (even privately) that Tibet is another country—or, for that matter, than Tea Party Americans will risk conceding that Barack Obama was born in the USA. For scientists to tolerate any exogenous source of consciousness would be to forfeit their eminent domain and vested legitimacy—the basis of their declaration of power. Consciousness must finally be either illusional or imaginary—a spinoff of thermodynamics and neural stacking. It cannot aspire to any higher status. If it ever gets a foothold outside entropy, their goose is cooked.

Everything that follows in my book is either an extension or ass-kicking of this dilemma; that is, it either addresses the crazy-making ploy of acting like conscious dudes who *de facto* deny their own existence, or it portrays consciousness operating *sui generis* under its own authority.

I mean to kamikaze rather than skulk into this snafu, avoiding "tao of physics" settles or stale resorts of quantum-mechanical metaphors that re-locate science and spirituality at layers of the same general paradox (though I think that these models are valid in their way). What I seek instead is an *actual* convergence of scientific and psychic attunements—very, very different birds that stick out hard beaks and sharp claws in trying to bash the other into nonexistence. I force them into *co*existence and frame their meanings in terms of each other's. I can't think of anyone else rager enough to operate at this frequency; yet I believe that it is precisely the dial tone of modernity as well as where we have to go.

It is also *déjà vu* for me. Decades earlier, as both an anthropologist and a psychospiritual seeker, I tried—more than 1,400 pages' worth in two books—to identify the pleat along which biological embryology and esoteric anatomy meet. Going back and forth between morphogenetic and occult incarnations, I hatched my own rubric: "The embryo is the universe writing itself on its own body."[2] It honors Darwinian logic insofar as the "writer" can be the equivalent of monkeys on typewriters, but it implies as well that the universe has (a) a discrete body, (b) something writable, and (c) a way to inscribe meaning on burlap in another medium.

What starts out in anatomy ends up in anatomy: to look for mind is to look at body—but you have to decide which "body." If it has a merely literal basis, it cannot bring context, identity, and meaning into presence. Mind cannot be scriven onto its tissues.

There are no two ways about it. Either human experience is a "perceptual window" through which the material world is depicted, an emergent, evolutionary state of the brain and nothing more, creating just a mirage of consciousness; or it is a transpersonal event that has always existed in the universe, a form of structured information that does not have to be physically embodied and can transmute from one state to another; *or it is organized by* a transpersonal event that has always existed in the universe in the form of structured information, does not have to be physically embodied, and can transmute from one state to another.[3]

Either this affair is a temporary excrescence of bubbles, which is already dissolving toward permanent erasure; or we are a phase shift within a larger tonation and consciousness is less an accident of molecules colliding in each other's soup than a rune painted in their fluvium with its own brush.

●

Consciousness cannot be addressed either solely rationally or solely intuitively. It requires increments of both, each a deep, committed breath. I have tried to draw *Dark Pool of Light* out of such breaths. I hope that they're mindful and worthy of your attention rather than the sort of indulgent logorrhea of which I have been accused, most recently by a fellow Amherst College alum who posted on our class website: "God bless you, Richard, but you really do need an editor!"

Would that it had been simple or straightforward for me. The long count of books I have written over almost fifty years to get to *Dark Pool* downright shocks me—the scope of them, their progeny in stacks and ragged rows, cartons cluttering successive basements and attics of my life. From *Solar Journal, Spaces Wild and Tame, The Long Body of the Dream, The Slag of Creation,* etc., through revised editions of *Planet Medicine, The Night Sky,* and *Embryogenesis,* to *On the Integration of Nature, The Bardo of Waking Life,* and *2013,* I have been on an inexplicable mission throughout which I have worked in indifference, isolation, oblivion, and inadvertence of the mess I was making (not to mention the forests I was destroying). These imperfect, weird, often rambling volumes—thirty-seven or so of them—shelve like an undiagnosed and unclassifiable heresy in my living room (and in the libraries of my readers). Even the key ones alone make motley piles at talks I give—so many of them by now, too old and dusty, many of them too fat, covers scuffed or bent, hard for anyone to grasp the intent or drift of this project.

In this later phase of my life I recognize the undeniable yet inscrutable fact that, page by page, Richard Grossinger spent a good chunk of his Earth-bound hours keyboarding testimony (from manual typewriters to Apple laptops). It is hard for me to grok quite why. All I can say is that it speaks to the daily power of the ineffable to sustain a human life.

I am mortified that I had to write so much and at such length to come to the simplest thing, but I have, and there it is. My books mark something like a snail's trail over bumpy ground—wonder and epiphany converted into hieroglyphs on papyrus that will eventually mulch with this whole civilization and planet. I could have traveled and adventured instead or spent more time just hanging out, but an unseen universe kept calling to be fed, and every day almost (it seemed) its gods were asking for a new prayer.

●

From childhood I have known at some level that the modern world, the world I was born into, doesn't believe in anything and also doesn't know *how* to believe in anything. Oh, lots of people believe in science and prog-ress or money, or they believe in God or Allah, but not really. God is just the stand-in for something else that they don't believe in either but are willing to accept as a substitute for belief because it serves some other desire or fear. Check out Émile Durkheim's *Elementary Forms of Religious Life.* There's much in the way of getting to God, and those folks who are most fervently sanctimonious and work themselves into the heaviest raptures of biblical fundamentalism, jihad, emu dances, speaking in tongues, and the like often are worshipping not a divinity but a clan, a gang, or their ancestors:*

> When the Australian [Aborigine] is carried outside himself
> and feels a new life flowing within him whose intensity sur-
> prises him, he is not the dupe of an illusion…. Religion …
> before all … is a series of ideas with which the individuals
> represent to themselves the society of which they are mem-
> bers, and the obscure but intimate relations which they have
> with it.[4]

The fact that those emotions and their intensity are not an illusion does not mean that they are then merely (or only) social facts represented in sacred metaphors and ecstatic corroborees. The universe is represented at multiple levels simultaneously in itself and in us.

By calling on Durkheim here, I am also introducing a series of internal contradictions that will take all three volumes to settle. The above is a famous passage in which (from my standpoint anyway) the French sociologist is *nailing one subterfuge*—that much of what passes for religious or spiritual activity on Earth is actually bonding with the parishioners' own moiety or tribe and its footprint in an ineffable realm—*while missing another* (especially when trespassing among indigenous chantries): that many non-Western sociological ceremonies are magical and psychic operations too, which is what makes them both social and spiritual. For now I simply want to identify the misplaced concreteness of religious passion when it is political patriotism rather than identification with the Divine

---

*I know that this is a bit condescending and precious for so early in the book, but it came up, and there's no way I'm ever going to get back to Durkheim, yet as an anthropologist I feel that I have to touch his base once—so here it is.

(or compassion for Creation and its sentient beings)—though of course (yet another paradox) it must draw on actual sacred energy in generating its chauvinistic tropes. It sure feels like God, but it's several octaves short of the actual overtone. Getting to the Divine takes unremunerated commitment and devotion to an absolute mystery. If you can package and sell it facilely, it ain't God.

Lots of people believe in science too, or at least in the world created by progress with its machines. But there's nothing to believe in there except trained appearances and a skein of distractions and entertainments ending in a funeral. Even if science believed in and taught a universe of consciousness, Earth under its sole rule would be a vacuous and unhallowed place. Where there is nothing but artificially sweetened objects to sample and their actual insides are surrogate, second-best options, you are going to come out believing in nothing anyway. Consciousness is not the sort of eccentric uncle who can be paraded out occasionally to entertain the troops and guests. Consciousness is the here-and-now portal to everything.

Science doesn't believe in consciousness anyway, so that leaves me a clean slate to take my best shot at its riddle—a double enigma that creates beings with identities as well as a body-mind situation that ties their destinies to the morphologies it generates. For all that physics and biology finally say (or don't) about life and existence, something happens, we are inside it, and it has shape and trajectory. There is an absolute event, beyond systems or beliefs, that does not defer to ideology, etiquette, or consensus: how to tell *its* story?

The universe takes, and can take, nothing away from us, but we don't know that because we do such a thorough job of taking it away ourselves. We are in ignorance of who we actually are.

●

This first volume of *Dark Pool of Light* tackles the psychophysical and physiopsychological aspects of consciousness, but it also sets the stage for the larger work, introducing consciousness as a sphinx and oracle. Consciousness is thermodynamic, neurological, cybernetic, phylogenetic, and maybe quantum-entangled, but it is also neuropsychological, neurolinguistic, behavioral, developmental, morphophonemic, proprioceptive, and cognitive (all of the above toured in Chapters One through Eight). Out of these qualities arise the reflections, meanings, and ontologies of Chapters Nine and Ten.

In the second volume I move from scientific definitions and circumscriptions of consciousness to psychospiritual and psychic exercises. I try to tease consciousness out of itself—to go into its fire and become the conflagration. Since obviously I cannot write about every aspect of consciousness, which would be to write about everything including that which cannot be formulated into language, I chose to tack the outer banks or at least to initiate reconnaissance there. I almost said "the margins and boundaries," but I do not have any idea what these might be.

From assuming the existence of consciousness at the start of Volume Two, I proceed into a rendition of consciousness's mesh, replacing science, philosophy, and psychology with shamanism, theosophy, and Buddhism. I offer peeks into systems as diverse as tarot, Maori psychology, astral projection, Gurdjieffian cosmology, meteoromancy, Teilhardian noospherics, and the Vedic cosmos.

Volume Two is also a how-to book: an introduction to psychic practice and lore. I woo all agnostics, hardcore skeptics, and atheist materialists to go beyond their prejudices and ideologies into the raw energy of themselves—to ask existence what it is. This becomes not only a book *about* consciousness but a celebration of consciousness and a parade of consciousness's most exigent children: self, identity, mortality, free will, empathy, fear, sacredness, creativity, dreams. Yet Volume Two falls under the heavy shadow from Volume One, so it is by no means a New Age free-for-all. Science has squeezed mind, identity, and meaning down to a nubbin and then tried to drown it in its own bathtub or scrunch it out of existence (of course, I am riffing off the early twenty-first-century pledge of Grover Norquist and the fiscal conservatives to starve and then geld government). So then what is left, and how and where do we begin looking for it? How do we validate what we actually feel?

At the end of the second volume I peel to where the Tibetan Dzogchen cutting edge meets the neuroscientific cutting edge. At that convergence of total transparency and total opacity the glass of materialism, scrubbed clear of itself, becomes an eerie dark pool of self-originating light. There consciousness is the ground luminosity and foundation of its own being, transcending category or space. True, that!

Volume Three is a herd of wild horses trampling over this ground. Coming out of the cauldron of the first two volumes' landscapes, it confronts us with the stark and inextricable fact of our own existence, again and again from different angles and on variant stages—each a stampeding mustang. What does this life mean and what is our destiny? Do we still have one?

Topics in the third volume include terror and its transformation, the role of evil in the universe, the ecospiritual crisis of humanity, suicide, death practices, family constellations, communication with unseen entities, shamanic initiation, synchronicity, near-death experiences, and the genesis of personal meaning. I provide a combined scientific/occult unified field theory of the universe, a so-called "Cosmic Eternity System," but I do it under a rock 'n' roll score.

Both the second and third volumes close with "hyperlinks" to the rest of the text: among them Sandor Ferenczi's cosmic trauma, Jared Loughner's Tucson binge, Paul Simon's "long-distance call," the hunting of mourning doves in rural Missouri, astrological Pluto-Charon, God as a personalized entity, sacred ballads, and the nature of time.

●

Finally consciousness just *is*. Its deposition continues to transcend all else like a forest fire burning and consuming whatever falls in its path, no matter the aegis or authority that put it there or recants it. Then the same blaze creates the same garbage all over again.

We didn't choose or define this manifestation; we apparently created and create it by our existence. The philosopher Alfred North Whitehead identified that core circumstance in his book *Process and Reality* (of that, the poet Ed Dorn noted, the big word is "and"[5]—getting from P to R without losing the baggage en route).

The process creating reality is invisible to its actors, camouflaged in the background of their unexamined conceits and body-mind kinesthesia: "A society does not in any sense create the complex of eternal objects which constitutes its defining characteristic. It only elicits that complex into importance for its members, and secures the reproduction of its membership."[6] That is the precipice on which I am trying to set this book. My argument is, in part, that we are so mesmerized by the decisive splatter of materialistic science that we don't see the actual universe.

What are Alfred North W's eternal objects? "The things which are temporal arise by their participation in the things which are eternal. The two sets are mediated by a thing which combines the actuality of what is temporal with the timelessness of what is potential.... By reason of the actuality of [the] primordial valuation of pure potentials, each eternal object has a definite, effective relevance to each concrescent process. Apart from such orderings, there would be a complete disjunction of eternal objects unrealized in the temporal world. Novelty would be meaningless and inconceivable."[7]

And clearly we have made it into the temporal world, absolved of disjunction and not entirely betrayed by the meaningless and inconceivable either.

In other words, the numinous requires the temporal, as the Cosmos requires worlds. Unnetworked meanings must be cuffed. Novelty, as Dorn's mentor Charles Olson teased a mostly uncomprehending audience at the Berkeley Poetry Conference in 1965, *is* Creation. He was sourcing Whitehead: "'Creativity' is the principle of *novelty*.... The 'creative advance' is the application of the ultimate principle of creativity to each novel situation which it originates."[8]

So how does novelty come flying out of the background to astonish us? Is novelty the prop and pillar that hold up the foreground, riveting us into compliance? Or is novelty the joker that brings the castle down, pillar and prop, all the king's horses and all his henchmen, again and again, each civilization and universe?

Whitehead's conceits drift at the outer fringes of my ken, but I get their general gaze. Eternal objects must descend into self-thinking thoughts.

What lies beyond novelty, like ultraviolet light from beyond our ambit, is a taste of, to echo Ray Davies and the Kinks, the "real reality." I rest my case on this: the strength of our species—our wake-up call and imprimatur—is, that whatever is in existence, we must weigh in on, and the weight with which we strike must be equal to the weight of the thing with which we strike as well as the thing we are striking against, and the strike itself must hold the cumulative substantiality and must measure the eternal time through which it is coming into being.

When an unanswerable question is being posed, the answer is always the same: what are we *not* talking about?

CHAPTER ONE

What the Fuck *Is* This?

How did it all get here?

And who are we? How did we get here too?

What is going on?

Why a universe at all? Why stuff? Why stations of consciousness? Why a pebble, an igloo, a croquet ball? Why anything?

Why space? Why shape? Why gravity? Why ground? Why heat? Why worlds? Why time? Why matter? Why? Why? Why?

How did all this trash and treasure get dropped on everyone's doorstep? How does anything emanate from anything else? How and why did it become this?

And, while I'm at it—why not something else? Something else entirely? *Anything* else entirely?

Why not nothing? Why not nothing forever? Why creatures? Why private views? Why ego identities? Why now? Why should anything wake into radiance?

Anteaters, shrews, snakes, wasps, and all the rest—to what purpose?

Everything in this world has a context, in fact many contexts. We deal only in contexts. There are big contexts: hunger, pleasure, survival, sex, shelter, profit, America, Christ died for my sins, romantic love. These drive not only behavior but meaning. Then there are small contexts: putting together a chair from parts, following a soap on TV, playing a chess match, supporting a candidate, an uncle's birthday bash, yoga class, stylish clothing, downtown, the sales and marketing team, the gun collection, plans for a holiday, being a hottie, tickets to a play, losing weight, the World Series, O. J. Simpson, Donald Trump, Kim Kardashian. Money is context, war is context, bribery is context, police are context, the Bloods and Crips are context, jihad is context, mathematics is context, public transportation is context, trying to find a soulmate is context. There are mega-contexts too: mortality, the dead, the universe.

But there is no context for the whole, for the entirety, the state of existence (at least in contemporary American culture). There is no context for us. The closest to a context is God, or matter and energy, or DNA, but that is all outsider buzz. "Being" comes down to what "being" feels like.

Since the human species manifested in the Stone Age, each of its members has been confronted with the same astonishing blaze. Reality in its naked presentation, shining and bristling from within and across proximal space while penetrating absolute space, is flat-out shocking and profound. Realer than a motherfucker! It is more profound than all the profundities conjured by science and philosophy. Along its most deepening seam it is subtler than anything in it. Cars traveling down the street on some planet, probably but not necessarily this one, not even cars, are not profound when viewed by everyday mind; however, in the vast unacknowledged scheme the fact that they exist at all and are piloted in orderly fashion is profound and weird beyond conception.

Scientists now explain the existence of nature (and mind) by equations of heat, entropy, surface tension, binary coding, and differential survival. They scan substance to where its gauze is most distended (the sky), most discrete (the subatomic nucleus), and most quantifiable (the algorithm), as they try to excavate condition and origin. Fat chance!

Philosophers buy this prognosis hook, line, and then some; they extract "being," meaning, and values from it.

Psychologists overlay ego, psyche, personality, and behavior—thermodynamic and chemical vectors traveling inside membranes. They replace the philosophical mind with the biological mind and neurotransmission.

Shamans, priests, and clerics set nature under sacred sovereignty.

Psychics tune to energies and planes not measurable or acknowledged by science.

None of these gets to the bottom of the weirddom.

●

Among depictions and rationalizations of reality, twenty-first-century upper-tier denizens are most familiar with the West's sanctioned brand: the survival-of-the-fittest, you-only-go-around-once market economy. Their lives occur on its mean streets amid its hemorrhaging urbanization, in progressively more acute cycles of crisis and cataclysm, clinical anxiety and depression, plus urgency in the context of ever dwindling time and possibility, incessant craving for more, endlessly more: more life, more goods, more thrills, more validation, more anything.

In towers and operating rooms of the corporations and academies, professional scientists continue to address reality as a riddle in forensics, a cold trail left in the galactic sky and in the cyclotron of matter, evidence quasi to a crime. Dismissing its phenomenal aspects, they stalk it to the Big Bang and subsequent fusion, fission, and differentiation in stellar cauldrons ignited by the blowout. Comparing indices and refining their assays, they dowse and test the "splatter" in hopes of exposing the weapon used, the nature of an unwarranted slash on the void.

But there is no such smoking gun. The stuff that broke through from beyond time and space is out of play, forever. This is a spill zone not a construction site—everything in it has been used before and as something else. Or not: same difference.

The universe is simply too deep, too old, too frayed, too insouciant to be explained. That is why grand unified theories of All That Is are, to a one, pretexts and vanities. Inquiry is limited to what came after the Big Bang, which is all that we can get at. Just about every item, every primo seed is missing from dossier and file.

Science supposes that creation was merely statutory—no design behind it, no rationale or impulse, no hint of an absentee landlord, only the absence of sufficient obstacles to prevent or impede its splay.

Imagine a malefaction without a motive, that begins with its commission—absolutely—no assets or adjuncts of any kind.

Materiality is the present idol of our manifestation; it guards Entry and Egress; it decrees: "Thou Shalt Have No Other Gods Before Me." And we don't.

●

Creek and Ainu philosophers, Australian Aboriginal elders, Tibetan shamans, and the Aegean cosmologists understood (and still understand) the engine better than do most citizens of modernity—and that includes sophisticated particle physicists. They understood it in the moment and bowed to its omneity: a light arising from darkness, a wind from stillness.

Once upon a time, the universe was sacred and unfathomable by simple emanation. Humans accepted the operations of nature as the mirror and counterpart to their own existence, surrendering to its primacy and innate dignity. They ceded a vast and absolute design and conducted a ceremony whose goal was adoration not interrogation. Before quarks and Big Bangs, they called it Spider Woman and Corn Mother and zoned its tiers by Chameleons, Swimming Turtles, Bouncing-Stick-Player-Toads, and Hyenas' Eggs. These are neither contrivances nor mere fables; they are not raw primitivisms either. They are hard-won intuitions of something before form:

"Verily at the first Chaos came to be, but next wide-bosomed Earth, a disk surrounded by the river Oceanus and floating upon a waste of waters, the ever-sure foundation of all the deathless ones who hold the peaks of snowy Olympus and dim Tartarus in the depth of the wide-pathed Earth, and Eros, fairest among the deathless gods, who unnerves the limbs and overcomes the mind...."[1]

Eros before matter, always. Listen carefully and you will hear the rustle and trickle of an actual universe, an inviolable presence, not a working factory.

"The Ground Squirrel said, 'I think day and night ought to be divided like the rings on the Coon's tail.'"[2]

Contrast and discrimination—on fur as among the rings of Saturn.

"A very long time ago there was nothing but water. In the east Hurúing Wuhti, the deity of all hard substances, lived in the ocean. Her house was a kiva.... To the ladder leading into the kiva were usually tied a skin of a gray fox and one of a yellow fox. Another Hurúing Wuhti lived in the ocean in the west in a similar kiva, but to her ladder was attached a turtle-shell rattle."

How was this possible before there were either foxes or turtles? It is because these stories encapsulate construction of a universe of events inside a prior universe of meanings.

4

"The Sun also existed at that time. Shortly before rising in the east the Sun would dress up in the skin of the gray fox, whereupon it would begin to dawn...."[3]

This is it! It might slip by as a pretty-boy metonymy if you overlooked its ontological cred: Everything arose from nothing. Concretely and explicitly. This is what it looks like if you peer inside this very minute: gray foxes and self-emanating light.

●

Viewing electrons, atoms, and chromosomes in the scientific manner as they shape-shift and deliver payloads doesn't alter or encroach upon their identity. For being exposed like a burlesque dancer, a mitochondrion is no less or more immaculate a riddle than it was inside Stone Age hunters. Western reality has no prerogative or supremacy over other brands. It may be the present operating system for modernity on Earth, but its roots are no more rooted, its arising no more fundamental or absolute. No one species's or planet's deposition has primogeniture or is endorsed by the universe. The same claims are made *implicitly* by the spider and the mouse.

Through the entitlement of its birth, each entity places its lien on existence. Albert Einstein and a 1930s sea squirt each expressed a sincere and desperate truth, equally confronted the fact of their being and rendered a coherent paradigm of it. They fed the universe's eyes, ears, and brain.

There is Bushman reality, Navaho reality, Aranda reality, Cherokee reality, Xhosa reality. Within each of these sprout countless personal realities. And these barely scratch the surface. Cat reality, snake reality, whale reality, wolf reality, worm reality, bacterial reality all are "real" too.

The osprey with its wingspan and talons, the owl with its judicious eyes and motion-detecting granules, geese with their star- and sun-maps, were knighted long ago by vanished gods. Currents of air, below and above feathers, fins resisting waves through rippling flow—these are sciences too. "Even the trodden worm...," declared philosopher William James, "contrasts his own suffering self with the whole remaining universe, though he have no clear conception either of himself or of what the universe may be."[4]

Amen, and God have mercy on us all.

●

Science as we know it is not science anyway, not by standards of worlds or-biting Rigel, Antares, and the Dog Star—or, if not there, then somewhere. The Big Science of the Milky Way provides an impartial jury for claims of truth by experimenting parties on separate worlds. The Meta-Science of the Universe alone knows everything (or anything) about any thing. Earth Science, endowed by private and corporate interests, offers only space-time audits from the perspective of deputies on one planet in one small capillary.

Alligator crocodile reality, dragonfly damselfly reality, realities on the billions of inhabited planets in the Large and Small Magellanic Clouds—there are more stars and skies, more heavens and earths than are dreamt of in our philosophies and operas.

Each entity gets born, lives, and dies on the universe's terms, and the universe is one serious mutha. We don't get to choose our own operating system or paradigm indefinitely. The universe owns all paradigms and sys-tems—and it is running a far bigger game than science.

So get off your high horse! Physics is not king of the universe. Earth is not the only game in town. Three dimensions are not the sole platform. Stranger realities arise continually on worlds in other solar systems, close to here and unimaginably far. We know their presences intuitively and un-consciously because, like hydrogen, consciousness is singular—we know them *as something else*. We know them *at all*.

Reality is the state in which we participate with everything else in the universe, a living fire that keeps emerging. And again at this next moment, and so on … in every creature in every crack and cranny, every tidepool and volcanic vent.

Yet scientific laws operate with impunity, as if official, as if someone other than us made them up and enforces them, as if they were cast in something more than the breccia of metaphor.

In its act of establishing a jural reality, science has detoured from honest inquiry into institutionalized ideology, using a bogus authority to enforce its sponsors' products. Our bodily existence and minds are now arbitraged in a futures market. Queued into motor pools, creatures are encouraged to trade in existence for algorithms, to refute their own beingness.

What used to be pure *scientia*—neutral knowledge—is a combo dicta-tor, morality squad, and hanging judge. When doctors confer cures, they must do so under a regime of terror, unacknowledged and reduced to muzak, falsified documents, and profit-and-loss statements. The Fates still decide how, when, and why each creature is born and dies. Clotho spins the thread onto her spindle. Lachesis measures it by her rod. Atropos cuts it with her shears, Charon receives it with a coin in its mouth. By usurping this province, by making DNA the oracle, a Taliban-like authority commands and deludes us (and itself) into thinking it is rolling the dice and cutting the cloth. Meanwhile it recruits us for its jihad: consumers all.

Body-mind is not even the sole frequency of intelligence. Beyond the charm of a matter-on-matter universe, other entities coalesce in untold dimensions of hyperspace. However divergent from our embodiment and shibboleths here, they are sordidly and viscerally real wherever they are because they are rooted in primordial awareness of their own existences and the common substratum from which they are arising. From their perspective today's local blue sky is the ultimate surreal backdrop.

For that matter the Earth is a planet that even *we* should never have seen, one that we were—yes—forbidden to see.

So I come back to my original question: Why us? Why here? Why now?

Why this gaudy manifestation, each granule, bump, and surly or succulent intent of it? How could you ever take it—your own existence, the warrant of "life"—for granted?

Just look around you at what has formed and stuffed itself into every gap. Witness pure existence arising, creating space and direction, lighting its own canopy, pouring through its own portal, filling the void with objects, shading its own light!

Empty yourself of preconceptions. 'I don't know what I am. I don't know what this is.' Like the gentleman songsters of the Whiffenpoof, "*We are poor little lambs who have lost our way./Baaa, baaa, baaa!*"

Let this confession fill your mind, roll across your skin, dilate into your chest and sockets, sink down below your shoulder blades, open your diaphragm, reverberate in your belly and lungs, drop into your genitalia. Answer the unanswerable question by an affirmation at your core.

Sense how deep and thick and omnipresent and sensational the universe is. Feel its silent stream of semblance. Hear its gurgle at a frequency so immediate, scrupulous, explicit, and snug that it is nonexistent. Watch its liquidity flowing from and to everywhere—the ground of yourself filling with a fulgent gleam. How is this possible? How is such an impeccable state of being and knowing allowed?

The moment you let go of your habit addiction, you explode in all directions. An intimidating audit, but not half-bad. At least it is happening at all.

●

Staring at surf, I am struck by the interplay of gravity, mass, and cohesion under lunar pressure, as rocks carve waves into glyphs.

We are sustained by foam as wide and precisioned as gravity, written by styluses as fine and hieroglyphic as air. What is spelled in our own minds is what was once written dumbly in the sea, in the calls of seabirds, welling up through ganglionic stations into sequestrations of self.

Mind is in constant dialogue with the intelligence of its own formation.

An imperative had to begin somewhere. Each motif indicates a source; otherwise there would be nothing at all.

What Sigmund Freud posited *vis à vis* dreams—that every entry and instance has an energetic prerequisite necessitating and providing it—is true as well of the waking dream, the simmering fog. Each item exists because it must. And there is no bottom or break to the ring of proxies engendering and sustaining it.

Where else would or could it come from?

Beavers gnaw down trees many times their size, pile up mud, dam rivers, store vegetation in cold houses under snow, patch holes in their dams. The semi-aquatic rodents permit muskrats to co-occupy their underwater huts and eat from their larder—why? From where does the symbol come to render and allow the gift?

Muskrats pay a "rent" of grasses and vines as they swim into and out of the communal refrigerator. Under what compact do the beavers monitor this transit?

What future and eternal meaning is synopsized in the screech, the caw, the yowl? Barking seals, baying hyenas, chittering moles, shrieking gulls— these metabolic packets don't merely provide meanings prior to language. They *are* meaning. Wild turkeys crossing a field at sunrise are screeching raw existence, intentionality, wonderment, and individuality back to the universe.

North American squirrels, though color-blind, discern a dissimilarity between acorns from red and white oaks, consuming the white ones which, by sprouting before spring, quickly lose their food value, while burying the slower-sprouting red ones for sustenance during late winter.

How does such information, at its every level of designation, get through the cables into molecular space?

In years when there is a shortage of red acorns, those same squirrels munch just enough sludge out of the white acorns to disable their sprouting capacity, and *then* they bury them.

Australian lyrebirds imitate car alarms, doors opening and closing, men with chainsaws cutting trees. Urban crows drop nuts into traffic in order get them cracked; they select streets with red lights because movement periodically stops there, allowing them to fly down and retrieve the meats unscathed.

Various species of birds pick up twigs in their beaks, then poke with them at grubs in tree trunks, agitating them to move in their dream of succor, to come out and be consumed.

Standing in shallow water, other birds make their wings into shade to trick fish to come to the surface.

The symbol is always and ever being born.

CHAPTER TWO

The Scientific View of Reality and Consciousness

I am looking to give you a quick and deep toke of the scientific world-view. Though it is not the *raison* of this book, it is my starting point, and I want to nail it as accurately as I can, stretch it as far as it will go. I want to track its feelers toward reality, to see how it gets itself there, how it subtends or protracts, and what it jettisons along the way. Then I want to come back the other way, from the exigency of our own existence with its counter-claims on the universe, and conduct its deepest gaze too.

The mainstream approach to reality/meaning is tied to a twenty-plus-centuries-old patent, its gathering thread sourced in a pancultural mega-project. This project is neither incidental nor opportunistic; it is genuine, earnest, and brilliant. Mankind has kept its faith and continuity and brought it to fruition through stringent inquiry and analysis despite everything else—and there has been plenty else along the way.

Science has been such a transforming revelation that civilization has forfeited possible alternative views. The technological domain shines so brightly that other paradigms are invisible.

Even philistines and reactionaries who despise highbrow endeavors—who strive to reimpose barbarian gods and patriarchal superstitions on the planet—embrace a scientifically enhanced world. They make entitled uses of its tools and weapons as if these were their dowry without anyone having bothered to study nature or carry out impartial experiments. Appropriating technology's fruits, they dine on its produce, commandeer its Volvos and Toyotas, travel in its planes, broadcast along its networks, tote its laptops to missions, and assemble its explosives as if they were anything but what they are. Cargo cult supreme! Fundamentalists (of all denominations) are devotees of materialism, yapping contrary slogans.

Matter is simply energy, and energy is itself an orphan of unknown stock, restless and short-lived. As long as matter and energy are adventitious, everything that roiled out of them, every form that they have taken, is happenstance and contingent, with no agency except circumstantial ones and no meaning except its own open event-field. It not only means nothing, it *is* nothing—rank, passing forms.

It means nothing, it is nothing, and it will all be obliterated down to the last tress of whatever, leaving no sign that anything ever existed.

We exist in a vast, profane, and drear neighborhood that arises because "stuff" from somewhere happened to aggregate and bundle in globs and clumps; then scrabbled into networks, grids, and membranes because (and only because) it had some innate penchant to gobble energy, the more the merrier. As membranous bundles cultivated agency, creatures awoke to assert and enact their sense of separateness. But only because egos were effective energy vampires. There is no way for consciousness to sneak into such a universe except as an intruder or artifact of something else.

It is not that science does not view consciousness or reality as real. There is plenty of hard stuff in consciousness: electrons, atoms, molecules, neural nets, neurotransmitters, brain tissue. But all this fury, while "real," is happenstance, and even if it does one day get located and assigned, it will still signify nothing, have no ontological status beyond its role as plaster and daub. From a scientific perspective thoughts are not even as real as heat or mass—interlopers too, interlopers all.

Where science and philosophy parley nowadays, consciousness is not a game-changer; it is not even consciousness; it is a hallucination of something "like" consciousness. It goes no more deeply into the basic makeup of the cosmos than its absence would; it has less meaning or value than a rock, or nothing at all.

Science is committed to the disenfranchisement of consciousness: why should living machines require individuation or privatization of consciousness, e.g., *self*-consciousness? In other words, even if there had to be higher-order functions to run metabolic programs, why do they also engender internal informational states and awaken first-order components? Why did representations, symbols, and internal contents go rogue? "[W]hy is the performance of … functions accompanied by experience…? Why doesn't all this information-processing go on 'in the dark,' free of any inner feel? … We know that conscious experience *does* arise…, but the very fact that it arises is the central mystery."[1] Indeed!

Why do hierarchies of monitoring systems turn on an interior light and take a look at themselves? And keep on looking (of course, that first look was addicting—you can't eat just one potato chip). Couldn't nature have settled on robotic versions of its worms, snakes, and squids? Could not octopi govern chains of subsystems in their stupor of many-tentacled motors?

Of what use is subjective experience anyway? Wasn't reality good enough by itself?

The High Court has recently ruled in favor of "inessentialism," a decree that any conscious activity in any cognitive system can also be performed just as well nonconsciously: "Consciousness did not have to evolve. It is conceivable that evolutionary processes could have worked to build creatures as efficient as we are, even ones more efficient and intelligent, without those creatures having experiences. Consciousness is not essential to highly evolved intelligent life."[2]

By this opinion, mindless automata, if assembled over millions of years by natural selection, despite their lack of introspection would be just as competent and resourceful as phenomenal creatures because they would *do* the same things, then transmit their genes. A relentlessly consuming robot lion or shark is as welcome in Darwin's universe as a minded one that is equally relentless. Hypothetically at least, the system could run fine without consciousness. Animals could prowl like aircraft flown by computers, adjusting for turbulence and conducting maneuvers on the basis of representing signals and contents and mediating them interactively with the outside world in input/output fashion *without actual experience*.

But just because this is theoretically arguable (and logically compelling) does not mean that it is the way that things happened. It clearly is not. And the issue is moot because there is only, so far as we know, reality *with* consciousness. Our heritage is legitimized *ex post facto*. We run world-tracking, memory-storing monitors in order (apparently) to compete with other programs running equivalent monitors, and the information-processing design of those monitors (in both cases) entails consciousness.

Some scientists file a sympathetic brief for the consciousness effect, though to them consciousness is just more efficient software running on more sophisticated hardware, another court-assigned client:

"Our species' success and our personal survival depend on successful commerce with the external world. This requires being in touch with the present state of the environment and drawing inductive inferences based on the past. In theory we could have evolved so as to succeed in being in touch and drawing the right inferences without phenomenal consciousness. But the fact is that *we did evolve with phenomenal consciousness.* Furthermore, phenomenal consciousness is hooked up to exactly the systems processing the information that, from the point of view of wise evolutionary design, it has most need to know about if it is to make a significant contribution to the life of the organism in which it occurs."[3] (My italics.)

In either opinion, pro or con, consciousness as a mode of subtle contemplation is thermodynamically superfluous. Only consciousness as a programming language has any operational or evolutionary value. The debate is not over whether the universe had any intrinsic interest in "knowing" or "being" or becoming a better universe, only over whether creatures with the *secondary* potential to know and be were better predators, more elusive prey, and more successful procreators. If they happened to develop knowledge and a sense of private awareness too and got interested in stuff like science, that was an accident, an unintended bonus—which (again) is why consciousness means "nothing" and can never advance its status.

Could there in fact be a universe without experience, without us (or anyone) to reify and validate it, to spare it from its own—if not loneliness and despair—vapidness? This is the *"hideous question someone is always asking/Egypt after Egypt."*[4] Hideous because it flirts with our already fragile existence and threatens to take our company and comfort too, while estranging us from "that with which we are most familiar...."[5] It is not survivable—or, if it bubbled up anyway, would lead only to wars and death camps.

We walk that fringe precariously enough, and not just because consciousness is under assault from physics and biology but because it is under assault by its own tenuous proclamation and inconstancy (and always has been).

The second law of thermodynamics, supervising energy flowing between objects of diverging temperatures, states that heat *only* travels to a colder region, degrading as random movement of particles toward disorder and becoming (ultimately) unobtainable energy. The measure of the internal state of a system at any point is its entropy: its relative order or disorder. In any system left alone, entropy has but two options: to remain constant or (most often) to increase. From the standpoint of creatures who rely on beating Murphy's Law, this is like things only being able to go from baddest to worst. The universe should by rights be cold and dark and vacant (or colder, bleaker, and emptying faster than you can say Jack Rabbit, or than Jack himself).

Although a machine, whether horse-drawn or petroleum-ignited, may do useful work for a while by changes in pressure, volume, and temperature, it *always* consumes more energy than it converts toward its goals. Living machines (eggs, tadpoles, and the like) are no exception. Usually some energy is squandered in friction or drag from the wearing out of parts (in locomotion or metabolism) but, even if it were not, the available mechanical energy to a machine cannot exceed that supplied to it from its heat source—the directed random motion of particles into it. In fact, it must always be less, even if, in the case of the absolute best machine somewhere in the universe, only fractionally so. Heat can never be converted into an equivalent amount of mechanical work; hence no creature has limitless (or even self-sustainable) energy to draw on; every motor runs itself down—in fact long before it is able to do anything much and certainly before it achieves something as outrageous as consciousness.

So how, despite this fiat, do privileged organisms arise from the dust? Why do their particles and parts not grind down and wear out or bolt the regime long before they engage in higher-order functions like stalking and evading stalkers?

There is no answer beyond the fact (again) that *we did evolve with phenomenal consciousness*, we are here and also know that we are here. The world is intrinsically comprehensible, and our existence is intelligible to itself. That's important. It's where meaning begins. Life means something to its disciples, even the simplest mudpuppy or snail.

Scientists assume that, if life and consciousness are operating at large in a strictly physical universe (as they seem to be), they must be somehow remitting their dues to entropy. So entropy is (yes, at first and even second glance) a rigid paymaster, a dictator of chaos, mayhem, and dead heat—the opposite of information. But information is also conveyed in pure molecular streams; hence entropy must not be *quite* the opposite of information.* There has to be another level at which higgledy-piggledy gets outfoxed. Bits of structure do find a way to steal from an unfettered romp of heat and gravity, generate and close metabolic systems, embed them interdependently in one another, then run their monitors in coordinated designs. Enter Germ World. Enter DNA.

Genetic programs, once written first as runes in alphabets derived from silt rills and wrinkles of shear force, dramatically augment the biosphere's capacity. Of course, DNA's cache and zipper must be primordially invented, must be fabricated and then maintained by entropic principles too.

So how does chaos turn into biological information and get organized in forms? Then how do those forms convert their informational basis into evolving chains of more complex forms?

The conveyance of information in binary molecular streams according to thermodynamics was demonstrated in a thought experiment by nineteenth-century physicist James Clerk Maxwell. He proposed a "wee creature" (later dubbed "Maxwell's demon") who perched inside a box of gas particles in

---

*I use the word "information" provisionally throughout this book. The *modus operandi* behind the term is less straightforward than it might seem. Meaningful differences can be traced back to their initial appearances in phenomena that can be studied by information-theoretical accounts of the *behavior* of physical substances in communicational settings—a mouthful. At the same time in complexity theory's context, "communication" delegates how complex phenomena appear, develop, refine, and evolve. In considering these together, I take information for granted and assume (whether as an implication of my own analysis or a good "spot" for communications theory) that intelligent ontological processes must be at the basis of everything. A quasi-scientific alternative, developed later in this volume, is a vast, information-producing-processing hyper-event, having no substantiality other than its computing.[6]

front of a valve that he manipulated in order to funnel faster, hotter molecules into a chamber at his right and slower, colder ones into another at his left. Thereby a gradient was introduced, information was created *ex nihilo,* and mechanical work was underwritten, like driving a piston (by heat) or turning the blades of a tiny windmill. This hypothetical creature permutes entropy into information (energy) in direct challenge to the second law of thermodynamics. At least in theory….

Maxwell was saying: we can reverse entropy. All we need is the equivalent of a little bobbin perched at an orifice between two chambers of equal temperature. If it sends high-energy molecules one way, halting them from returning, and directs low-energy molecules the other, likewise impeding their return, this baffle will separate hot from cold, using a nascent quiddity. Once primitive information is created, it interpolates itself; ultimately it can do all sorts of useful work (like nest-building and digestion), thereby increasing the free energy of nature. Eventually it generates networks that transform its energy into subjective information and relay it to even more intricate networks.

We don't need to conceive how elaborate and intricate designs and coalitions of terminals arrived on the Earth; all we need is a bare bobbin.

To get beyond gas particles in boxes to the Rock and Roll Hall of Fame by the bobbin route may be a stretch, but it is the kind of run the universe can go on when it shifts its momentum and density in a particular direction (because the universe has a lot of momentum). Somehow enough constituents of cellular sets, and then nervous systems, were tamed by enough wee creatures positioned strategically and funneling heat and gravity into information that they arrived at cobras and frogs. A link between the physical world (as molecules) and knowledge (as design) requires mucho hardware in between: helices, stem cells, precursory mesenchyme, invaginating membranes, sheaths of axons and dendrites (or their equivalents), not only out the gate but at each successive phase, converting entropy and energy into smart bits and managing to circumvent the Second Law over and over without actually breaking it. By this work-around—at the crossroads of matter, energy, and information—primordial motors convert stellar debris into living machines. In the process they fashion near limitless channels for the incursion of information. For some reason—and we will explore that mystery on different levels for the remainder of this book—the chemical reactions and potential chemical reactions making up the total system seem to want to flow toward worms and mosquitoes. This is no mere or minor anomaly.

The antecedents of information may have been furrows in evaporating puddles, mud in suspension, gas in gradients, serendipitous sieves; their successors now incorporate astounding amounts of lossless data compression in manifold algebras of bit flow. They recruit and transmogrify molecules from gross disturbances of gases and liquids into slim computers (or trillions of tiny cybernetic devices procreating, source-coding, and executing themselves through soil, air, and seas). Good morning! These apparatuses are the repositories of so much intelligent jabber that we have come to think of them as *transcendent of ordinary matter*. Even the single wayward beetle on a log is more complex by far than an iPad, for it is self-organizing, needs *no* programming, and was *never* programmed. A swarm of insects is a swarm of entropy-defying computers.

Glimpse back for a moment at Maxwell's demon as he tries to figure out which piece of dust to start with, which gas molecule to measure. He is always flummoxed. When he tries to measure them all at once, he has to put in a gargantuan amount of work only to find that there is no displacement from equilibrium. Although he is creating neurobiological information that can later be extracted, he is dissipating it at the same time and rate (or faster). There is nothing from which he can extract profit or even compensate his labor. He is ultimately so frustrated that he dissolves back into gas particles; in fact, he could not have existed in the first place.

But there he is! Eddies form out of entropic streams and sport in zones absurdly far from equilibrium. In these unbalanced regions, amazingly complex systems—specialized, low-entropy structures—congeal spontaneously. A cell is one such structure; the collection of cells we call an embryo is another. Energy pours into these vortices, deviating from equilibrium albeit at the expense of increased entropy in their surroundings. Signal-processing and process-collocating, they redefine the universe. Fresh molecules are then recruited from splashes, streams, and dust devils, conscripted as well into the same taut, functional designs. Invaginating disks with guts induce (and become) fissioning bubbles in floating, swimming colonies.

Once "alive," cell lattices go through the appearance (to our backward look) of operating like man-made appliances, stealing work from entropy and somehow paying their bills on time. While mechanically conserving energy, the biosphere converts quintillions of separate thermodynamic functions rooted in entropy into trillions of higher-order enthalpic topologies flaunting it. Energy seeps into their cells—chemical energy as glucose—and, in the same reaction, carbon dioxide and water are released. And that's the least of it.

As these bubbles outsmart their own algorithms, they transcend the belabored and grim attitudes of their source thermodynamics. Various bats, blowfish, and baboons shamelessly and nonchalantly promenading could care less about their dim prognoses. They are here, and it is nationtime. They breathe the air, propel gaily through water, and swing imperiously on the branches as though these were made expressly for them.

What a universe! Regions of RNA aggregate, dissolve, and send messengers along hieroglyphic creases, promulgating novel syntax and machines; proteins catalyze one another's alchemy. The fabric of life reaches out into raw heat dynamics and acquires more and more territory. This kind of stuff is off the charts.

Science's final "best guess" explanation for the emergence of consciousness out of life is (more or less) that bound molecular clusters assembled themselves under natural selection, closed entropy, took on agency, fed off each other's energy, and evolved into ever more efficient machines. Inside the uncertainty states kindled by their neurons, heat dispersal was converted into binary flows that became informational templates and ultimately got run up the ladder as synapses into relay stations aggregating to receive them. Axons and dendrites conducted these streams of cumulative data-flow into an emerging central ganglion, which then gave off the hallucination of a weightless, antientropic interior: the apparition of thought as a chemical mirage. Its hallucination bred better trackers, killers, and evaders—winners of natural selection's "Survival of the Fittest" Xbox. But again and even so, there is no explanation for why this grid should have become cognizant of its own being.

●

18

Entropy into data is a game that we will let rest for now, but it underlies the consciousness enigma. After all, to get from a barren universe to existentiality, you need a sturdy bridge across matter (chaos) to information (design), another from design to first-person portals. There are no shortcuts or alternate boulevards; the road trip is: matter to mind along series of hypothetical bridges.

From Maxwell's day, science has tried to ordain the validity of spans across unspannable gaps, those of information and self-organization under Darwinian selection. Though clever and serviceable, they are suspended from sky-hooks, grounded in thin air. Creatures crawling from equations based on jerry-rigged assumptions are less plausible by far than anything conceived by Rube Goldberg, but they do seem to hold with resilience, even insolence. They flow toward being. They eventually rap about what it means to be "something" without intrinsic substantiality. They even sponsor their representation at conferences on their own state of consciousness.

This is one level of work-around, but what about other levels, above and below metabolism? And how did the whole venture/scheme of working around entropy get started in the first place?

Behaviorism is fundamentally "a refusal to talk about consciousness"[7] or, more diplomatically, to do away with it by dismissing its primacy or by feigning to account for it otherwise. Behaviorists "believed, at least as a working hypothesis, that whatever might go on 'inside' an organism was irrelevant to a scientific explanation of that organism's behavior. They proposed to treat all organisms, including humans, as black boxes, hoping to discover objective laws relating the box's inputs (stimulus) to the box's behavior (response) without ever having to include the box's 'experiences' as a factor in their calculations," ignoring "what seems to be the most important feature of human life—what it feels like from inside...."[8] They preferred behavior dry—hold the mayo!

Functionalism, a close colleague, provides the tautological view "that mental states are definable [solely] in terms of their causal relations to sensory inputs, behavioral outputs, and other mental states."[9] Animals are machines run by sophisticated autopilots.

Functionalists and behaviorists "charge the defenders of phenomenal consciousness with believing in a fiction ... and [then] creating a philosophical problem out of it."[10] Behaviorism's first commandment is that "the world is made up of nothing over and above 'physical' elements, whatever their nature (waves, particles, etc.) might be"[11] and "persons have no more of an inside than particles."*[12] Complex relations of those physical elements ignite the phantasm "consciousness."

The attack on "consciousness as consciousness" has a long, distinguished pedigree. In a turn-of-the-century polemic entitled "Does Consciousness Exist?" pre-modern psychologist William James summarized the nineteenth-century's indictment in anticipation of a quick twentieth-century execution: "For twenty years past I have mistrusted 'consciousness' as an entity; for seven or eight years past I have suggested its non-existence to my students…. [Consciousness] is the name of a non-entity…. Those who cling to it are clinging to a mere echo, the faint rumor left behind by the disappearing soul…."[13]

That vanishing soul has been functionalism's scapegoat as well as its sacrificial lamb for many moons. Science requires a supporting material source for everything real, so mind cannot promulgate itself without a fully accredited sponsor in matter at every level, not just by information theory. If it exists, as it apparently does, it must map onto hard stuff, for in nature surface features are always "caused by … underlying microstructures"[14]—consciousness is generated proximally by the brain and nervous system; only dolts or Jesus freaks would consider otherwise. In fact the orthodoxy of neuroscience—and the coda for most of this volume—is that "conscious mental states [must literally] supervene on brain states";[15] they must reduce to a cerebral-hemisphere "topographic saliency map."[16] There can be no thoughts without thought-generating oatmeal, and we know where the urn of that is kept.

The brain has not always been a solitary candidate for the seat of consciousness. The ancient Greeks, for instance, considered the heart an auxiliary organ of mindedness, and diverse cultures have treated the belly, the larynx/throat, the kidneys, and other organs as containing or contributing to consciousness. Seventeenth-century French philosopher René Descartes

---

*The idealist counterargument is that particles can have no less of an inside than persons.

anointed the pineal gland as consciousness's gateway, the touchstone through which traffic from the senses converts from ciphered signals into the performance of mind.[17] He chose this tiny pinenut-shaped nodule because it is "one of the brain's few unpaired structures,"[18] is opportunely located "near the intersection of three major ventricles of the brain,"[19] and has no other apparent function. That kind of reasoning wouldn't work today.

In modernity the scavenger hunt has come down to the brain itself, "the place where it all comes together,"[20] the remaining issue being whether the *whole* brain is needed to confer the mantle of mind or whether a tipping point lies inside one or more of its modules.

The human brain consists of three concentrically arranged shells, their anatomical configuration developing atop the old invertebrate brain stem as if an exotic extension of the spinal cord—or the most elegantly coordinated and designed malignancy in the universe. As this growth protrudes, its spiraling motifs coil out of the apex of a relatively simple sensory ladder that defines the "mental" status of worms *et al.*

Homologous to the control ganglia of crustaceans, insects, and marine polyps, the mammal brain is a super-ganglion, assembling itself on the templates of less complex ganglia embedded in it and historically linked to one another in its genesis within neural design.

Imagine a hard drive of concentrated custard in which chips and circuits are so thoroughly miniaturized and mushed together that you can't find them even with the best electron microscope. The only hypothetical way to back-engineer a computer into a brain is to take an entire hard drive with all its programs, condense them into a tiny sector of the nucleus of one cell and then miniaturize that object to nano-scale without losing any of its content or functions, then repeat the process billions of times in billions of cells until all the components, logic gates, chips, casings, PCBs, resistors, polymers, batteries, wires, circuits, transistors, solder, copper, zinc, and silicon disappear into meat. Don't expect Apple to hold a press conference any time soon.

The integers of the brain comprise not a *more primitive* recipe for a computer but an *unimaginably more advanced one.* Then the virtual and ineffable equivalent of a metallic data processor is cloned billions of times and linked wirelessly inside self-replicating holograms of networks. Its units are so small that they are undetectable. In fact they may turn into quantum objects before they hit the limits of matter (see Chapter Eight).

> The human brain has been described as the most complex object in the universe. Certainly a lot goes on in this warm fist-sized ball of meat. Various exotic fluids pour, soak, and trickle through its channels and crevices. A veritable drugstore of chemical substances is synthesized there, put to strange uses, then broken down and recycled for further use. Legions of brain cells are born (in the early months of life), connect up to other cells, and carry out their mysterious cellular tasks in various neural communities before they die. Trillions of electric signals travel through the brain's wet electrical networks, each impulse inducing a weak electrical and magnetic field that races across the cranium at the speed of light. Torrents of electrically charged ions escape through suddenly opened cellular gates only to be captured one by one and sequestered again inside a brain cell....[21]

This is a lot of activity, but it is not close to an adequate precondition or qualification for mind. Where and how does consciousness arise in a burgeoning tuber? "With so much activity going on all at once, it is difficult to tell which brain functions are important, which irrelevant, for producing the phenomenon known as ordinary awareness."[22]

Within the context of such a self-arising object, "important" and "irrelevant" don't even mean what they usually do.

●

As the brain's outermost shell to which all lower cerebral centers run inputs for integration, the cortex is the most likely applicant for a center of consciousness: "a crumpled layer of gray matter the thickness of an orange peel with the consistency of tapioca pudding..., mak[ing] up seven-tenths of the entire nervous system, containing perhaps 8 billion nerve cells interconnected by almost 1 million miles of nerve fibers."[23] Since the cortex is the last-developing part of the brain in both primate evolution and the *Homo sapiens* fetus, it has been surmised that it somehow provided extra oomph or caucusing to jump-start mentation. Science-fiction cartoons of brainy aliens with bulging cortices honor this trope—an obligatory pre-requisite for "Thomas Edisons" and "Pablo Picassos."

Yet creatures without significant or *even any* cortical development have some form of awareness. The cortex might run philosophical and social extensions of consciousness, but it is not its fountainhead or baseline.

Subsidiary to the cortex, various subcortical structures form a second shell, while a centrally located thalamus comprises a third, older module that is nested inside the cortex and subcortex. As if created by a self-similar wave, these amalgamated structures are fissured down the middle into left and right hemispheres linked by soft cables, connectors, and a prominent band of tough tissue, the corpus callosum (literally "calloused body") consisting of 20 million of its own nerve fibers. I discuss the twin cerebral hemispheres and corpus callosum in Chapter Five.

In the "theater of consciousness" spirit of Descartes, theoreticians have identified "mind" in various cerebral components within these shells. Among the leading candidates are the anterior cingulate cortex (located along the corpus callosum and the apparent controller of blood pressure and heart rate as well as various cognitive, decision-making, and emotional processes, including empathy and reward anticipation), the basal ganglia (located below the cortex and controlling voluntary muscle movement and, in that capacity, adversely affected in Parkinson's disease), the claus-trum (a vertically curved sheet of cells adjacent to the basal ganglia with a somewhat mysterious role in harmonizing and timing inputs), the hip-pocampus (a paired twisted structure under the cortex involved in the for-mation of long-term memories and navigation as well as where fibers from diverse areas in the brain converge near gridlock), the recticular formation (a diffuse netlike pattern of neurons arranged like a stack of fuzzy poker chips and extending from the spine into the brain stem and the thalamus), the brain stem itself, and zones in the prefrontal lobes.

The cells of the reticular complex, making up as many as ninety-eight distinct clusters, have earned their status as an initiator of cognitive atten-tion and reasoning power. Their ascending influence has been described as wakefulness, readiness, a general "take note of." If the cortex introduces advanced reasoning functions of mammalian cognition and behavior, the reticular formation and brain stem, which are present too in simple, pre-cortical animals, might activate the flow of baseline conscious awareness itself: the cosmic flip from matter to mind.

Even though spread-out and undifferentiated, the reticular complex apparently activates "motor will," serving as a relay and guardian of the pathway to the cortex, also gating external signals to the outer brain during deep sleep and dreaming. By highlighting aspects of sensory flow, the reticular formation calls the brain's attention to what it should or should not be involved in—the biological state of an existential passage between nothingness and being. When the reticular cluster is stimulated, animals react more alertly. The descending influence of the reticular formation both facilitates and inhibits motor activity.[24]

Whereas the ape's primary attention is cortical, the lizard's is reticular—and the reptile came first. In many ways we are still lizards, governed by electrical bursts of fear and appetite in our brain stem. Everything loftier is a projection of those emotions onto the cortex's puppet theater.

None of these subcortical and thalamic structures has been granted the primacy of Descartes's pineal, but each of them has been elevated at one time or another to a threshold role in mind. Yet they all fail coronation. Experiments show that lesions to or removal of each or any of these sectors do not do away with the flow of consciousness. The loss of the hippocampus, for instance, only impairs learning. (It is one of the first structures to show the effects of Alzheimer's disease, but Alzheimer's sufferers are still conscious.) Lesions to the reticular formation will cause a victimized cat to go to sleep, and it is impossible to arouse the animal no matter how disruptive and discordant the attempt. At best, it will briefly stir. Yet it is still a dreaming cat. Damage to the frontal lobes, speech centers, and left brain lead to cognitive disorders but not to the total demise of consciousness. I read online that one graduate student in mathematics had "barely any cerebral cortex at all, thanks to water on the brain suffered in early childhood"[25]—an absence which, if true, speaks for itself *vis à vis* a the purported top-heavy installation of consciousness.*

---

*I know, this seems absurd, makes no sense at all, you can't believe most of what you read online, and the link is now gone—but my point is, anyway, that the brain has a complex nonlinear and indeterminate relationship to its own properties and functions.

Clearly we have no saliency map for the mind, whether it is considered an indivisible whole or a homeostatic merger of synaptic components—whether it has ignition at a tipping point or represents a coalescing aggregate of incremental streams. We cannot track "mind" back to *any* single moving parts of cerebral anatomy. "[I]t is not even clear how ... changes at the neural level relate to those at the psychological level."[26] Mind as we experience it neither resembles the brain nor has a cause-and-effect relationship to it, at least one that we can discern or conceive. There is no resemblance between the brain as a morphology and consciousness as an existential state, nor any way to connect the subjective components of consciousness with molecular structure: "Between these two, perceiving mind and the perceived world," there appears to be nothing in common.[27] Yet "together they make up the sum total for us; they are all we have."[28]

This is such a remarkable and redolent paradox that it will pop up again and again in different guises through all three volumes.

In order to correlate consciousness to the brain and solve the brain-mind problem under classical mechanics, one would have to pinpoint surges of cerebral activity that are prompted by relatively slight triggers and that contain "structural or logical information that is present in the associated conscious psychological surge ... a 'function' that generates a kind of beingness, namely inner light, which appears to be absent from the underlying physical matter";[29] in other words, we need to demonstrate consistent relationships between thought- or behavior-generation and sites of cerebral activation.

Physicist Nick Herbert characterizes the possible brain-mind link as a hypothetical "purple glow" emitted by an identifiable portion of the brain "in association with conscious experience.... [A]lthough the purple glow is physically identical to ordinary light, its production cannot be explained by normal electromagnetic mechanisms. In order to fit the purple glow phenomenon into physics, scientists must invoke a previously unsuspected new force linking light and the inner life...." Now all we need is that glow! Not so fast. "It goes without saying that experimental evidence for purple glow or any other special physical manifestation of awareness is nil."[30]

On the other hand, scientists can posit and even prove experimentally that the brain is a biological trigger for effects of consciousness and that by-products of consciousness are its activities' main outcome. These secondary surges of "purple" are not mirages or mere hypotheses. Neuropsychologists have confirmed links between schizophrenia and brain-lobe activity that are based also in genetic complexes. Oceanic, mystic moods among epileptics and migraineurs are uniform and reliably enough correlated with regions of brain activation to be declared substantially neurological in origin. On a grander scale, palaeontologists can track how gross cerebral capacity in the fossil record correlates consistently with the emergence of more and more intelligent lifestyles as indicated by the contexts of the bones.

But even if we grant all of those, proven and unproven, in full we still cannot hook or tag consciousness itself—the liqueur inside which events take place and which seems to precede them ontologically—to anatomy. "Everything that lies between these two terminal points [brain and mind] is unknown to us and, so far as we are aware, there is no direct relation between them."[31] They are separated by not only a gigantic gap of essence but a fundamental difference of nature. If that gap were to be sustained, it would be a major bummer for science.

The fact that "nothing in the functional neuroanatomy of the brain suggests ... a general meeting place [with mind]" may be a quandary for functionalists, but it is not the full dilemma, for positing such a center would not even hypothetically solve the problem. It "would merely be the first step in an infinite regress of homunculi. If all the tasks that Descartes assigned to the immaterial mind have to be taken over by a 'conscious' *sub*system, its own activity will either be systematically mysterious or de-composed into the activity of further subsystems that begin to duplicate the 'nonconscious' parts of the whole brain."[32]

This is critical. It means not only that nothing has been found but that there is no way to look, no direction in which to proceed, no place to bottom out. Back to mindless pudding—plenty of bridges, all sturdy, but leading nowhere except back into each other. There is no actionable gradient of subsystems, no "deeper headquarters, any inner sanctum, ar-rival at which is the necessary or sufficient condition for conscious experi-ence"[33]—no superchamber in Oz. A deferment to infinite subsystems leads in a circle, paralleling the regress of matter and energy to the Big Bang and ending at the same sterile periphrasis. I pick up this discussion of infinite regress in Chapter Six.

When you think about it, how can a mere locale or gruel of carbon compounds be the origin or operating center of something as ontologically fundamental as consciousness? Look at a brain; then consider consciousness. Where's the fit? It is like showing a can of tomato soup to a tortoise and expecting an intelligent response. A thinking bio-contraption would have to show tracking capacity well beyond a holographic laser or a Zeiss projector of the universe. Trying to fit consciousness into the brain is like trying to fit a city inside an acorn. Yet intricate cosmologies, autobiographies, and universes sprout from each of our private acorns. Even a lizard without much of a cerebral cortex navigates the landscape ingeniously and seems to know in some fashion that a cosmos and even itself exists.

Plus there is this: the relationship between consciousness and the brain does not have the same fundamental or categorical basis as that of other assignments of cause and effect in nature: gravity and mass, combustion and fire, bacteria and disease, fertilization and ontogenesis. Those are each and all erewhile tucked inside reality inside consciousness. They fit without pleat or spillage. "[T]he passage from the [magnetic] needle to the current, if not demonstrable, is conceivable, and ... we entertain no doubt as to the final mechanical solution of the problem. But the passage from the physics of the brain to the corresponding facts of consciousness is inconceivable as a result of mechanics...."[34]

The case for consciousness as an artifact of brain activity at any level or by any specific function inevitably runs into the rebuttal that the "mind" effect originates in a manner fundamentally different from anything else in the physical world *despite* its apparent partial congruence to particular neural patterns and activities of the brain or its interim obedience to thermodynamic constraints on information.

Consciousness just doesn't look like the brain. As a maverick entity refusing circumscription of context, it continues to expand illimitably, encompassing everything it is supposed to supervise and confine as either cause *or* effect. It literally puts *the city inside its acorn* again and again.

What else breaks such fundamental rules of space, position, and scale? How can mindedness even have a single designated cause or be a cause's singular effect? How can it be pinned down to *any* topology, geometry, or algebraic map?

Dualists consider that attempts to implant a city in an acorn are a waste of time, for mind and matter are two essentially different modes, "each with its own laws and manner of existence."[35] Mind exists for reasons independent of any possible reduction to body. Even scientists who disdain dualism are stuck ultimately with the dualists' conundrum:

"[T]his is what seems so shocking—a perfect science of the brain would still not lead to an ontological reduction of consciousness in the way that our present science can reduce heat, solidity, color, or sound…. [T]he irreducibility of consciousness is a primary reason why the mind-body problem seems so intractable. Dualists treat the irreducibility of consciousness as incontrovertible proof of the truth of dualism. Materialists insist that consciousness must be reduced to material reality, and that the price of denying the reducibility of consciousness would be the abandonment of our overall scientific world view."[36]

Yet abandon it they must, either overtly or surreptitiously. The enterprise of siphoning the mind into the brain as such is futile. By their own admission, scientists know next to nothing about mind. Its remainder appears in none of their equations, and it would seem that the world would behave pretty much the same without it. It is the upstart in the quagmire as well as the poltergeist in the machine. They are up that famous creek without a paddle unless they cut off their own heads. They may be able to track *particular aspects and trails of consciousness* by gerrymanders of brain anatomy, but they cannot strike the match itself—nor can they imagine how to strike it. That is neuroscience's Catch-22. Where in tarnation does a new quality just hop on board and kindle matter into a private interior theater?

Yet the lack of a transparent connection between consciousness and the brain (or between information and matter/energy) does not crimp heroic attempts to reduce intangible mindedness to ordinary matter. The house assumption is that it is only a matter of time before the manufacturer's guide is found. Here are some diverse and sundry briefs from science's side of a century-long discovery phase headed for a trial somewhere north of *Bleak House:*

- "I do not believe that we can ever specify what it is about the brain that is responsible for consciousness, but I am sure that whatever it is it is not inherently miraculous."[37] Innocent until proven guilty. Or is it guilty until proven either?

- "*[S]ome* property of the physical world explains [the phenomenon of consciousness]—though we cannot ever discover what that property is."[38] "Ever" is an awful long time for a property to go missing.

- "There must *be* some property of the brain that accounts non-magically for consciousness, since nothing in nature happens by magic...."[39] This is a matter of faith for not only science but contemporary logic: no magicians or magic allowed in this old house. Exactly why?

- "[T]he human body, like all living bodies, is a machine, all operations of which, sooner or later, [will] be explained on physical principles. I believe that we shall, sooner or later, arrive at a mechanical equivalent of consciousness just as we have arrived at a mechanical equivalent of heat."[40] This is a major ontological assumption posing as a mere physical extrapolation.

- "[T]here exists some property of the brain that accounts naturalistically for consciousness; we are cognitively closed with respect to that property; but there is no philosophical (as opposed to scientific) mind-body problem...."[41]

- "We do not completely understand how humans can be conscious, but neither do we understand how they can walk, run, climb trees or pole-vault. Nor, when one stands back from it all, is awareness intrinsically more mysterious than motor control."[42]

- "[P]henomenal qualities ... are physical properties of the brain that lie within the scope of current science [even though] ... they are conceptually irreducible..., [and] neither *a priori* imply, nor are implied by, physical-functional concepts.... [It is not the case that] they are irreducibly non-physical-functional facts, forever mysterious, [or] pose an intellectual problem different from other empirical problems, or require new conceptions of the physical."[43]

- "The explanatory gap, as it appears to me, is an epistemic or conceptual phenomenon, without metaphysical consequences.... The illusion is of *expected transparency:* a direct grasp of a property ought to reveal how it is internally constituted, and if it is not revealed as physically constituted, then it is not so. The mistake is that a direct grasp of essence ought to be a transparent grasp.... [T]his rather elementary and unremarkable conceptual fact [has been] blown up into a metaphysical problem that appears to require an extreme solution."[44]

His objection is that we are confusing disclosure of reality with reality in the raw, while the universe may be a little bit more torqued and underbellied than that. Nonetheless wouldn't you concur that consciousness is pretty damn extreme by comparison with everything else?

- "[N]o one ever said that consciousness was diaphanous, that we could see right through to its underlying structure and nature."[45]

- "We cannot plausibly take the arrival of life as a primitive brute fact, nor can we accept that life arose by some form of miraculous emergence. Rather, there must be some natural account of how life comes from matter, whether or not we can know it. Eschewing vitalism and the magic touch of God's finger, we rightly insist that it must be in virtue of some natural property of (organized) matter that parcels of it get to be alive.... Presumably there exist objective natural laws that somehow account for the upsurge of consciousness."[46] This is science's favorite, plaintive dirge: 'Get me out of here with my boots on!'

- "[T]he human mind may not conform to empiricist principles, but it must conform to *some* principles—and it is a substantive claim that these principles permit the solution of every problem we can formulate or sense."[47] Every one but this one! Isn't a single exception allowed, especially if it happens to be the elephant in the parlor? And if the elephant is the same one that had been squatting there for millennia before it was even noticed by Heraclitus, Aristotle, and Aquinas?

30

- "[T]he brain is apparently made of quite ordinary materials; there is not the slightest evidence that our body uses any supernatural processes to produce the phenomenon of mind. The brain as a bodily organ seems from the outside to be no more or less remarkable than the heart or lungs."[48] But let's not conflate the clothing with the horse.

- "We grant that there are spooky possible worlds where Materialism is false, but we insist that our actual world isn't one of them."[49] Cute, but I'd like to know where and under what conditions this guy would accept Planet Spooky.

- "[T]here is nothing eerie going on in the world when an event in my visual cortex causes me to have an experience of yellow—however much it *seems* to us that there is…. We confuse our own cognitive limitations with objective eeriness."[50]

- "At the present time the status of physicalism is similar to that which the hypothesis that matter is energy would have had if uttered by a pre-Socratic philosopher."[51] That depends on what you mean by "matter" and then what you mean by "energy." I come back to this proposition under the parasol of Claude Lévi-Strauss's totemism in Chapters Six and Seven.

- "[I]t would be a mistake to transform the fact that there is a limit to our capacity to describe things we experience into the supposition that there are absolutely indescribable properties in our experience."[52]

- "It *may* be that every property for which we can form a concept is such that *it* could never solve the mind-body problem…."[53] Failure to discover the mechanical cause of consciousness would then simply mean that the system is more complicated than our mind's capacity to decode its elements. There are "'problems,' which human minds are in principle equipped to solve, and 'mysteries,' which systematically elude our understanding…. Some of the more arcane aspects of cosmology and quantum theory might be thought to lie just within the bounds of human intelligibility."[54] Hold that thought for the "quantum brain" in Chapter Eight.

- "[M]onkeys are not able to comprehend the concept of an electron, and armadillos are not even up to the task of doing elementary arithmetic."[55]

- "[T]he character and three-dimensional distribution of complex stellar objects in a volume of interstellar space hundreds of light years on a side [is] made visually available to your unaided eyes from your own backyard, given only the right conceptual framework for grasping it, and observational practice in using that framework.... The cast of inner space is potentially the same."[56] I don't think that the percipience and resolution of external galaxies is actually the same as the percipience and resolution of the city inside its acorn, a city that includes galaxies as well as squirrels and is not located externally or, in fact, anywhere.

- "When the human science community was comparably ignorant of such matters as valence, electron shells, and so forth, natural philosophers could not imagine how you could explain the malleability of metals, the magnetizability of iron, and the rust resistance of gold, in terms of underlying components and their organization. Until the advent of molecular biology, many people thought it was unimaginable, and hence impossible, that to be a living thing could consist in a particular organization of 'dead' molecules. 'I cannot imagine,' said the vitalists, 'how you could get *life* out of *dead* stuff....'"[57] But molecular biology hasn't conferred life on dead molecules; it has benefited from the fact that they are already alive—close but no cigar!

- "'Fire' was used to classify not only burning wood but also activity on the sun and various stars (actually fusion), lightning (actually eclectically induced incandescence), the Northern lights (actually spectral emission), and fire-flies (actually phosphorescence). As we now understand matters, only some of these things involve oxidation, and some processes which do involve oxidation, namely rusting, tarnishing, and metabolism, are not on the 'Fire' list."[58] Quite so, but the multiplicity of an apparent phenomenon does not negate the actuality of each apparency.

- "Currently, we have several similar mysteries on our agenda: consciousness is one, but not the only one; the problem of the freedom of the will, and the interpretation of quantum mechanics are also equally mysterious. It is possible that we may never have a solution to the mystery of consciousness, but it would be unintelligent to give up the effort, because we have no reason to suppose that consciousness is inexplicable."[59] I would repeat: there *is* a reason for consciousness's exemption: unlike fire, it holds the focal point, agenda, and reality basis of everything else. Also, aren't free will and quantum mechanics merely subsets of consciousness?

- "Meanings *change* as science makes discoveries about what some macro phenomenon *is* in terms of its composition and the dynamics of the underlying structure.... In the case at hand, I am predicting that explanatory power, coherence, and economy will favor the hypothesis that awareness just *is* some pattern of activity in neurons. If I am wrong, it would not be because an introspectively-based intuition is immutable, but because science leads us in a different direction. If I am right, and certain patterns of brain activity *are* the reality behind the experience, this fact does not in and of itself change my experience and suddenly allow me (my brain) to view my brain as an MR scanner or a neurosurgeon might view it."[60]

This is a significant point that scientists assume but mostly forget to bring out in their own defense: identifying a mechanical source of consciousness need not diminish the power, enjoyment, or meaningfulness of it. To assign a purely physical provenience to consciousness does not make our experience of it any less incredible or wonderful or reduce our humanity.

Finally these arguments come down to four key points: 1. We don't have to identify the material substratum that causes consciousness (transparency is not required); we only have to confirm that consciousness is materially based, and we can do that by showing cause-and-effect relationships between aspects of consciousness and the activity of neurons, synapses, and particular components of the brain, the actual link to be delivered later; 2. Other scientific problems once considered intractable or irreconcilably dualist have been resolved satisfactorily without dualism (progress is its own default posture); 3. The limitations of the human brain are not the limitations of physical systems or of scientific explanation; 4. No discovery about the source of consciousness can eliminate or trivialize our experience of it, nor should it.

Let's say, for argument's sake, that someone discovers a consciousness process composed of certain neural structures. As I am headed towards a newsworthy public demonstration of the purple glow, I begin to wonder how I can be shown *anything* that will convince me. I don't understand how the presentation of a process *within* consciousness and simultaneous with the meanings arising from an experience of consciousness can be contained indefinitely inside any simple physical object or relationship of physical objects. What am I supposed to observe in the coupling: dialogue bubbles, dancing elves, celestial music? How is consciousness to be pinned down, captured, and shown to itself?

What happens when the physical parameter is then shut off or elided? Will something else not identifiable with any neurophysical structure disappear too (the "purple light")? If not, what does it do next? You can't have consciousness alone in a box without a conscious entity, can you? If, as is more likely, the effect disappears, then the riddle will remain. It will be no different or less enigmatic than death by any other means.

Let us for now concede to functionalism the relation between the brain as a structure and consciousness as not only its main effect but its evolutionary function. Instead let's focus on the prior problem of what consciousness flat-out is. After all, it's a bit weird to look for a purple matrix or glow if we don't even know what the thing glowing is supposed to be. As per fire (above), "it is not even clear that the word 'consciousness' stands for just one sort of entity, quality, process, or whatever."[61] Consciousness may be a complex of related phenomena converging on a single-seeming apparency.

While, according to most neuroscientists, "the existence of consciousness can be explained by the causal interaction between elements of the brain at the micro level," those same scientists agree that "consciousness cannot itself be deduced or calculated from the sheer physical structure of the neurons *without some additional account of the causal relations between them.*"[62] That's basic to the point of axiomatic: consciousness is not

34

a "thing" but a process or confluence of processes. We can't find it perhaps because there's nothing discrete to find.

A simplified model of causal interaction goes something like this (and it's a long, serpentine journey with many changes of vehicle and discontinuous modes of transportation along the way): Billions of ectodermal (outer-layer) cells in multicellular organisms specialize into nerve cells (neurons) that also penetrate ectoderm and mesodermal and endodermal tissues. Neurons acquire distended membranes (axons and dendrites) that do not quite touch, so they communicate by dispatching electrical signals across the brief watery gaps between each other (called synapses). These transmissions either activate (discharge) or inhibit (repress) the neighboring neuron. Once activated, a neuron sends the message to the next clone of itself, and so on.

The filaments' signals, however, are so weak that they can't cross even this tiny moat, so the synapse serves instead as a clearinghouse, converting one kind of currency (electricity) into another (chemistry). The filament (axon) of a transmitting neuron releases a neurotransmitter into the gap, which crosses the synapse and attaches itself to the receptors of the recipient neuron (usually a dendrite), from where it travels into the cell body. Among dozens of different preformed packets of chemicals for neurotransmission are classes meditated by dopamine and headquartered in the midbrain and classes mediated by serotonin with their headquarters in the brain stem.

But it is not nearly that simple a system. Any two adjoining neurons have multiple synapses between them, which produce a composite signal that must rise above a certain threshold voltage for the axon of one neuron to fire across the synapse of another and cause a neural event like a muscle contraction. Then the first neuron reabsorbs the chemical, or the neurotransmitter is metabolized by enzymes so that its prior standing state is reestablished.[63]

As these discharges occur continuously among the neurons of an organism wherever they are located in its body, the cumulative flow of information cascading upward toward cerebralization and cognition is inconceivable. Each eye alone transmits "on the order of 100 million bits per second of information along the optic nerve through certain midbrain relay centers into the occipital lobe located at the back of the brain. From every square inch of the skin, the largest organ of the body, tactile messages pour into and up the spinal cord, then pass through the brain stem into the thalamus, where they are relayed to the sensory cortex.... The quantity of information streaming into the sensory cortex may be as large as 10 million bits per second. From the ears, along the acoustic nerve, about 30,000 bits per second of auditory information [pass] into the brain stem, where it is relayed to the primary auditory cortex located near the border of the parietal and temporal lobes."[64] Smell, taste, and other channels provide less pure information but obviously contribute to the overall flow.

"[I]f we assume that each neural synapse corresponds to 1 bit of information, then the cerebral cortex considered as a communication channel is capable of dealing with about 10 trillion bits per second—100 billion synapses firing at a maximum rate of 100 per second. Viewed strictly as an unconscious data processing machine, the human sensory/motor system consists of relatively modest input and output data flows linked by an enormous amount of computational power. (The human brain by this estimate is 10,000 times faster than the largest supercomputers.) Of course, almost all of this neural data processing goes on below the level of awareness."[65]

We have jumped quickly from simple chemical reactions to a complex, multi-tiered phenomenon, e.g., from a raindrop to rainfall to rivers and oceans. This is why, as we shall soon see, scientists tend to lay their unfinished theories of consciousness on the sacred altar of quantification. We are a "reign of quantity" civilization, an idolatry for which spiritual philosopher René Guénon indicted his entire century.

The stream of consciousness (which disburses such things as "how to track caribou" and "how to split atoms") is at base a "yes" or "no" (activation or inhibition) toggle rather than a steady-state mist (but hold that thought). Insofar as information in digital systems of this sort obeys discrete marks and values, the binary basis of neuronal firing might seem to confine our net neurological capacity to bland pixels of yeas and nays (like the dots of Claude Monet's abstract impressionist lilies).

Yet there are no such inherent limits on the nuancing of gestalts in biological consciousness. A thought or image is not just the result of a series of neurons firing (or not firing) in long on/off chains; it is the complex outcome of numerous firings and degrees of composite firing of many related neurons during which each one changes its frequency along a continuum, increasing or decreasing its strength and thereby qualifying its message. Meanwhile other neurons in its vicinity (as well as elsewhere) participate simultaneously in exciting or inhibiting signals along their own continua, and each of them have multiple synapses with one another too, which operate at different degrees of distance or tightness. There are so many neurons and signals firing at any one site that choppy digital information effectively turns into a smooth analogue flow. Of course, it is still a Monet; there are no other known pathways to landscapes or meanings. What we get, though, is a more nuanced palette. How it nuances its nuances is the forest of symbols wherein we dwell.

If the brain were a computer, it would be a hybrid of digital (binary), analogue (gradational), and parallel (manifold simultaneous) processors. But, as we shall see in subsequent chapters, the brain is not even a machine. For all its garish meat and explicit wiring, it is a mystery object of uncertain origin, the mechanism of which defies analysis or description. It only "models" as if it were an Apple product because we have invented and assembled computational machines on its prototype and through its operations. Computers are not plans for a brain; they are primitive facsimiles or subsidiaries—by-products of already existing activity.

Once inside neural networks, consciousness operates according to ordinary rules of cybernetics, communications theory, neurolinguistics, and the like. The dots and strands in consciousness's fabric (attention, memory, and perception) are no longer points, threads, or projections like motion pictures; they are binding products—collectivized synchronies of synapses linking back and forth until they come together in multidimensional messages pulsating into and through a central node. The brain turns digital, analogue, and parallel signals together into images and thoughts. The unity of the perceptual and cognitive field as against its fragmented and staccato basis and binary status must be established moment to moment by serial mechanisms—continuous firings of neurons that the brain binds together across different sensory modalities.[66] I discuss this loom of consciousness in the following three chapters.

In addition, "conscious experience involves neuronal groups locking into synchronous or semisynchronous oscillatory patterns in the 40-hertz range ... with cross-cueing keeping [them] smooth and predictable. Getting the right neuronal groups into synchrony and keeping them in synchrony for a short time may be sufficient for the experience."[67] I'd say "necessary" rather than "sufficient"; they may be sufficient for *the cohesiveness of the experience,* for they ground an electromagnetic trope in a biological system.

As the strength of various synapses is increased or decreased, neural cells become both pre- and post-synaptic, and unified records of these events are logged in the short-term memory of the brain. "[T]he neural system subserving a sensory modality provides the *content* of an experience, and the non-specific system consisting of resonating activity in the thalamus and cortex provides the *temporal binding* of such content into a single cognitive experience ... *'It is the dialogue between the thalamus and the cortex that generates subjectivity.'*"[68] Subjectivity then is little more than self-referentiality: memory.

That too is three guys' extrapolation: a shot in the dark and a mega-oversimplified one at that. By this rendition, neurotransmitters are effectively the medium of subjective experience. Neurons alone activate synapses, synapses flow through firing stations, composite neural traffic self-censors at benchmark thresholds as it integrates upstream, neural nets accrete flows into ganglia, and ganglia coalesce it in cerebral fields in some fashion—these systems evolve autocatalytically, chemico-mechanically, nonergodically. The experience of mind is generated, in effect, by bioelectrical mucilage, neural salvos, parallel reciprocal fibers binding independently originating grids and causing them to activate as hallucinogenic mirages. But how? Why?

All scientific vindications are, in principle, reductively quantitative: the more neurons, neurotransmitters, and cortical mitoses there are, the more alert the organism should be—the more synapses, the more mind—even though (remember) maps of the brain and nervous system in no way suggest a map of consciousness, and the ingredients for mind do not reside componentially or collectively in tissues.

It is unclear why any number—even a near infinite set—of empty shells, spools of wire, continuous serial mechanisms, binding products, neurons, cross-cueing, pool balls, or electrified chambers melded and strung together in gradients should ignite consciousness *ex nihilo*. A television, yes. A voice synthesizer, an artificial parrot, why not? A locomotive, an iBook, okay. A chess-playing robot, yeah. But awareness itself—no way!

Cells and neurons are, for all intents and purposes, no better suited *than ping-pong balls* for fashioning thoughts. Nothing times nothing is still nothing; nothing to the power of nothing does not increase. Carbon or silicon circuits are of the same order (and ontological status) as plastic-enclosed air. There is no phenomenological universe inside a ping-pong ball or a dendrite (or a brain). Consciousness and identity do not arise from or accrue through neural infrastructures, nor is the city whooshed inside its acorn, no matter how densely consolidated the carbon-based wiring, how deep the binding and cross-cueing—no matter how much time is allowed for synapses to flirt with their own algorithms and analogues and find their golem.

The paradox of a conscious mind rooted in matter continues to be fêted, poked at, and venerated by flabbergasted scientists and philosophers of science as they struggle to tame it:

- "How can technicolor phenomenology arise from soggy gray matter?"[69]

- "How could anything material have conscious states? How could anything material have semantical qualities? How could anything material be rational?"[70] "[H]ow does any physical mechanism give rise to any kind of phenomenal experience?"[71] "How could the aggregation of millions of individually insentient neurons generate subjective consciousness?"[72]

- "Neural transmissions just seem like the wrong kind of materials with which to bring consciousness into the world, but it appears that in some way they perform this mysterious feat."[73]

- "It is not that we possess bad, partial, or flawed theories of the inner life. We have no theories at all, even bad ones. Instead we possess only vague fantasies, philosophical hunches, and speculative, untenable guesses…. Kurt Vonnegut's fictional Tralfamadoreans *(Sires of Titan)* accurately assessed the current state of human awareness research: 'They could not name even one of the fifty-one portals of the soul.'"[74]

- "[U]nlike solidity, consciousness cannot be redefined in terms of an underlying microstructure.... *[T]he appearance is the reality.*"[75] "Appearance as reality (end of story)" is an axiomatic feature of consciousness that could easily slip by. No matter how refined and impartial our devices become, there is no way to get at reality apart from our consciousness of it and no machine that is not created out of and operating as consciousness itself. If the appearance is the reality, how do we get out of the loop or turn over the broken record? How do we objectify or externalize the purple glow?

- "What makes the bodily organ we call the brain so radically different from other bodily organs, say the kidneys ... body parts without a trace of consciousness?"[76] I am not so sure of the latter's totally unminded status (see Chapter Five).

- "[T]he supervenience of the mental on the physical is marked by the fact that physical states are causally sufficient but not causally necessary for mental states...."[77] Take away those ping-pong balls and all hell breaks loose.

- "There the brain is, an object of perception, laid out in space, containing spatially distributed processes; but consciousness defies explanation in such terms. Consciousness does not seem made up out of smaller spatial processes."[78]

- Even though certain visual regions of the cortex light up, respond to retinal input, it "is, strictly speaking, not an isomorphism of any order. It is simply a topographical relation of a very crude sort, preserving only order, but not size and not really shape.... [N]o part of the occipital cortex turns red upon perceiving a red object!"[79] Nowhere in the brain is there an abstract thought of a triangle—or relativity theory or romantic love. Thoughts are not "square any more than they are literally blue."[80]

- "[T]he relationship between information processing explanations and accounts of experience [is] analogous to two different maps of the *same* part of the world; neither map need be totally deducible from the other."[81]

- "[C]onsciousness is not perceivable by directing the senses onto the brain ... [in fact consciousness] is not perceivable by directing the senses anywhere...."[82]

- "I hereby invite you to try to conceive of a perceptible property of the brain that might allay the feeling of mystery that attends our contemplation of the mind-body link: I do not think that you will be able to do it. It's like trying to conceive of a perceptible property of a rock that would render it perspicuous that the rock was conscious."[83]

- "Would anyone deny that Tolstoy's 'The Death of Iván Ilych' is paper and ink or that Giorgione's *The Tempest* is canvas and paint? And yet, how far does that get us in expressing what these works are? Am I wrong in feeling that 'a vast assembly of nerve cells' is an inadequate description of *me* or that those words fail to answer the question, What happened to me? Am I looking for a narrative, a confabulation, to interpret a debility that is no more and no less than synaptic wiring and firing?"[84]

- "Were our minds and senses so expanded, strengthened, and illuminated, as to enable us to see and feel the very molecules of the brain; were we capable of following all their motions, all their groupings, all their electrical discharges, if such there be; and were we intimately acquainted with the corresponding states of thought and feeling, we should be as far as ever from the solution of the problem, 'how are these physical processes connected with the facts of consciousness?' The chasm between the two classes of phenomena would still remain intellectually impassable."[85]

- "It is surely a striking fact that the microprocesses that have been discovered in the brain by the usual methods seem no nearer to consciousness than the gross properties of the brain open to casual inspection. Neither do the more abstract 'holistic' features of brain function seem to be on the right lines to tell us the nature of consciousness. The deeper science probes into the brain the more remote it seems to get from consciousness.... Advanced neuropsychological theory only seems to deepen the mystery."[86]

"... only seems to deepen the mystery...."

Though we try to overwhelm the paradox of consciousness with rubrics of neuroscience, we cannot resolve how cognition could be dependent on an organ that is itself dependent on cognition for its own cognizance. It is possible that the problem of consciousness is insoluble in minds such as ours, e.g., conscious minds that have no access to information or meaning apart from their conscious states. How can the referent also be the referrer *and* the referee? Plus, how could consciousness be hardwired if the consciousness-bearing parameters of mind are not "wires"? How do cells distend into membranes, membranes into synapses, synapses into synopses and syntaxes, syntaxes into syncretisms—abstraction into concretion? How does the universe get inside an acorn and illuminate infinite spaces of astronomy, philosophy, mythology, mathematics, and poetry there?

The whole affair collapses under the weight of its own absurdities, tautologies, and gerrymanders. The only way around them is to talk as fast as if imitating Abbot and Costello performing "Who's on First?" except with neurons, synapses, and brain anatomy.

We are stuck with existing only insofar as we *don't*, e.g., as series of monitors talking to other monitors and their secondary representations and control hierarchies, while at the same time we make a case for our nonexistence (e.g., the "illusion" of consciousness) from the only place where such a case can be made: ego-based awareness. For all the operational expedience of neural maps and territories, introspection remains an ineluctable factos. It is where the myth of materialism confronts its own face in the mirror and turns away, always.

CHAPTER THREE

Consciousness: Everything and Nothing

We cannot explain consciousness—how it got here, how it made a nest in matter, what its meanings mean.

Though science kicks the tires on consciousness, more or less, it does so conditionally and peremptorily. It doesn't fetter or source it—but then it can barely get from matter to life without being singed down to its last short hairs. And prior to that, astrophysics squeezes nothing at all into matter by the *deus ex machina* of the Big Bang, a drupe spraying the juice of nowhere everywhere. As we saw in the previous chapter, binding energy and agency, let alone subtle awareness, inside membranes inside zooids leaves us with, at best, a prodigal contraption of gum, baling wire, string, and lint.

There is nothing that consciousness is "like." Not even close. Nothing else cores reality outside-in, makes continuous replicas of it, transferring sound, color, feeling, taste, smell, etc., while converting it into a field with a private framework of experiences so implicit, immediate, and convincing that the embedded-in individual becomes the hub of all identities and meanings. Each creature feels unique, complete, at the center of its own reality—not only that, but as though being alive and conscious is the only game in town or that ever was in town: the center ring of all creation:

- "This consciousness that is myself of selves, that is everything, and yet nothing at all—what is it?"[1]

- "The phenomenon of consciousness does not have clear-cut boundaries, and its complex structure does not admit any easy formulations."[2]

- "[H]ow are we to understand our insistence on the existence of consciousness given that we cannot find a place for it in any reasonable theory of the world?"[3]

- "We do not yet have the final answers to any of the questions of cosmology and particle physics, molecular genetics, and evolutionary theory, but we do know how to think about them. The mysteries haven't vanished, but they have been tamed.... With consciousness, however, we are still in a terrible muddle. Consciousness stands alone today as a topic that often leaves even the most sophisticated thinkers tongue-tied and confused."[4]

- "[T]he relation between conscious experience and subvening physical state ... is absolutely unique in nature."[5] And then some! Meteors and clouds do not run private cinemas.

- "Conscious mental states and processes have a special feature not possessed by other natural phenomena, namely subjectivity. It is this feature of consciousness that makes its study so recalcitrant to the conventional methods of biological and psychological research, and most puzzling to philosophical analysis...."[6]

- "Nobody has the slightest idea how anything material could be conscious. Nobody even knows what it would be like to have the slightest idea about how anything material could be conscious."[7] Putting "anything material" in a position to recognize both its subjective content and objective context can't be done: "[W]e have at present no conception of what an explanation of the physical nature of a mental phenomenon would be."[8]

Let's pause here a moment. You have to get this "no conception of what an explanation ... would be" clearly and in the plenary way in which it is stated. *We just don't know. We don't have a clue.* We can lead the horse to water (the brain), but we can't get it to take even a single sip. And that's a lot of work for nothing, to drag the horse there if it's not going to do anything.

From the hint of a mirage, an entire universe (equal to the other one) springs into existence from sea to shining sea, galaxy to illuminated galaxy and, as if that weren't enough, that phantom universe is the only thing that allows us to be in the official one. Meanwhile we can't even get one spark or sparkler to connect them or turn *itself* inside-out.

Remember, everything else in the universe (except consciousness) is not only tangible but subject to segregation and inspection. Every sample can be submitted to microscopic and/or spectrographic analysis. A scraping from an alien spacecraft's crash-site can be flown to a forensics lab and bombarded until it divulges its componential bar graph. Even a spoor from a transdimensional UFO can be analyzed by liquid-chromatograph mass spectrometry. The light of the remotest nebula is reducible to laboratory-familiar wavelengths and evidence. All stuff is elemental, made of atoms. Atoms have common protons, electrons, nuclei, etc., while every force affecting them resides somewhere in the ample thermodynamic-gravitational spectrum. These manifestations and effects are the same throughout the universe.

Not consciousness. Consciousness has no elemental band or spectral shift; no atoms, no electrons. It is not thermodynamic or gravitational either.

- "Turning a tool on itself may be as futile as trying to soar off the ground by a tug at one's own bootstraps. Perhaps we become confused because whenever we are thinking about consciousness, we are surrounded by it, and can only imagine what consciousness is *not*. The fish, someone has said, will be the last to discover water."[9] One can't take a picture of a camera with that same camera or look through a telescope at itself.

- "[N]othing *in the visual field* allows you to infer that it was seen by an eye."[10] All that we are left with is an infinite regress of internal mind's eyes watching their perceptions of each other's renditions, and dead-reckoning that not only are their images sourced in a visioning device but its collocation of sensory data (such that it becomes sight to an eagle probing the lower zenith for moles) is set in another, higher-order device that turns it inside-out while making copies of it in its own medium.[11]

You find yourself imbibing an aqueous reflection in your five-plus senses on a monitor that is both inside and outside—around you, through you, wrapped gravitylessly around every particle of you, generating interest and urgency while activating instinctual behaviors. It also makes you *you* in a real-life theater with your own exclusive, advancing-while-receding show.

- "No matter how large the frontal lobes of our biological descendants may become, they will still be stumped by the mind-body problem, so long as they form their (empirical) concepts on the basis of perception and introspection."[12]

- "[C]onsciousness] is impossible to define in terms that are intelligible without a grasp of what consciousness [already] means.... It is impossible to specify what it is, what it does, or why it evolved. Nothing worth reading has been written about it."[13]

Heavy-duty stuff! Consciousness alone in the universe is its own justification, its own imperative, its own rationale. It has a monopoly on the basis for its own authentication: a strike is not a strike until the umpire calls it so. And there is no ump.

After all, 'consciousness as opposed to what?' Everything we know and do begins as consciousness, nothing exists for us outside of consciousness; yet that consciousness has no discernible root, no frame, no explanation. It creates itself instantaneously without a proximal mechanism (forget the brain; it is merely a platform and conduit of consciousness).

In trying to give consciousness a context, we might say that it is a mode of embedding reality inside another medium, while representing its objects in space apprehensibly to itself. Despite its lack of access to actual landscapes or any capacity for extrinsic molecular transport—without any mass, momentum, or leverage of its own—consciousness miraculously sticks itself onto reality in order to capture its moving replica. This tangible annexation is immeasurably more pervasive than 3-D, Dolby, Smell-O-Rama, or Philip K. Dick's most extravagant "virtual reality" parlor.

While imprinting a fluctuating point-by-point facsimile of reality, consciousness scans, maps, simulates, retrieves, and binds itself in semi-synchronous analogues of synapses. We become aware of the world as a series of projections that wrap their epitomization around our sensoria in such a way that they exude it from within. As a proxy syrup flows into us, consciousness touches and tastes and even swallows its massless reflections, translating things and events into aliases and archiving them in its soft mirror without budging or affecting the objects themselves. It literally seizes reality without altering it.

What the bulky ganglia and lobes of the brain can't accomplish, their minion does nimbly, adroitly, and matter-of-factly. It shoots palps into every blackbird and pie. These appendages *have no demonstrable existence;* yet they capture and assimilate the world and matter indiscriminately. "Phenomenologically, we feel that the mind 'lays hold' of things out there, mentally 'grasps' them, but we have no physical model for what this might consist in…."[14] It is like cartoon bubbles extending from Elmer Fudd and Bugs Bunny and eating the frame. "[W]hen I see a stick, it should not be supposed that certain 'intentional forms' fly off the stick towards the eye."[15] What allows [the mind] to "reach out to objects of experience … to *arc out* into the world"?[16] "What determines the direction of its throw?"[17]

"[T]he brain cannot incorporate the external in the way the mind can. We have no physical model of how consciousness can lay hold of the physical world in the peculiar way it does."[18] It "can reach across vast stretches of space and time; it has a universality that transcends the particular actions and experiences of speakers."[19] "[C]onsciousness has the power to 'lasso' things…;" it "seems to extend an invisible hand into the world it represents…. [H]ow on earth could my *brain* make that possible? No ethereal prehensile organ extends from my skull!"[20]

And it is not just you, my friend. Mental intrusions penetrate the world from all its myriad creatures: individual "Earths," each at its own scale and vantage, millions of them passing one another on an average city street (those of birds, squirrels, and insects too).

Nor is thought *ever* turned into a hard or durable representation: "[C]onsciousness … appropriates the objective while holding itself aloft from it; it takes the physical in but refuses to be ruled by it."[21] It is astonishing for its cohesion, speed, instantaneity, and thoroughness:

"[W]e monitor so much at any given time and achieve so high a degree of integration and control…. We have elaborate and remarkably non-gappy visual models of our environment; we have our four other main sense modalities, which supplement the blooming, bursting phenomenological garden already furnished by vision; we have proprioception of various sorts that orient us within our surroundings, and (most important) we have almost complete freedom of attention within our private worlds; we can at will attend to virtually any representational aspect of any of our sensations that we choose."[22]

Though we don't scan every contour and granule, the model we reconstruct is good enough to create absolutely convincing holography in the meadows and lakes of our minds.

Despite this virtual transfer, inside consciousness there are no direct links to meaning; everything is made up of internal messaging: precedents and paradigms, signals and networks. Like life itself, the "consciousness system" organized itself over at least three billion years, so carries three billion years' worth of chips, transistors, wiring, source codes, data, internal references, and hyperlinks*—all wrapped around its own emerging semantic components and operational properties. It is mediated, species by species, through nets and tiers of deep proto-symbolic matrices: "recognitional concepts … formed against a conceptual background."[23] Thought is built always on former thought, new behavior on old behavior. Prior to designation, content is on the level of the buzzing of insects that fills this planet unperceived.

System noise is turned into reality or, more properly, runs back and forth over its own congeries so many times through so many different output channels that it finally "makes sense." Yet it was always making sense out of the sense of prior systems—from initially arbitrary chains of signals, synapses, and static converting the world into its medium, to refined cognitive fields as the system evolved. The basic operation was as seamless for an antediluvian worm or trilobite as it is today for a possum or leopard, and just as passionately real. Consciousness was in place long before humans began witnessing it.

Ultimately consciousness is sourced and furbished in what it is conscious of: the installation of reality. Reality alone makes consciousness real and

---

*It starts with intracellular endosymbiosis and the possible quantum states of protoneural microfilaments and microtubules (see Chapters Seven and Eight).

gives it a rationale. Without phenomena (and phenomenology), mind would not exist: Without a world of which to be conscious and a way to understand our consciousness of it, a sensory flow and computing mechanism (no matter how many gigabytes it holds) would be meaningless. If there were only reality *without* consciousness, no tree would ever fall in any forest.

"[W]hat makes a mental state conscious cannot be our consciousness of it. If we have conscious experiences, beliefs, desires, and fears, it cannot be our introspective awareness of them that makes them conscious."[24] That (again) would be like looking through a clear glass at nothing. "[W]hat makes [things] conscious is the way they make us conscious of something else—the world we live in and (in proprioception) the condition of our own bodies."[25]

Though consciousness may have other exceptional qualities, its defining performance is as a copula between organisms and their external world. It is, "in effect, a mysterious medium in which something relatively mundane is contingently embedded."[26] But how mundane can the world be if the embedding which brings it into existence is fathomless and phantasmagoric? How mundane is reality if consciousness can get inside it and take it inside itself? Operating inside-out or outside-in, consciousness sticks *two* mundane realms (neural flow and reality) together in a decidedly nonmundane way.

Biophysical consciousness literally invents biophysical reality. A dark pool of light is sucked into each organism and creates itself while creating an existential organism. The match that lights its flame is immediate and yet predicated on an archaic moment when it hovered between inertia and provisional sentience, sputtering untold times before catching. That first ember is still burning, transferring cell to cell, instilling a quality not previously in nature: a city-bearing acorn, *another* universe reflecting trillions of times through itself while reaching into its own spatiality and inverting it.

The plot continues to thicken: consciousness has no "central viewing room" with a "single monitor to do the viewing."[27] It is neither a multisensory cinema projected out of synapses nor an output scroll of neurons; it is not even a continuous, nonlinear performance of neural states. It consists of multiple, intersecting levels of information, interrelation, interpretation, and reinterpretation in the context of a continuous flow of fresh information. But there is no one watching the movie, in fact no "immaterial audience in the brain ... no undischargeable homunculus ... no such locus at all ... no single Boss Unit or even CPU [central processing unit] ... to serve as chief executive" (unless you want to elevate the reticular formation to that status). There is likewise no internal or executive scanner or audience in the theater, no "turnstile of consciousness nor one central inner stage on which the contents of consciousness are displayed in one fixed temporal order...."[28]

Tiers of representation alone elevate neurons above the plateau of input/output machines, but even those are evolving genealogies of analogues and consociations. "*Thinking is spelling* (and transformations thereof),"[29] codes and subcodes in queues of glosses.

Each experience, even if specified in the brain, does not even cognitively exist until it is represented in the context of other representations—which are equally transitional to yet other rudiments and figments:

> Part of what makes the world a world ... is the density of
> relations and connections among the objects which are simul-
> taneously the *constituents* of which the world is composed and
> also *constituted* as the objects that they are by their relations
> within that world. Put another way, the (experienced) objec-
> tivity of phenomenal objects consists in part of the enormous
> diversity of the perspectives from which they are accessible
> within the representation. Any phenomenal object is delim-
> ited and defined in part by the relations they bear to it. It is
> in part the density of these interdefining relations that gives
> phenomenal objects their "thickness," their objectivity. The
> fact that phenomenal representation of objects involves such
> dense and interdependent relations might help to account for
> the high degree of semantic transparency; any phenomenal
> representation of an object would of necessity also be a repre-
> sentation of its myriad relations within its world....[30]

The "world representation" is a mirage generated by a bundle of cables wrapped around themselves. At another level it is their generalized congruence and vestigial remainder: the sheer thickness of the neural system's representations cancels out the fluctuations and interference patterns of its own individual threads (and broken threads), turning subject into object and rinsing its quagmire until it functions as a pellucid glass. We see through the glass into something else, even though it is anything but a glass (more like a seething welter of snakes too close to the portal to notice). Plus, consciousness's semantic properties are *also* diaphanous. In fact, consciousness requires semantic transparency—"the extent to which [it] can be said to understand the content of the internal symbols or representations on which it operates." This explanation then interpolates: "The basic idea is that the greater the extent to which the system's use of a representation reflects an understanding of its content the more the representation can be said to be *semantically transparent* to the system. The relevant behavior by the system will consist in parts of behavior connecting the symbols with the outside world through input and output relations, but will also include behaviors relating symbols to each other in ways sensitive to their content."[31] The pain in one's leg "is the intentional object of your experience; it is not the experience itself. The content of your experience is that there is a disturbance of some sort in your right leg. The intentional object of the experience is located in your right leg. The experience itself is not located in your right leg. If the experience is anywhere specific, it is in your brain."[32] Maybe. But it is located also in language, belief syndromes, instincts, habits, and personal history. You feel the pain through all of these filters as an interpretation, not a diaphanous fact. Somewhere in the convergence upstream, the interpretation gets inside the acorn: ouch!

So reality is neither an event nor a replication of an event; it is a cascade of relational memoranda: "more-or-less serial virtual machines implemented—inefficiently—on the ... hardware that evolution has given us,"[33] comprising a matrix of information-processing and behavioral modules and a complex hierarchy of parallel specialized processors and circuits integrating one another's inputs and outputs,[34] culminating in "multitrack processes and streams of interpretation"[35] and the "simultaneous interaction of many ... representational systems" with a "combinatorial explosion"[36] of network states and properties.[37] "Who's on First?" again.

To exist at all, the mind "must cope with rapid sequences occurring at the limits of its powers of temporal resolution,"[38] winnowing, sorting, and making a universe out of them. "[A]ctual ... simultaneities and sequences ... [are] of no functional relevance *unless they can be accurately detected by mechanisms in the brain....*"[39] The rest just flows into Neverland.

Let's take a closer look at the indeterminacy whereby the actual sensory stream is lost. It vanishes as it is distributed within its own network, which itself is fluid, conditional, perpetually under revision, always in a phase that no longer exists a moment later: it is "*not* a single definitive narrative ... but a parallel stream of conflicting and continuously revised contents...."[40] States of consciousness are the outcome of underlying computations and discriminations, some promoted to higher functional status, others abandoned. It is because we have *too much data* flowing in at once and there is such substantial repetition, mosaic distribution, and overlap that we recognize only the dividend, discarding far more than we keep. Even the dividend fissions through multiple unspecified and inchoate shufflings of its own drafts. No canonical version is ever adopted, and no reality is filed in the brain to corroborate the reality generated by the mind.

Like the figments of dreams, the ciphers behind working reality defy logical consistency. Chronology is anachronistic, parachronistic, prochronistic. Not only does space in the external world not translate neurally into space in the mind, but no particular content is ever exactly itself (or a faithful copy). Representation itself is circumstantial: "[A]ny candidate [sensory] code can represent any perceptual dimension; there is no need for an isomorphic relation between the neural and psychophysical data. Space can represent time, time can represent space, place can represent quality, and certain nonlinear psychophysical functions can represent linear or nonlinear psychophysical functions equally well."[41]

Sensory and proprioceptive onsets do not mark equivalent onsets of representation; instead the "narrative stream or sequence [remains] ...

subject to continual editing by many processes distributed around in the brain…";[42] "incorporations, emendations, and overwritings of content can occur, in various orders."[43] The "arrival time of signals themselves … [is] irrelevant to the information they carry,"[44] as backwards referral[45] and memory revision fill in gaps and blanks, not accurately but credibly enough to produce an illusion of cohesiveness and meaning.

Stated otherwise, as content is built up from sensory and proprioceptive inputs and outputs, unconscious and conscious meanings continuously shift toward each other in episodes of ambiguation and disambiguation. They don't mirror reality but *invent it* out of their feedback. "[P]erception turns imperceptibly into memory, and 'immediate' interpretation turns imperceptibly into rational reconstruction."[46] The process is similar to states of Heisenbergian uncertainty, perhaps for the same reasons and on more than one level of uncertainty.

Despite consciousness's equivocation, its underlying relativity, its reliance on gears of different historical and biological vintage, and its continuous reconstruction from multiple drafts—it works; it creates the seamless fiction of an anchored and personalized existence out of mosaics and disjunctive representations. The brain is "a massive parallel processor with many billions of operations going on at the same time, [yet] our inner experience seems to possess a single center: whatever is going on seems to be happening to only one being."[47] We construct and cultivate farms and villages with enough acumen and accuracy to manage them. We transform our habitats according to our changing constructs and social orders. We eventually assemble gigantic cities and networks of roads, write detailed software for video games, send rockets to the Moon, and guide satellites beyond the Solar System, all with the justifiable confidence that we are imaging and navigating actual external space somewhere, in fact (through our telescopes) to all of irradiated infinity. That we have *consensus* reality becomes seeming proof of reality *per se:*

We are "brainwashed" by the mind's demonstration and display (even as we are "mindwashed" by the brain's display). But there is no choice in the matter: if an animal weren't totally impressed, indoctrinated, and on board, it wouldn't survive. The same representational process as well as its resolution in self-identity imbues arthropod preconsciousness as it does human philosophy. Consciousness is not just another thing: "The world contains subjectivity as a rock-bottom element...."[48]

Grok it right now! Look at it closely. It's yours for the noticing. It has coherence and unity. It is not going to relax its grip on reality; it is not going to vacate under persuasion or duress. It cannot be filibustered away by lawyerly or semantic sleight. "[T]he term 'subjective' refers to an ontological category, not to an epistemic mode."[49]

This is crucial. Where ontology (the nature, existence, and essential reality of a thing) meets its epistemology (how we know what we know about it, or know anything at all), consciousness sits preternaturally like Lewis Carroll's Cheshire puss. "The ambiguity between the appearance versus the reality of consciousness" is a dilemma that "presumably nothing else in the world suffers from.... As such, it tends to uniquely blur the distinction between the epistemology and ontology of consciousness: if all there is to (the reality of) conscious states is their appearing in a certain way to subjects, and if they have no existential status independent of their so appearing, the ontology of consciousness seems to collapse into its epistemology."[50] Wow! That's even sicker than an event horizon being sucked into a gravitationally collapsing neutron star; it's the whole proxy universe vanishing into itself like an ouroboros devouring its own body and leaving nothing behind. We see a thing and think it, even as the seeing and thinking are swallowed into their own representations: what they are is how we know about them at all.

"How we know" is the final basis of a thing's reality—but, as science dudes, we crave better than that, we want the cattle to have real meat, independent existence, give off a nitrogen scent, whether anyone else believes in it or not. We suspect that things darn well *do* exist, but mind throws a monkey wrench into the equation—and a twisted one at that.

If we discount either the epistemological or ontological aspect of consciousness (it doesn't matter which) in hopes of salvaging an objective link otherwise between it and the physical world, the circular nature of the conundrum restores itself by default, routing us back inside the tautology with which we began: the appearance *is* the reality.

There is finally no way around the relationship between the content of consciousness and consciousness itself, between reality and the experience of a reality—and that is the rub. We cannot separate ontology from epistemology, so we might as well not try: reality behaves like reality, even though it is not—or maybe reality does not behave like reality, even though it is.* "Consciousness receives the world such that the density of relations and connections among the objects ... are simultaneously the *constituents* of which the world is composed and also *constituted as* the objects that they are by their relations within that world."[51]

And that's true from moment to moment, experiment to experiment, before even the reality-bending observations of quantum physics. Scientists have no tools, no precedents for measuring and contending with anything without first taking it (and themselves) out of the barrel with no name (consciousness). But then how do you take the barrel out of the barrel, especially if there is no way to get at either, no independent modes for distinguishing them from each other, no space between them, no yardsticks of verification, and finally no barrel?

Instead what theoreticians do is set up naming conventions for the receptacle such that it is included in its own contents, and then they pretend that the warehouse in which it sits has a special dispensation to operate outside the commerce that has constructed it:

"[T]here would be no content in a world without consciousness. Accordingly, we labor under an illusion if we think we can *complete* the theory of content without even mentioning that contentful states are associated with consciousness.... When we think we are conceiving of content in the absence of consciousness, we are really treating a system *as if* it were conscious, while simultaneously denying that this is what we are up to."[52]

This is the intractable mess in which consciousness has landed science. It approaches the ostensible real (a snake) as if it weren't what it seems to us (an object of consciousness) but (instead) something else that *is* real. Then what's the *real* reality? Where does the rubber meet the road, the snake its mirage? When does the damn illusion coil and bite?

Unwilling to sanction either phenomenal reality or abandonment of its own devout materialism, science must straddle an unstraddleable line,

---

*An entire domain of Hindu metaphysics rests upon the ultimate disposition of a piece of rope mis-seen initially as a snake: what *is* it before its erroneous "snakeness" is rectified?

using the temporary fig leaf of dualism as an operating placeholder while clinging tenaciously to its own nondual ideology. Yet any "theory of consciousness requires the addition of *something* fundamental to our ontology ... [because] the subjectivity of consciousness is an irreducible feature of reality—without which we couldn't do physics or anything else."[53]

The unspoken verdict is that consciousness is simultaneously a temporary, unassigned noncategory and an honorary hallucination in which we are permitted situational belief, even absolute allegiance—a statutory artifact of the brain as well as a withheld option to allow and facilitate discussion of reality otherwise (dinner and banter afterwards).

It is like the artifice of a boy playing Viola in Shakespeare's *Twelfth Night* during the era when women were forbidden to act on stage publicly. When Viola disguises herself as a boy, her actor cannot simply play a boy; he has to play a boy pretending to be a girl pretending to be a boy. A pure behaviorist must portray the role of consciousness pretending not to be conscious in order to demonstrate that it is really *not* conscious.

Clearly consciousness is either a super-ontological event or the illusion of one.

●

In the seventeenth century a new breed of scientist-philosophers went looking for "mind" as something distinct from a spirit or soul. René Descartes proposed a baseline demarcation of consciousness as "everything which we are aware of as happening within us, in so far as we have awareness of it."[54] Right off the bat, that shows how hard it is to declare anything that is not fundamentally either vacuous or tautological. I mean, Descartes was no dummy; he said famously, *"Cogito ergo sum."* A later seventeenth-century philosopher, John Locke, proposed: "Consciousness is the perception of what passes in a man's own mind."[55] Yeah? And?

But early on, just to pin that much down and brand it was a big deal.

Eighteenth-century philosopher Immanuel Kant described mind as "continuous sensuous manifolds of time and space within which phenomenal objects and events are presented,"[56] "whether [these] be the spatial manifold of perception or the merely temporal manifold of inner sense." This creates the "'thing-liness' ... and concreteness ... of phenomenal experience ... the construction of a model of the world that in some sense itself *is a world*."[57]

Consciousness is, in modern philosophical shorthand, "whatitislike-tobe":[58] "the self-awareness that attends perceiving, thinking, and feeling."[59] But what exactly does that mean, again as opposed to anything else? Like falling in love, wise men talk, but no one really knows.

Continuity is a signature feature of consciousness: "A conscious mind is an observer who takes in the information that is available at a particular (roughly) continuous sequence of times and places in the universe."[60] William James compared this quality of thought to the fluidity of water and emphasized that mind was not only an undisturbed flow through individual focal points but the medium itself: "this free water of consciousness that psychologists resolutely overlook. Every definite image in the mind is steeped and dyed in the free water that flows around it."[61]

James also identified consciousness's focalization or "attention ... [as] the taking possession by the mind, in clear and vivid form, of one out of what seem several simultaneously possible objects or trains of thought."[62] Attention is becoming *focally conscious*[63] such that awareness and attention form a cone of convergence.[64] Elsewhere James gave this property a Darwinian explanation as "exactly such as we might expect in an organ added for the sake of steering a nervous system grown too complex to regulate itself."[65]

We have come a long way since Descartes but, at the same time, we have not travelled at all. "By consciousness," declares contemporary philosopher John Searle, "I mean simply those objective states of awareness or sentience that begin when one wakes in the morning and continue throughout the period that one is awake until one falls into a dreamless sleep, into a coma, or dies or is otherwise, as they say, unconscious."[66] Back to Locke!

Modern philosophy, using information theory, arrives at another, purely operational definition of consciousness—"monitors monitoring monitors": "the scanning of one part of our central nervous system by another.... In perception the brain scans the environment. In awareness of the perception another process in the brain scans the scanning."[67] Sound familiar? It's those Boeing flying robots again. The overall frame of reference has shifted from "what consciousness is," an impenetrable proposition anyway, to "what consciousness does." Consciousness no longer stands for an immaterial entity; it performs (and is) a utilitarian function of nature with thermodynamic input and output—a service module that does its browsing way high up on the synaptic ladder by comparison with the entry ports of data.

In other words, as perception gets routed inward and upward, it runs its own representational networks on monitors, which "experience" themselves as they look into each other—wherever they are located and however they orient their "I" in relation to their act (let alone how such a process originates in the universe at large). Someone is probably peeking, so the resourceful materialist has a fallback position:

The "consciousness module" does not exist passively in relation to information; it imposes its higher-order commands in the form of interpretations and attitudes: "A 'second-order' intentional system is one that not only has beliefs and preferences by virtue of obeying such regularities, but in particular has beliefs and preferences about beliefs and preferences."[68] Whether that turns out to be second-order or fifth-order is arbitrary and moot. Who's really counting?

By this definition mind becomes a consensus reached by multiple-order digital-analogue systems dialoguing inside us. As their neuronal colloquy proceeds, the brain, nervous system, and incipient properties of consciousness exchange content and "gossip" with one another to create states of self-organizing mentation until those states somehow become subjectively aware.

To one neuroscientist "our sense of subjective awareness arises out of our dominant left hemisphere's unrelenting need to explain actions taken from any one of a multiplicity of mental systems that dwell within us.... These systems, which coexist with the language system, are not necessarily in touch with language processes prior to a behavior. Once actions are taken, the left [hemisphere], observing these behaviors, constructs a story as to the meaning, and this in turn becomes part of the language system's understanding of the person."[69]

But how does a mere monitor construct a story? How does a light get turned on inside? Is it by those ping-pong balls again?

Memory has to be the "computer" module that binds the rest together—an absolute system criterion and a characteristic of all creature consciousness on Earth: "I am who I remember myself being."[70] And I exist only "as far as this consciousness can be extended backwards to any action or thought."[71]

Memory fits the cybernetic paradigm of "mind" like a glove: its central monitoring module a sort of sensory-retrieval vault. If you can't log the events that you experienced in your successive states of awareness (if they vamoose without a trace), consciousness is zilch. Without memory, computation is pointless. However significant or vivid the stuff it registers, there is still nothing to keep track of and no one to keep track of it. A count (ac-count) is the whole ballgame. If you are not going to keep score, then wind blowing leaves off trees or waves depositing what-not on sand are calculators too. A waterfall is a viable hard drive.

In that sense functional consciousness is little more than "the name for all the different types of awareness attached, as it were, to each distinct sensory module and to the declarative memory system."[72] If we try to wire various modules and components together as if we were generating biological consciousness from scratch, we get something like an internal attention-firing mechanism "in which information about multiple individual modalities of sensation and perception is combined into a unified multidimensional representation of the state of the system and its environment, and integrated with information about the most important elements, the retention of the experience thus gained in the memory according to the needs of the organism, generating emotional reactions and programs of behavior to adjust the organism to its environment with the creation of a pattern or model of required action, and production of the appropriate program (plan) to control the selection of necessary actions; and finally the comparison of the results of the action with original intention ... with correction of the mistakes made."[73] This is out of control, and no one's going to catch Jack Rabbit (or put "Who on First") anyway.

But the entire spectrum of human consciousness is not logically finite in this way. The analogue-digital nature of brain processes cannot be represented by any series of scientific statements about the brain (or modules in a brain simulation) or any finite logic-based system. Even memory in a brain is not the same sort of module as in a computer: "Conventional computers consist of a small central processor plus massive amounts of memory storage space, each word of memory stored at a particular address where it can always be accessed. The brain differs from computers in that there seems to be no space at all allotted to memory. The brain's memorization facility seems to be diffused, in some ill-understood way, into the brain's sensory, motor, and emotional processing networks."[74] The miniaturization and placement of the drive is global and nonlocal. Hence, it is not a module at all. But then what is it? How does it operate and confer a "memory storage" property?

Lacking such skinny, can we even build a computer that becomes "conscious"? Yes, but only in the sense that it can simulate a conscious human brain to any arbitrary accuracy within the limits of its computing power, which will likely be more than adequate to fool us soon (if singularity hasn't been reached already). At that point, a computational system will demonstrate the complex mixed logic and extralogical behavior of human brain function. Computers will produce outcomes that cannot be wholly predicted in advance from their inputs.

In 1950 cybernetic pioneer Alan Turing foresaw the coming dilemma of artificial (machine) consciousness and proposed a straightforward protocol that any robot or computer would have to enact in order to be considered conscious: "A computer in a box passes the Turing test if it can convince a human being that there is another human in the box."[75]

A Turing-certified system will be able to "simulate" the output of our neural processes and in that sense *be* an extensive "theory" of the way brain produces consciousness. Its successors may eventually be able to clone human brain activity, including the apparent effects of awareness or consciousness, but convincing us is all that is necessary.

Though future machines will be experienced as "thinking" like humans with "consciousness," I suspect that they will never be reduced to a series of finite simple logical statements about the brain producing consciousness. There is no way either for humans to convince robots that they aren't really people (the alternative referent for "they" works here too), and there is likewise no way to convince the proud architects of these machines that their golems aren't "real."

The differences between Turing machines and creatures supersede transphysical or quantum effects; the sheer presence of random events and recursiveness in complex systems is enough. Just as a Turing machine cannot predict all the possible computational outcomes of the device it has logically defined, logic-based efforts to formulate all the possible outcomes of brain activity including expressions of consciousness can never produce *the experience of consciousness.* Even the most advanced current programming models of brain activity are merely Turing machines in disguise.[76] They cannot join a pack and hunt with the other wolves.

This would point back to Cartesian dualist explanations of mind as a "nonphysical substance." If consciousness is immaterial but real, where are you going to store memory in it? And how are you going to introduce nonphysical components into a physical machine? First you have to "find" such components (in nature or wherever); then you have to figure out how to jam them into the console. Talk about "the emperor's new clothes"!*

By "imagining first a nonbiological machine, and then a program that would realize various human mental abilities,"[77] we pretend to retro-engineer ourselves. The principles of nature that we appropriate in reconstituting computational functions and installing a form of our own mode of knowledge in computers are assumed to be the ones blindly and randomly chosen by nature in designing *us,* but only because *there are presently no others.*

As to the drop-dead complexity of living machinery arising as well from mere adventitious concatenations of water, wind, dust, sunlight, and the rest—well, nature could afford to work slowly, methodically, extravagantly, by trial and error. It could consider just about any and all options, digital, parallel, analogue, and who knows what, at every crossroads, because it effectively had unlimited time, unlimited resources, and no agenda. It

---

"The Emperor's New Clothes" by Hans Christian Andersen ("Kejserens nye Klæder"* in Danish) is a story (as most of you know) about two weavers who promise to make their Emperor a suit of clothes invisible to those unfit for their positions, stupid, or incompetent. But there is no outfit at all. When the Emperor parades before his subjects naked, thinking he is wearing magical duds, a child cries out, "But he isn't wearing anything at all!" So it is with Turing machines: they are not delivering first-person consciousness at all.

made up for its lack of innate intelligence by the discursive intelligence of a three-billion-year sorting cycle. That's a lot of time for even the most diffuse computing process, even one without a hard drive (or using eddies and shear force), enough apparently to be the matrix and womb of all being and experience. No much more discrete machine has come close to replicating this feat of entropy and chaos.

CHAPTER FOUR

Degrees of Consciousness: Protoconsciousness, Preconsciousness, and the Freudian Unconscious

i. The Spectra of Consciousness

The mystery of consciousness attends its incremental emergence in nature as well as its loss or degradation in organisms. What qualifies a creature for consciousness? Are there degrees of it, or is consciousness an on/off switch—you have it or you don't?

Entities or states that may or may not have full consciousness or (alternately) lack our human version of consciousness comprise a spectrum from protoconsciousness to consciousness. It is ontological and phylogenetic (evolutionary): pretty much a one-way street.

Even the simplest and most primitive life form has a "belief system" that connects its inner and outer worlds, whether you consider that "alpha state" a spirochaete, protozoan, flatworm, newt, or mouse—and whether you deem its landscapes ontological or just biochemical tropisms.

Another, epistemological and psychodevelopmental spectrum locates states within consciousness in which it or some aspect of it has been temporarily or permanently shut off, or is dormant or otherwise compromised. I will postpone that till Chapter Five.

Neither spectrum generates a firm line between what is conscious and what is not. You can't even have unconsciousness or systemic nonconsciousness without consciousness first because these are also consciousness-based modes of data.

A related issue is whether consciousness is solely a transitive ego function (an individual creature's consciousness *of* something) or can be an intransitive transpersonal phenomenon: state consciousness of an abstract mode of knowledge prior to individuated creatures.

Neuroscientists believe that neural pathways constitute themselves algorithmically without prior content or syntax; they are novel, discrete, and generate their sole meanings out of their emerging anatomy and functions. But many philosophers consider consciousness and knowledge a general characteristic of systems and the universe.

There is a difference between an operational function inside biology that provides information only from its contingent operations and a transcendent domain of preexisting archetypes. Does matter kindle consciousness *ex nihilo,* or does consciousness supply reality out of a storehouse of *a priori* content, or might these modes be concomitant?

An absolute barrier separates consciousness as a product of brain activity from consciousness as an aspect of the universe elicitable by any felicitous activity. In the former case "mind is like the light that goes on when you throw the (matter) switch. The switch controls the light; the light never controls the switch."[1] In the latter case, consciousness, by definition, *had to exist biologically too* (once it existed at all).

Situational (functional) consciousness is arbitrary and circumstantial; it did not have to exist in the universe but came into being anyway. If the algorithm had played differently, this very same universe would be nonconscious forever.

Twentieth-century French philosopher Jean-Paul Sartre proposed that consciousness is always *consciousness of,* for "a consciousness which would not be consciousness (of) something would be consciousness (of) nothing."[2] There is no consciousness separate of worlds and events of which to be conscious, which is not quite the same issue as whether consciousness is absolute and fundamental or circumstantial and developmental. Either mode must be "of" something, but one mode is innate and inevitable and the other is circumstantial and provisional.

The two modes converge where the enlightened nihilism of Western existentialism confronts the nihilistic enlightenment of Zen Buddhism. To the individual ceding his or her biological consciousness at death the difference between permanent obliteration by shutting down the neuron stream and obliteration by the absorption of personal consciousness into a transpersonal stream is not as great as it might seem. Polar rubrics conceal alternate interpretations of how the universe actually works, so the antithesis is more semantic than experiential. I will shelve this conundrum till Chapter Nine.

One other note: some of the accounts in this chapter have leisurely sidetracks and diversions. Bear with me. It is all in the service of consciousness—its masks, negations, ploys, and the masks people put on to evade, exploit, or pretend to collude with it.

ii. Protoconsciousness to Consciousness

- Protoconsciousness operates "nonconsciously" in amoebas, rotifers, rhombozoa, sponges, tardigrades (water bears), and simple worms (goblet worms, arrow worms, mud dragons, etc.). Insects and spiders carry out complex activities through inherited, hardwired behavioral patterns—taxes, tropisms, and general movements toward and away from stimuli—but "there is little reason to think that they have a general and central representation of space."[3] They don't know much of anything, but they do "exist" in the sense that consciousness itself exists. If they were the only existential beings in the universe, it would still be a conscious universe.

Bacterial chemotaxis and one-celled attraction-repulsion without neurons are different from but not *that* different from certain self-organizing inanimate chemical cycles; an average observer could not distinguish metabolism from pattern-repeating bubblings. Are protozoa only consciousness-like chemistry or true protoconsciousness? The treachery of this borderline state led scientifically trained members of the *Viking* Martian lander team in the late 1970s to declare (in opposition to each other) that their experiments with soil samples had discovered biosignatures of bacterial metabolism on Mars and that they had merely encountered a virulent chemical reaction.

An incalculable divide exists between bare functional consciousness (a centipede seeking food in soil) and self-consciousness (a chimp looking at itself in a mirror and attempting to rub a red mark off its forehead surreptitiously painted there by a scientist). But what is its measure? Are sentient tingles enough to ignite true mind or do we require conceptual overlays to advance a candidate to the consciousness program? Do our qualifying imperatives include (1) intuition of continuous time, (2) self-awareness, (3) a world map, (4) and/or the construction and maintenance of a belief system of self?[4] Can "consciousness" be any motivated action or imagination of *any* universe, however depleted or sparse: "to be" against "not to be"—and it's all downhill from there? Well then, are chemicals any less "motivated" than volvoxes?

Whether you officially deem it "consciousness" or not, the truly miraculous thing happened long before humans or even primates came on the set. Tentacles of jellyfish medusae and statocysts and pressurized canals of comb jellies are "minded" at core, not only moment to moment but once upon a time, creating image-complexes at the bioelectric threshold of thought. "Phenomenal representation probably predates the advent of meta-cognitive self-models in evolutionary development."[5] Protoconsciousness is ripe with potential meanings, many of which it has since given rise to, most of which it probably has not. In fact, self-models may have been "recruited" for their semantic transparency.

Rotifers and sponges have awareness prerequisites, an ember of internal light. As the ingression of novelty into space-time spiraled from worms to crustaceans and insects, from fish to amphibians and reptiles, cognitive strategies replaced taxes and mimeses, behavior became subtler and more complex. Birds and mammals have far more awareness than flukes or fleas—clear apprehension of their own beingness and a certainty about the world and their intention to act in it. Hawks could not hunt without full-blown representations, each an *Imago Mundi*. Yet insects likewise interpret food sources and carry out martial acts in protection of their bodies, nests, and young. What property transmogrifies from a wasp to a falcon, a bumblebee to a shark? What does a cobra "know" that a termite doesn't? And what does the knowing, in either instance, mean?

What makes a bobcat or groundhog such a tenacious defender of its beingness? Would we deny the same solipsism to the roundworm or anemone? While the crab's commitment to its own actions seems to qualify it for selfhood, pretty much everything else in its imaginal life is up for grabs.

Mammals (chimpanzees, mice, dogs, and the like) have operational awareness and problem-solving ability. But again, where did these come from if not simpler animals? Are not our own thoughts the developmental equivalents, 600 million years later, of a sponge's autonomic pulses, or of 120 million years of accumulated mollusk-level inquiry?

The functionalist definition (while we are at it) valorizes agency *sans* awareness: "[A] thing is conscious, at all, if it is conscious to any degree at all—that is, if it has at least one internal monitor operating and contributing.... The more monitors, the more integration, the more control, the more richly or fully conscious."[6] Ping-pong balls matter! Mr. Crab is invited back in. A functionalist's crayfish and geese are relieved as well of burdens of self-reflection and self-awareness; they just need to run. But, guess what, so do we.

A-Consciousness or Access Consciousness is defined as follows: "A state is access-conscious if, in virtue of one's having the state, a representation of its content is (1) inferentially promiscuous, that is poised for use as a premise in reasoning, (2) poised for rational control of action, and (3) poised for rational control of speech."[7] It tops out at Turing-level. "Access Consciousness has to do with information flow.... [T]here are things to which we are informationally sensitive but not experientially sensitive."[8] The basic operations of our body and metabolism are clearly informational without being experiential.

P-Consciousness or Phenomenal Consciousness has "experiential properties of sensations, feelings, and perceptions ... distinct from any cognitive, intentional, or functional property."[9] These "monitors" operate systems, but their possessors also experience them. Whether that amounts to anything marketable in an Access universe remains to be seen.

Knowledge without awareness (A-Consciousness) is common throughout nature.[10] When scallop-stage shellfish evade predators in strategy-like patterns, they likely do not "know" (in our sense) what they are doing. Do they grok why they flee the shark's shadow or do they just get out of Dodge? Do they experience anguish or pain when they are caught or about to be eaten?[11] Do they have a flicker of regret *vis à vis* their existential situation? Do they "care" whether they live or die?

Some scientists take the position that invertebrates do not feel pain or grief—or it doesn't count as suffering because they do not have representational or cognitive faculties by which to experience or witness it.[12] This argument has been extended disingenuously to mammals: their pain lacks moral weight, which excuses any sort of harsh or antipathetic treatment of them in the service of human use.

The animal-rights movement rebuts this from the premise that *all* animals have historically intersecting nervous systems, minds, and feelings. Kinship in protoplasm gives any sentient creature an equal right to compassionate treatment. Elephants and manatees experience anguish from the same basic synapses as we do. Why not the same meanings? Why shouldn't the same ethics apply to them?

"[I]t is absurd to say that the rat [in a Pavlovian electroshock experiment] isn't conscious of the tone and the shock. The rat is alive, awake, and remembering. What he doesn't possess is the highest level of human self-consciousness. He no doubt has some kind of me-ness, a subliminal sense of his own organism and its drives to fight, play, flee, have sex, and eat, and he surely recognizes his fellow rats and can distinguish them from predators. He does not have an internal narrator telling the story of his adventures in the lab with those gargantuan scientists in white coats delivering tones and unpleasant bursts of electricity."[13]

Our story is just that: a story. It cannot simply privilege itself on the basis of white coats, but of course it does, even as the lion privileges itself by growl and mane.

Of course, humans treat other humans as objects, so unexamined extension to other zooids is inescapable. If wise and judicious Thomas Jefferson could rationalize owning and impregnating slaves, is it any wonder that procedures carried out routinely in animal slaughterhouses nowadays go largely unchallenged? Workers carve up live heifers as they shriek, toss sick piglets into dumpsters like sod, and crate live and dead hens together. That's consciousness too, crying out that it is suffering.

- We cannot conceive of "other" consciousnesses. We haven't a clue as to what it is like to *be* an owl as opposed to, say, a lobster; a moose as opposed to a tortoise. They are existential states.

We might imagine the palpability of a bird's "feel" of flapping its wings or its up-and-down soars by analogy, for instance, with hang-gliding, but the spackle of a bird's surf-ride and full phenomenal dashboard through Earth's magnetic field would be as unfamiliar to us as sight to a blind person.[14] Likewise the rich and varied odors of a sewer to a rat and the sonar landscapes of a bat or dolphin are excluded from the human experience. So is the cold scorch of photosynthesis, however it might register chemotropically in the cells of a plant.

"Bats perceive different secondary qualities from us when they employ their echolocation sense; it is not that they perceive precisely the same qualities and embed them in a different (non-representational) medium."[15] They are not humans pretending to be bats; they are bats. To fancy being a bat, one can "imagine that one has webbing on one's arms, which enables one to fly around at dusk and dawn catching insects in one's mouth; that one has very poor vision, and perceives the surrounding world by a system of reflected high-frequency sound signals; and that one spends the day hanging upside down by one's feet in an attic. Insofar as I can imagine this (which is not very far), it tells me only what it would be like for *me* to behave as a bat behaves. But that is not the question."[16] After all, to a bat it is hardly "poor" vision or "upside down" or "yuk! a mouthful of insects."

"I want to know what it is like for a *bat* to be a bat.... I cannot perform [this] either by imagining additions to my present experience, or by imagining segments gradually subtracted from it, or by imagining some combination of additions, subtractions, and modifications.... [N]othing in my present constitution enables me to imagine what the experiences of such a future stage of myself thus metamorphosed would be like. The best evidence would come from the experiences of bats, if we only knew what they were like."[17] Good luck! "[W]e can never become bat-like enough to have those experiences and still be ourselves."[18]

Though there is an empathic continuity between "batness" and "human-beingness," they are unique states of being without admittance to each other.

It is even harder to imagine invertebrate consciousness. "What we succeed in imagining, if we try to get inside the mind of a spider or a notebook computer, is either an implausible cartoon (with anthropomorphic talk balloons) or something that hardly seems to us to deserve the title 'consciousness'.... We are not well placed to receive the idea that there can be very low degrees of consciousness."[19]

Likewise "a Martian scientist with no understanding of visual perception could understand the rainbow, or lightning, or clouds as physical phenomena, though he would never be able to understand the human concepts of rainbow, lightning, or cloud, or the place these things occupy in our phenomenal world.... Members of radically different species may both understand the same physical events in objective terms, and this does not require that they understand the phenomenal forms in which those events appear to the sense of the members of the other species."[20]

You cannot look into someone else's brain and "see" the nuances of a rainbow or lightning there; you cannot feel their exact happiness or anguish. You cannot wrap your senses around their experiences and perceptions in the way that you do your own interior "feels." I'll come back to those Martians in just a moment.

- Are plants conscious? How about crystals? Both show coordinated growth, patterning, and design "memory." Plants respond to inputs tropistically. There is, in fact, an entire bibliography of outsider testimony regarding the "secret lives of plants," including their "feelings" and capacities for recognition and even love.[21] A separate quasi-scientific literature addresses the ostensible "memory of water."[22]

Quite apart from metaphysical speculations, indigenous cultures on Earth have recognized that plants entheogenically transmit practical, cosmological, and medicinal knowledge—from where, it is not clear. This is not even—or

70

not *just*—subliminal realms of knowledge that botanicals awaken in animals who ingest them; it may not be unconscious creature awareness (plant or animal) either. There may be modes of "state" or alien consciousness that entities in nature share at radically different "neural" frequencies and levels of recognition or even that they somehow co-create with each other.

Plants could embody the transpersonal, objective intelligence of the Earth or of certain molecular configurations, or they may be the conduits through which otherwise-disembodied intelligence passes. Ayahuasca and peyote visions are totally up for grabs as regards both their ontological and epistemological status—and the degree to which either is human, botanical, archetypal, or some other exotic thing.[23]

- Are air conditioners, vacuum cleaners, thermostats, automobiles, or jet planes conscious? Don't be so quick to dismiss these guys before you have a working definition of consciousness. They *do* operate multiple-order internal self-monitors and feedback loops. They behave like willing servants (and even occasionally get scolded or kicked like them).

The underlying question here is whether or not there is such a thing as artificial or machine intelligence (so-called AI). Robots and digital computers function *as if* they possess criteria of consciousness, but from another perspective they borrow their consciousness from us. We embed counterfeits of our own minds—complex information-processing grids—into their networks, polymers, and logic gates.

The answer may depend on whether we go from the bottom up, extrapolating consciousness from machines to life forms, or from the top down, taking our own consciousness and transferring it to machines.

- Is consciousness as we know it limited to carbon-based animal brains on the Earth, or is this neural chauvinism?[24]

If we should encounter living creatures that evolved on other worlds, would they be conscious by our definition, or would we need a new definition for them? Does our mode of phenomenology and knowledge translate off-planet? Is physical consciousness itself singular or plural?

Remember, a machine-like representation for brain activity that will satisfy the Turing test does not similarly resolve the question of whether *human* consciousness can be embodied in some other sophisticated information-processing system. Do dolphins (with a water/hydrocarbon system like ours but organized differently) have the potential of not just consciousness but human-equivalent consciousness? Are there information-processing entities in the universe with some other biochemical and/or metallurgical matrix that is not related in any way to Earth's biomatrix but that has "consciousness"? Would we call it Turing-style consciousness? Would we recognize an alien biomatrix's attempts to communicate with us? If it came in attack mode with weapons or, conversely, bearing gifts, I am sure that we would regard it, rightly or wrongly, as conscious in our genre.

The pictorial messages inscribed on the plaques sent into outer space aboard *Pioneer 10* and *Pioneer 11* (1972–1973) were designed to be read by conscious beings anywhere, made out of anything. The recipients were supposed to be able to decipher, if not the precise rebuses, at least the fact that an attempt was being made by another sentient species to convey information to them.

I am guessing that consciousness is singular, as I cannot imagine how a creature from another world would be conscious of some other reality or conscious of the universe in such a different way from us as to require a whole new definition. Consciousness is consciousness is consciousness—hegemonic, mysterious, and tractable enough to encompass just about anything (and anything else). In this regard a minded creature in a remote galaxy is no more or less alien to us than a hypothetical nearby Martian, Callistan, Alpha Centaurian, or Pleiadian.

- Is consciousness exclusively restricted to properties of carbon-based tissues, or could silicon or some other element develop its own different pathway to consciousness?

Silicon may not allow the necessary psychological states and functional relations to occur rapidly enough for innate mindedness, but there is nothing in principle, at least that we know of, that confines consciousness to the sort of organic carbon-based materials endemic to human neuron chains and brains. Yet it remains an open question. Would a silicon creature be able to communicate by the same signs and mathematical logic as

72

we do, or does silicon "think" differently? Without knowing what "causes" organic carbon to "think" (and if it's unique to carbon), we cannot answer this question decisively.

Probably consciousness has universal properties and qualities that transcend the discrete arrangements of protons and electrons in atomic orbits and could install itself as an independent function similarly in any elemental compound—same issue as E.T.s. But while we're at it, what about the difference between carbon-based and silicon- (or other-) based Pleiadians?

• In the case of cyborgs made up partly of human tissue and partly of machine parts, do their artificial elements transmute integrally into a part of their consciousness of the whole? Do chips that have been inserted into a brain themselves become "contact conscious," and, if so, how is the sapient quality transmitted or conferred?

Again, we are stymied by not knowing how *anything* becomes conscious in the first place: "Even if it is in principle possible to invent a 'consciousness' monitor, a device that would detect the physical signs of the presence of consciousness…, no such technology is anywhere in sight, because it is not even known what exactly is to be measured."[25]

In a thought experiment suggested by John Searle, your own cortex is replaced neuron by neuron by an artificial computer, as silicon chips that perform the same functions are "progressively implanted into your dwindling brain…."[26] Searle posits that three outcomes are conceivable: the first is that nothing changes: you retain your inner life through your new silicon prosthetics, as each chip instantaneously takes on the function of the neuron it is replacing in every regard.

But then what is the ontological status of the silicon chips themselves (before and after)? Do they become miraculously conscious by absorbing an immaterial oscillation of mindedness, or are they prosthetic devices mimicking consciousness in a conscious system? Not even mimicking, but seeming to mimic it in parallel functioning.

A second possibility is that you are left "totally paralyzed but possessed of a normal flow of inner experience." You will be able to hear the doctors "expressing regret over the apparent death [of your brain] but you could not tell them that [you were] still alive."[27]

A third possibility is that "you find that the area of your conscious experience is shrinking, but that this shows no effect on your external behavior,"[28] which remains the same. "As the final chips were being exchanged for the last neuronal circuit, Searle would be dimly aware of a familiar voice, seeming to come from a great distance, saying to the doctors: 'Yes, yes. I feel fine. The operation was a success.' Then all awareness would cease: John Searle would be dead. An 'empty' Searle zombie would get up from the operating table. In this third case, after he returned to the universe, the Searle zombie would find itself in the unusual position of unconsciously giving lectures on the subject of consciousness...."[29]

At what threshold does somebody cease being a self-identified life-form and turn into a dead robot? What is the relationship between a "familiar voice at a great distance" and the "dimly aware Searle"? And, as long as there is one aware reference point, wouldn't ego consciousness itself remain?

However, once "the silicon chips are able to duplicate the input-output functions of the whole central nervous system, even though there is no consciousness left in the remnants of the system...,"[30] "Searle" himself is mentally dead, he has no mental life, though to an observer he is the same person and his behavior is indistinguishable from before.

As long as a person is cognizant of becoming a "mindless" robot, he is, by definition, not a robot, at least until he becomes unaware of his plight. So where does presence emerge from nullity, thought from simulation of thought? And how likewise does presence devolve into nullity, thought into zombiehood?

A related thought experiment can be undertaken using the concept of the "ghola" (serial clone) developed by science-fiction novelist Frank Herbert for his *Dune* series. Swordmaster Duncan Idaho is a ghola, recreated again and again from cells of his deceased body, initially soon after his death and then in succeeding generations extending over 3,500 years. Each "Duncan Idaho" picks up his own memory from the moment of the first one's death and proceeds to have a separate existence.

Even if we allow that every clone of D. Idaho is an independent human being with his own free will and set of priorities, albeit proceeding from the memories and personality structure of the original swordmaster, what is the status of Duncan Idaho himself? That is, do his life and experiences terminate at his own death (like the life and experiences of the "John Searle" cyborg upon completion of the silicon replacement above)? If so, then each subsequent "Duncan Idaho" is a different person having nothing to do with the inner life of the original Duncan Idaho but using his memories like silicon implants.

However, there is also the possibility that a transpersonal "Duncan Idaho" continues to incorporate the experiences of all his clones in a single meta-human network. I explore this notion in Volume Three, Chapter Seven.

- Is consciousness limited to the physical realm or does it transcend matter and hang out in other states? Do transphysical sentient entities exist, "creatures" such as angels, spirits, elementals, and gods?

Addressing that issue, if you haven't already guessed, is a signal agenda of this book.

iii. Sigmund Freud's Theory of Consciousness

Here we enter an entirely different domain of consciousness: unconsciousness. I do not mean that there is an actual unconscious mind; I mean that consciousness in its makeup has an intrinsic unconscious component. In fact most of systemic consciousness is unconscious.

Consciousness is squirrelly enough for us to try to analyze and delineate but, when we add unconsciousness to it, we encounter difficulties extending in all directions. Not only does the unconscious "mind" determine many of the qualities of aware mindedness, but it maintains its own measureless kingdom and basis—likely far older and more pristine than consciousness and cardinal to it as well. Though impenetrable to ordinary inquiry, it shares consciousness's etiology.

Unconscious mind is a game-changer, for it not only overthrows the tenets of positivist philosophy, it blows functionalism out of the water.

Well, not really, because functionalists can (and do) claim that the unconscious is system static, access debris, entropy at its finest. They treat it as such: machine noise and network chatter, even when it does obvious damage.

Despite rationalist dismissals, the unconscious "mind" drives the processes of consciousness. While not conscious (literally *un*conscious), it maintains fundamental properties of consciousness in a comprehensive sense. In fact, unconsciousness is consciousness's underlying developmental phase and remains an inseparable aspect of its expression. You can't exclude the unconscious aspect of consciousness from any inventory or complete any theory of content without it because the forces of nature (or the gods) didn't beget conscious mind separate of an unconscious basis— no doubt for good reason. "Systemic unconsciousness" is critical to the evolution of mind and the destiny of vertebrates—it doesn't get any more big-time than that.

We do not know if consciousness and unconsciousness use the same "software" or if they represent a convergence of systems and circuits; but, either way, they are inextricably bound in each other's neuropsychological loads and outputs. They deliver one energetic expression at different levels, while each appropriates and transforms both the operational and representational channels of the other, as they require each other to discharge their meanings. To track consciousness to its basis in unconsciousness is to invite the beast into the garden—yet it is an invitation only in the sense that New Orleans invited Hurricane Katrina.

For centuries, scientists and philosophers were so engaged in targeting conscious reality that they missed the ways in which their version of consciousness was not even "conscious" within its own paradigm. Before Freud, whatever was not overtly conscious was consigned, for the most part, to a vast undeclared nether zone ranging from nonmental and nonminded to nonexistent. Early theoreticians made the mistake of focusing on logical intention and motivated acts of reason; thus they fooled themselves into believing that irrational acts, slips of the tongue, and other "lesions" could be discarded as incidental by-products of the mental process rather than profound measureless states of meaning, while madness itself was deemed an exogenous state of possession.

The unconscious, however, is too blatant for whole generations to have missed entirely, for instance in the context of "invisible" animalcules in droplets of pond water or the imperceptible molecular components of matter—ours is a universe at every level composed of iotas and interiors.

But mind was much more elusive than paramecia or cells and of course did not yield to microscopy; the philosophical glass was the only tool. Folks from pulpits to royal courts to the battlefields of the Thirty Years' War suspected the unconscious enclave, but they either devalued it or were clueless about its extent and implications. They projected intimations of it onto supernatural rather than psychological agents. Suspects included demons, devils, witches, imps, and rival scions of sympathetic magic. Scholars and authorities of the era never considered a shadow phase of consciousness itself. It wasn't the right time to go native.

Yet there were some instances of prescient clarity. Anticipating Freud's theory of trauma and sublimation by more than three centuries, Descartes noted that "the smell of roses may have caused severe headache in a child when he was still in the cradle, or a cat may have terrified him without anyone noticing and without the memory of it remaining afterwards; and yet the idea of an aversion he then felt for the roses or for the cat will remain imprinted on his brain till the end of his life."[31]

Seventeenth- to eighteenth-century philosopher Gottfried Leibniz proposed: "There are a thousand indications which lead us to think that there are at every moment numberless *perceptions* in us but without apperception and without reflection…. In a word, *insensible perceptions* are of as great use in psychology as insensible corpuscles are in physics, and it is as unreasonable to reject the one as the other under the pretext that they are beyond the reach of our senses."[32]

By the nineteenth century, "English physiologist William Carpenter, the German psychologist Gustav Fechner, and the German physicist Hermann von Helmholz each maintained that there was a psychological unconscious, not just a physical one. Thoughts, memories, and ideas could reside outside of our experience."[33] The notion of subliminal drives in everyday life could not have been more explicit, yet it took Freud's perspective for them to get rummaged, scaled, and institutionalized.

As a scientifically oriented philosopher at the turn of the twentieth century, Freud took consciousness for granted. He didn't worry about "Let there be light" because he was too busy separating light from darkness. He did not even develop an ontology of consciousness; he conceded thought as antecedent biological energy bound in symbolic formats. Then he located its latency in a zoological substratum and developed his version of it from the link between animal mind, which is instinctual and unconscious, and human mind, which is ego-based and has emotional and intellective phases extending beyond its core animal heritage. He did not so much model or conceptualize the unconscious as become submerged in it as soon as he took his quest for the origins of the psyche in a biological direction.

Freud categorized mental life as "some kind of energy at work,"[34] meaning cellular energy—metabolic, neural streams converted cerebrally. He didn't fret over proximal source because he could leave that on the treadle of evolution (entropy to matter to energy to life to mind). He came to the same essential conclusion as Isaac Newton and Charles Darwin before him: all motion and form are kinetic and thermodynamic, and every mass that moves, including the infinitesimal masses of thought, solves a prior energetic equation and creates the conditions for the next, *ad infinitum,* preserving only what it can't refute or dispose of. The universe doesn't engage in superfluous recreation; it is a turbine of exiguous energy, the varying expression of its force resulting locally in structures of differential design, expression, and stability.

For its colossal kilowatts, nature remains cautious, conservative, minimalist; it doesn't make anything it doesn't have to: mind is not an ambitious project; it is what the living membrane releases as minimum discharge of its own entropy. It is the universe's prior inertial state.

As a neurologist, Freud would have liked to have created "a biological model founded on brain stuff—neurons."[35] During his years at medical school he spent a summer at the Zoological Experimental Station in Trieste, dissecting and studying the histology of eels in hopes of making a map of consciousness and behavior. Then he spent another formative six years analyzing neurons in a physiology laboratory. Yet he finally turned away from biology altogether and toward philosophy (in which he had taken college courses) because "not enough was known about neural processes to produce such a map." Yet he always assumed that his psychological theory of instincts would be converted to "actual brain functions ... sometime in the future."[36] A biologist slumming as a philosopher, he wrote a biophysical philosophy.

"We seem to recognize," Freud proposed, "that psychical energy exists in two forms, one freely mobile and the other, by contrast, bound.... [W]e speak of cathexes* and hypercathexes of the material of the mind and even venture to suppose that a hypercathexis brings about a sort of synthesis ... in which free energy is transformed into bound energy."[37] The binding of free energy is the piston whereby the mind converts zoological drives into psychological and social objects, and also becomes conscious.

Freud's human mind consists of three distinct elements or layers: an id, an ego, and a superego. These are neither hard structures nor even pristine emotional states; they are his intuitive map of interactive biological energies: late nineteenth-century myth-making at its finest. Yet the id-ego-superego triunity has held its own and established itself in both social science and pop culture, in the process changing our definition of ourselves. Taking on lives of their own, the "instincts" have not only stabilized as constructs but developed far-reaching affiliates, as their uses have deconstructed their original meanings to form new domains of nonlinear object relations. A series of id-like, ego-like, and superego-like quiddities now do seem to generate mental life and shape each individual psyche. These constructs have become so powerful and widely applied that they may by now be almost neurological in the sense that their application has modified our actual nervous systems. We don't know that mental waves can translate (in Lamarckian fashion) into cellular dispositions or induce actual cerebral structures but, if not, the Freudian instincts have at least altered our cultural interpretation of our nervous system and the way we regard ourselves, so they have a psychosomatic impact. Now we neuralize and act out patterns with shifting boundaries that function much like ids, egos, and superegos in our brains.

---

*Cathexis is the common translation of Freud's term *besetzen* for "bound energy."

Id is Freud's one indispensable instinct, ego and superego its transpositions or sublimations. Latin for "it," id represents our primordial and original energy stratum, below the threshold of consciousness and beyond nominal form. It is a reservoir of unbound needs, desires, and aggressions—untamable impulses and raw hungers. It cannot *ever* become conscious.

Prior to id, biological existence emerges from the interplay of two polar energies that are more like Greek gods than Newtonian vectors: Eros (libido), the function of which is to cathect, establish unity, and neutralize the other force, Thanatos (death), the function of which is to unbind, undo, and return to chaos.

More than psychological, the Eros-Thanatos dialectic is archetypal; it births gravity, heat, and matter, as bards from Hesiod to Wilhelm Reich have bruited. Invisible deities transcend dichotomies and flaunt their superiority by trading identities with each other and appropriating each other's expressions. In Freudian parlance they do not exist by attribution but "change their aim (by displacement) and also ... replace one another—the energy of one instinct passing over to another."[38] Eros and Thanatos are a Western manifestation of Taoist yin and yang.

Libidinal energy from the realm of the id streams into each organism, converting animal drives and excitations into mindedness and, through the development of ego (*Ich:* "I"), reconstitutes them as signs, concepts, and institutions, which camouflage their rowdy pedigree. In this way, instinctual energy is transferred into social meanings; impulses become phenomenologies. Language additionally binds them in ideational flows.

By definition the id's energy cannot breach awareness, so everything we know about it derives from how it discharges and represents itself in the ego. As a secondary libidinal phase the ego is not even as "real" a fiction as the id, of which it is an unstable half-life. As the ego is constantly recreated by the id's interaction with the world, its single role is to discover, in loyal service to the id's demands and desires, "the most favorable and least perilous method of obtaining satisfaction."[39]

Consciousness in Freud's system is the ego's resolution of libidinal flow. In its passage through an emergent egoic structure, a quantum of the id's pressure is released and made provisionally conscious. But unconscious charge is unappeasable, never satisfied with a limited expression or outlet.

Our tissues, somatic structures, and evolving sense organs cathect the id's assault into a beachhead of Eros against Thanatos, life against death. They process unconscious energy as metabolic functions and convey them into the outermost ring of the ego. It is like keeping angry lynxes or hyenas in a pen, while trying to route their entropy and fury into Hume's philosophy. At the same time, the ego is assimilating neural impulses from the outer world as well as integrating its own internal visceral sensations. These three streams conflate as "the body itself takes the place of the external world"[40] and the ego's perception of the diverse information reaching it cobbles a replica, which is actually the inside of its own body as it renders the inside of the world (or the inside of the body inside a representation of itself). Consciousness functions as if it is a buckled translucency of an unconscious flow, the ego being both the buckle and the force of buckling.

The genesis of our thoughts is unconscious, their meanings insensible until they hit the ego where they are transitionally represented. As a consequence of its own unstable state of development and its feeble structure and lack of true boundaries, however, the ego lapses continually back across the id's boundary. This creates a standing inertial charge beneath behavior and thought—an everyday provocateur of neurosis and anxiety and, in more extreme cases, psychotic breaks with reality.

Consciousness is under constant revision, the definitions of its cargo fluid and relativistic. Id and ego fluctuate, not only because they are fluctuating anyway but because we can't conceive them otherwise. Before anything enters consciousness, e.g., has developed the innate potential of exchanging its unconscious state for a quantum of minded disposition, it has borderline access to consciousness; at that moment it is, in the Freudian lexicon, "preconscious." Whenever its items erupt into awareness, we call these temporary states "consciousness."

Only the outermost rung of the ego can be stimulated or engaged consciously. Its deeper core, which contains threshold quanta of the same material, remains preconscious. As such, it has simultaneous access to conscious and unconscious layers as well as to the verbal residues and symbolic mechanisms linking them. For Freud, consciousness is merely a a transitional occasion. Meanwhile unconsciousness potentially contains *everything* that is unconscious, in not only ourselves but the universe:

"Everything else that is mental is in our view *unconscious*.... [C]on-sciousness is in general a very highly fugitive condition. What is conscious is conscious only for a moment."[41]

Tell that to the judge about to bang his gavel!

The instincts have metahistorical implications, for it is through mam-malian and primate evolutionary development that the id (as proxy for a collectivity of biological drives) got bound into more complex representa-tions in the service of evolving neurobehavioral capacities. Pure ids never roamed the Earth as forerunners of sharks, wolverines, or tyrannosaurs—egoic structures were always welling up to contain them.

Egos at a reptilian phase cathect very few objects, but dinosaurs are still in-cipient "people"; they have character formation, albeit troglodyte. Lizards and frogs have faint personalities. As the reptilian-mammalian container subse-quently evolves by ganglia toward centralized cerebralization, more and more libidinal energy is contained and more and more of the external world is cap-tured in egoic structures. Libidinal containment is consciousness's back-story.

The human mind continues to be sabotaged by its atavistic heritage in the form of pratfalls, missteps, and jokes as well as a broader spectrum of instincts, imperatives, and psychopathologies. The relationship between id and ego remains fluid to keep the psychological charge alive and real. Through emotional life, art, play, and authorized conflict, ego tries to con-vert as much of id's vital original charge as it can, thereby forging the human universe. Bards, clowns, and shaman-priests are among our spe-cies's emissaries on this pulse.

Just because a thing is unconscious does not mean that its effects are not conscious in powerful ways; the ego simply knows them as other things. It is motivated unconsciously by them while thinking it is prompted by *those other things*. Lost memories, repressed feelings, sublimated instincts, disguised wishes, and tabooed desires all impose themselves according to their respective charges. Through the preconscious stratum they determine the components of consciousness without themselves becoming conscious: "In Freudian theory, the *unconscious proper* consists of repressed processes, exerting stress on the conscious component of the subject's life and shap-ing his or her daily life in substantial ways."[42] Unconscious guilt and shame have powerful phenomenal effects while remaining unrepresented except in subliminated forms.

In order to cathect the energy of Eros stored in the ego's personality—to bind it into useful objects and free it from a fate of pure narcissism or psychosis—the ego must constantly rebuild itself by projections and objectifications from the world and its body. It must extricate itself from its own purely unconscious sources. It does this through a process of representation—though paradoxically both the world and the unconscious mind are similarly unconscious in their essence and both pre- and post-synaptic. For this reason, some psychologists, blending Freud's etiology with tropes from neuroscience, have posited that the inundation we experience as mind represents an oscillation in the brain between conscious and unconscious (or preconscious) phases of distribution. That is the kind of neurological "confirmation" that Freud sought (or imagined he sought) for his theory of instincts.

Freud's third mental construct is the superego *(über-Ich)*, and its one objective and task is to limit libidinal satisfaction, to resist any incursions of id into ego—which, once the ego is formed and has a structure, is like trying to keep children out of the cookie jar: a full-time job. The result is an interplay of primordial libido (id) through preconscious mediation (ego) against superegoic resistance—with a metastasis of biological energy into diverse aliases.

Again, this is not real anatomy or even neuroanatomy; it is nineteenth-century metaphor under an evolutionary paradigm: an intuition of how the emergent dynamics of consciousness play out in a Kantian/Hegelian mode that is also Darwinian or Haeckelian. Ontological metaphors get enlisted into thermodynamic uses. Since there are no MRIs or ultrasounds of consciousness anyway, you take what you get.

Like ego, superego is an artifact and by-product of id, though it expresses itself mostly through mores and taboos. Why the id would endow a superego to frustrate itself is a riddle lost in the unauthorized marriage of Eros and Thanatos.

As the superego is forming, "considerable amounts of the aggressive instinct become fixated within the ego and operate there in self-destructive fashion."[43] And it doesn't dissipate or go away. Superegoic suppression of this discharge betrays itself in wars, illicit wishes, anti-social desires, and erotic fantasies—in bursts and binges of lust, envy, disgust, corruption, rage, and sociopathy.

The crisis of suppressing and repressing instincts invests exceptional power and authority in the superego, which get represented by society's totems and taboos, initially among tribal people, then through the guilds and edifices of civilization—all upwellings of id cast into egoic landscapes. To Freud, art and religion are nostalgic vestiges—futile attempts to fulfill the hungering of the id under superegoic policing—but then so is everything else.

Civilization, in Freud's final despairing and pessimistic view, is a vast neurotic superegoic superstructure, created to frustrate the ids of its citizenry. "Life is to be not enjoyed but endured"[44] was his aria muttered at age eighty to his disciple Ernest Jones, who repeated it solemnly beside a bust of the master at the conclusion of a documentary on his life.

iv. Dreams to Psychoanalysis

Dreams are a canvas on which unconscious and preconscious phases paint their immediate contents, thereby betraying their atavistic loyalties and amoral ambitions and imperfectly resolving their conflicts and contradictions. Dreams are also the intervention through which the unconscious mind releases apparitions of uncertain origin that are too frightening or alien to encounter in waking life yet also too powerful to suppress entirely. Their compromise oneiric status is tolerable for the ego, yet transgressive enough to permit a minim of unconscious content—not only to prevent psychotic breaks but to nourish the ego.

Freud declared dreams the "royal road to the unconscious,"[45] which meant (of course) consciousness too, for mental contents are generated by ongoing thoughts, experiences, and their associations. From the first clue of a penitent's dreamwork, a psychotherapist unravels the discontinuous threads and layers of its structure. By free connection he or she lurches as if willy-nilly from one item (or symbol) to the next, following a serpentine course until the links behind the pathway of the dream's formation, usually a proxy wish fulfillment, are disclosed.* A dream is also a compressed enough file that it can cathect countless, even contradictory, narratives and agendas in its plot. I will get to that territory in the second and third volumes of this book.

The syntax and hyperlinks in a dream transpose recognizable features and memories (personnel, landscapes, and events recently familiar to the dreamer) onto primal traumas, unconscious drives, and suppressed and forbidden information. As an analyst deciphers each oneiric construction, he identifies relative influences and roles of both current and ephemeral events and deep-seated traumas and crises (and perhaps transpersonal visitations as well). Each of these supplies its own charge for the dream's theater (waking life's theater too, though the degrees of illumination are reversed, with the present far outshining in brightness, though not entirely eradicating, the muted past).

The more intense an experience, no matter how ancient, the more likely it is to work its way into a dream, but also the more recent (even if indifferent), the more accessible it is to the oneiric field. In this way contemporary events provide handy ingredients for combination with older, more profound elements to constitute an urgent narrative, one which convinces the dreamer that his or her dream is real.

Freud excavated the syntax whereby dreams bind, link, and transmute id into ego while casting oneiric trances: sublimation, condensation, displacement, conversion reaction, introjection, and reversal. These functions distort tabooed information and palliate terrifying events while repackaging them as safe adventures and credible narratives for the gullible dreamer. What allows dreaming to supplant "lifing" is its cathexis of an underlying unconscious charge during the relatively permissive regime of sleep, e.g., a state of permeable semi-consciousness unregulated (or less regulated) by the superego or by the descending influence of the reticular clusters.

---

*The functionalist's rebuttal, if you haven't guessed, is that "dreams are a kind of chaotic discharge, a churning nocturnal garbage dump that involves no higher-order functions, that dreams, by their very nature, cannot hold or reveal complex ideas [and have] 'no primary ideational, volitional, or emotional content.'"[46]

The fluid, symbolic, and situational basis of consciousness is openly flaunted by the weird juxtapositions in dreams: why else would one park his car in a dense forest or toss a fishing line into a bathtub and pull out giant cod? But dreamers assent to such unlikely scenarios and hijinks. Consciousness is subject to not only normative systems, interdependent symbols, tiers of representation, and overwriting of content, but that ultimate truth serum, the biology of belief wherein neuroanatomy (reticular diffusion, melatonin secretion, alpha-theta wave-shifts, etc.) flirts with Freudian psyche.

This is also is why waking consciousness can never be considered fully "conscious" even when it is. Any functionalist definition of mind sags under the weight of its unconscious layers. Likewise any phenomenological system must account for not only the dynamics of the unconscious but the pretexts by which the conscious mind deflects it. Reality is undermined not only by the gap between the brain and the mind, leaving it rootless, but by the mind's partial integration of its own states. Again, what we have been deeming consciousness is substantially *non*conscious and *un*conscious.

Freud and his successors adapted the theory of instincts as a blueprint to treat neuroses and mental disturbances, using symbolic "talk therapy" to elicit unconscious contents and then integrate them with conscious behavior. The therapist not only guides his patient through the deciphering of meanings that have become cathected in his cognitive and affective structures but participates sympathetically with him in the endeavor, substituting his own engaged presence as an alternative to an emotional state under treatment. As the patient projects his neurosis and conflicts onto the therapist, he experiences their empathic objectification, and he gradually learns to reincorporate them more neutrally in his own ego.

Modern psychoanalytic transference is self-treatment by alias, but then so is shamanism. In neo-Freudian terms, both shamanism and psychoanalysis operate through rituals of libidinal cathexis that "disturb" a rigid mental (or disease) pattern and its static symptom complex. In both modes, the origin of the illness is relived and abreacted (released) through emotional participation; e.g., acting out in words, behavior, and imagination a version of the nexus that caused the crisis. The psychoanalyst's recreation of his or her patient's story is no less imaginal and semantic than the shaman's, and in neither case is the account's power negated by its mere symbolic reenactment or the fact that its content has also been ritualized. A sand painting or a dance with chanting and rattles is as therapeutic and abreactive as an interpretation of a nightmare or cathartic revisiting of childhood relics. There are no literal or fixed events in the psyche anyway, only representations.

Freud's system was practiced therapeutically for decades before it became subjected to medical and scientific scrutiny whereby it was "exposed" as an elaborate myth, then delegitimized as an accountable treatment. Psychology's symbols and approximations had slipped by censors previously because "Freud" was a superstar, a recognized brand, and no one thought to try to corroborate (or refute) his methods by updated scientific research; talk therapy was left in its own time warp.

For classical psychotherapy the brain is subsidiary to the mind; in fact, in Freudian practice there is no brain, there are not even really neurotransmitters—it is all feelings, affects, instincts, and behavior. Stated otherwise, Freud's goal was to represent the mind, not the brain and not the neurohormonal system (except secondarily). Though he was a neurologist with scientific ambitions, he was first and foremost a shamanic (symbolic) healer. So he conducted "talk therapy" without reference to his prior neurological map (though without specifically renouncing it). It was a case of two utterly different paradigms occupying the same space, each pretending not to notice its bedfellow.

The shift from the mind to the brain began in the 1970s, as psychiatrists perceived that not only were they facing stiff competition from amateur counselors and nonmedical therapists of various ilk, but their medical colleagues were better paid and preferentially honored as true doctor-scientists. Awakening suddenly like Rip Van Winkle to an unforeseen landscape, psychiatrists hastened to trade in their old-fashioned Freudian paradigms and scoot from *de facto* brainlessness to *de facto* mindlessness—from consciousness as states of mind to consciousness as neurotransmission and cerebral discharges; in practice, from resolutions of unconscious conflicts through therapist-patient dialogue and ritual abreaction, to neuroendocrine-based diagnoses and draconian chemical solutions courtesy of psychopharmacology:[47]

By fully embracing the biological model of mental illness and the use of psychoactive drugs to treat it, psychiatry was able to relegate other health care providers to ancillary positions and also to identify itself as a scientific discipline along with the rest of the medical profession....

Nowadays treatment by medical doctors nearly always means psychoactive drugs, that is, drugs that affect the mental state. In fact, most psychiatrists treat only with drugs, and refer patients to psychologists or social workers if they believe that psychotherapy is also warranted. The shift from "talk therapy" to drugs as the dominant mode of treatment coincides with the emergence over the past four decades of the theory that mental illness is caused primarily by mental imbalances in the brain that can be corrected by specific drugs. That theory became broadly accepted, by the media and the public as well as by the medical profession, after Prozac [fluoxetine hydrochloride] came to market in 1987 and was intensively promoted as a corrective for a deficiency of serotonin in the brain. The number of people treated for depression tripled in the following ten years, and about 10 percent of Americans over age six now take antidepressants. The increased use of drugs to treat psychosis is even more dramatic. The new generation of antipsychotics, such as Risperdal, Zyprexa, and Seroquel, has replaced cholesterol-lowering agents as the top-selling class of drugs in the U.S.[48]

Stripped of its phenomenological basis, mental illness is now so commoditized and up for grabs that, in the United States alone, a prescription-drug boondoggle flourishes amid armies of lobbyists and influence-peddlers. The already treacherous brain-mind wicket has become a cash cow yanked back and forth between competing political factions and meanings. Yes, consciousness is a black hole that will suck in anything that gets close to its magnetic field or attempts to secure it—shrinks included.

Freudian therapists treated a proxy mind with proxy symbols, staying well out of science's overly concretized bow wave of neurons while substituting their own forgeries of its terms. They were oblivious to their self-contradictions: you can't claim the right to use consciousness as a crutch while decertifying it in principle. Nonetheless, psychoanalysis proceeded with exactly such a bait-and-switch under tacit protection of a social contract not unlike the one that privileges mother and child or teacher and pupil. No one in Freud's circle tried to wrangle consciousness itself into the ring, yet all of its personal effects were invoked while retaining their full authority *as consciousness,* including the official, warranted ego "self."

For this reason early- and mid-twentieth-century Freudians completed an unintentional conversion of shadow shamanic and alchemical practices into a verbal and semantic confessional grail in lieu of the more accepted medical coupling of mental and emotional styles with chemistry and neuroscience. I am guessing that many psychiatrists kidded themselves that they were doing outlier "physics" and "biology," using superegos and ids—medical-affiliated aliases—though they were actually doing semeiology and sympathetic magic.

Then in a brave new world of health-care commodities and insurance-company vigorish, postulates with no anatomical or biochemical basis (like ids and egos) lost their validity and standing overnight, becoming pricey metaphors. Where is an id located among axons and neurons with their terminals and synapses? Except by neurotransmitters how does brain turn into mind? Of what use is a theory of unconscious instincts in a world where chemical imbalances are the official cause of mental illnesses ostensibly curable by psychoactive drugs? If mind is a hallucination, then madness is too. If consciousness is redefined as chemical in nature, then pharmacy is its sole legitimate treatment, the variables being recipe and dosage. The great sleeping dog—the brain-mind conundrum—that had been left to lie for generations was now stirred into futile agitation. Then it was herded into one very tiny doghouse:

> When it was found that psychoactive drugs affect neurotrans-
> mitter levels in the brain, as evidenced mainly by the levels of
> their breakdown products in the spinal fluid, the theory arose
> that the cause of mental illness is an abnormality in the brain's

concentration of these chemicals that is specifically countered by the appropriate drug. For example, because Thorazine was found to lower dopamine levels in the brain, it was postulated that psychoses like schizophrenia are caused by too much dopamine. Or later, because certain antidepressants increase levels of the neurotransmitter serotonin in the brain, it was postulated that depression is caused by too little serotonin.[49]

The id-ego-superego menagerie may be lightweight and facile by comparison with the trajectories of hardcore neuroscience, but something crucial, even indispensable, is intuited through its glass darkly. As the Freudian model was blasted into ridicule and irrelevance and belittled as "pseudoscience," it was replaced with another "pseudoscience" without any sense of the former's actual depth or how it was earned over centuries of ontological inquiry leading to Freud, likewise without any clue as to the shallowness or provinciality of its replacement. Consciousness was never, even under the most excessively reductive regimes of materialism, mapped successfully into moods and mind states based on enzymes, hormones, and brain chemistry or their equivalents, not even close.

Because Freud translated neurological and brain events into only *symbolically* represented processes, patients and physicians were able to work with their objectifications under humane circumstances at a human scale. Transference (empathy) was the central tool of emotional medicine. But unfortunately everyone woke up, fell out of the wrong side of the bed, and declared in unison, 'How about something verifiable?'

Did you bright guys all forget, we don't actually know what consciousness is? And we don't begin to know that it responds better to objectively objective than to subjectively objectified medicine.

Add to this the clinical dilemma that experimental psychopharmacists can barely (if at all) discern a difference between placebo effect and psychotropic drug influence. In fact, there is a school of thought that *all* psychopharmaceuticals, even ones like Prozac and Xanax that "improve" depression, anxiety, attention deficit, hyperactivity, and obsessive compulsion, may be placebos at best![50] At worst, they are hallucinogenic disruptors that actually *increase* symptomatology (see pp. 115–116).

Understand, no one is professing that these psychotropics are placebos in the sense of sugar pills that cause no meaningful physiological change at all while somehow actuating a positive change through the psychosomatic effects of a change in belief (mind over matter) as when a patient in a control group believes that he or she got a "real" drug. Mental illnesses are not the same as physical ones; both the mental disease and the mental cure are located ultimately in the mind or behavior (even when the true cause is presumed to be a neurohormonal imbalance, virus, bacterium, or other physiological defect).

A positive placebo effect in psychotherapy does not necessarily entail an actual physiological change or, more precisely, the *same* sort of "miraculous" physiological change as a placebo effect on an explicit organ disease. The outcome is subtler but also more devious: the administered drug could cause a chemical change resulting in a positive psychosomatic change but not (if one could track precise molecule-mood relationships) the psychosomatic change proposed; however, the effect would be "real" (non-placebo in origin) in the sense that a real physiopsychological chemical event actuates a secondary "suggestive" effect that "cures" the targeted mood. Of course at a deeper level, it *is* a placebo—that is, I am guessing that it is near impossible to tell the difference between a pharmaceutical effect and a placebo effect in psychiatric medicine because all pharmaceutical effects have cascading placebo effects too, and these must eventually envelop and baffle the chemical locus.

The point is, consciousness is not obedient to hard designation or direct approach. Psychosomatic at more than one level, it has its own reasons and responds to its own mysterious bugle.

The greater disadvantages of poorly proven drugs as against unverifiable instincts are: (one) the inculcation of more serious long-term toxic side-effects than therapeutic improvements; (two) the failure to recognize that other, more benign substances might trigger the same placebo effect; and (three) the creation of a bogus link between a belief system and an event in the world, to the benefit of a single industry and the detriment of the overall psychological, physical, and economic health of the populace or the advance of knowledge in an instance in which the future values and sanity of the culture are at stake.

The whole post-Freudian clinical obsession with doping, medicating, and mood modulation may be less a breakthrough regarding the chemical basis of consciousness than a conspiracy of follies, fallacies, delusions, and biased experimentation* followed by too-big-to-be-wrong companies' aggressive marketing of their own products regardless of their meetness.[51]

To commoditize and capitalize depression, anxiety, and schizophrenia as *organic brain diseases* created new markets with billions of dollars' worth of business, while it also alleviated both sufferers and their families of responsibility, either for the disease or the cure—passing the former on to nature (mostly bad genes) and the latter on to Big Pharma, which happily accepted the commission and supplied its heroic drugs. As chemical metaphors took the place of ids and egos, antipsychotics were prescribed in lieu of "talk therapy." Exorcism returned from where psychoanalysis had banished it half a century earlier, to be performed by corporations like Pfizer, Roche, and Merck rather than the Church:

> [B]y emphasizing drug treatment, psychiatry became the
> darling of the pharmaceutical industry, which soon made its
> gratitude tangible…. [The drug companies] showered gifts
> and free samples on practicing psychiatrists, hired them as
> consultants and speakers, helped pay for them to attend con-
> ferences, and supplied them with "educational materials."[52]

Is it any wonder that the U.S. has a runaway epidemic of mental illness in the late twentieth and early twenty-first centuries? The more psychotropic drugs that have been administered, the more widespread and serious have become the psychopathologies and aberrant behaviors that they are supposedly treating. The more Ritalin (methylphenidate hydrochloride) is prescribed and consumed, the more kids "need" Ritalin just to show up at kindergarten and first grade. The more Prozac, lithium (carbonate, citrate, and sulfate), and Xanax (benzodiazepine) that are ingested, the more bipolar and dysfunctional the populace becomes. Unhappy people are being

---

*It has been widely documented that negative drug trials were routinely discarded until there were enough positive ones to garner FDA approval (see the articles by Marcia Angell cited in the Endnotes for this section).

turned into crazy people because happiness and fulfillment are viewed as not only artificial constructs but chemically imported mirages. But then scientists think that people are machines anyway. Brave new world indeed!

There is finally no limit on how much industrial commoditization can be billed to and poured into Medicare patients. In 2011 it rivaled the entire gross domestic product. The unsolved enigma of consciousness is about to eat the house.

v. The Unconscious

Quite apart from the validity of his theory of instincts, Freud made a quantum leap in conceptualizing the provisional status of consciousness in relation to its unconscious substratum. He redefined "unawareness" as not just *non*-consciousness but a many-layered hierarchy of psychodynamic preconscious and unconscious processes. He demonstrated that we can never know what is in a "mind" because of the depth of its unrepresented components.

Freud's crowning achievement was to wrest unconsciousness—or whatever it really is—from the oblivion into which philosophy had exiled it, and to compel science to reckon with its own pagan heritage. He sought to amend the overstatement of the role of consciousness in mental life. In the process he confiscated mankind's progressive agenda, its hopes for making the universe explicit and conscious, and human society perfectible. As psychology and the social sciences began to consider the mind's ontological basis in the animal kingdom alongside its epistemological basis in individual egos, they found themselves getting separated from their own rationalist legacies.

The "Stephen Hawkings" of the world are still hunting assiduously for a trope that Freud disposed of more than a century ago: explicit reality and a conscious cosmos. The founder said, in effect, 'There is no utopian realm, heavenly city, or grand unified field theory in your future because most of the energy in the system is not only inaccessible and unrepresented but unrepresentable, untamable, irrational, and incorrigible to boot—and a good half of that is a death wish, while the other half is not exactly jolly—Eros in name only. There's nowhere to turn for the constituents of a conscious philosophy, a just and socially responsible society, a rational universe, a closed physics, or even basic courtesy.'

Beyond ordinary awareness and phenomenology, gazing back at philosophy, is the antipode: Leviathan, Dementia, Freak, Bedlam. It changes *everything* about meaning and is determinative of much of what it is.

Abstract expressionist, quantum mechanical, relativistic existential, dada, punk, and hiphop tsunamis followed this pronouncement—genocidal ones too. Together they made Freud prophetic, the unanointed King of the Twentieth Century, the Cassandra of a civilization in crisis.

●

If you want to come back at the unconscious the other, existential way, you can check out the deconstructionism of French Algerian post-Freudian philosopher Jacques Derrida (1930–2004). Derrida was the master of assertion by elision. He exported the unconscious from the objective condition which Freud had reified and made it the absolute outcome of all therapies, all sciences. He exited in the other direction: back through nature into the Void.

For all the babble and busyness of us, the planet's high politics and theater, the extravagant tableau of creature aggression, the hoopla of hungry and horny beasts, there is finally nothing here, nothing at all.[53] Which is everything. Reality (wave-particle) annihilates itself at the formulation of its own being. Why not consciousness too?

The desire mill and compulsions of humanity conceal that, whatever we are doing, we are also doing its opposite, negating each act, erasing by asserting. Everything in the world, every engine of human will, contains deletion of the same thing at another level, as everything sinks toward whatever it actually is (electrons, positrons, bosons, quarks) but also into its own *innate* antithesis.

Underneath, the universe is an unrelieved blank, an expunction of itself in a mute, pagan rain that will fall here tomorrow and five million years from now, until the Sun dies. Nothing is not nonconscious.

There is no pardon from the paradox of being, the role of the unconscious and, of course, mortality, which elides every mark, every half-baked or memorable utterance: those fragments of Heraclitus (at least the few still left) as well as parables of Augustine, Hamlet's soliloquies, verses of the Apostles, sutras of Dzogchen masters, and the runic scribing of Maglemosian pundits on swan bones—every ideation, every expedient, every design and insignia, as well as the crater left by every eradication. As we each die, we take our remaining meanings and memories with us.

By claiming the gap itself, Derrida dug a hole around the already large hole that we thought we were excavating (post-Freud), and then he broke it off lock stock and barrel: the dirt, the digging, the hole itself.

But he was also a Talmudic scholar who placed Meaning before Zero or the Sign. What else could he have been doing under instantaneous erasure of his entire work except giving birth to the pleroma by feeding the eternal question: What Am I?

What Am I? What Am I Not?

Freud skirted only the edge of the shock wave of the darkness, the black mirror and the obliteration of the black mirror. His unconscious trope slit a gash into Western culture; then Derrida took the gash to the next level, adjourning its offices, schools, curricula, hospitals, and parliaments, all occasions and academies in which the Freudian unconscious ceremoniously held forth like some sort of professor emeritus. He said: it not only is unconscious; it is *unconsciousness.* "Is" is *not.* Finally and forever. By then it was the only way to salvage "is."

CHAPTER FIVE

Systemic Consciousness:
Nonconsciousness and the Loss of Consciousness

Very little of what is going on in the nervous system enters consciousness. Its unconsciousness is systemic from biological and historical architecture as well as its psychodynamic operation, a series of sublimations embedded by series building on other series, series that (as noted in previous chapters) had implicit "semantic" content long before minded recognition. Nonconscious sentience not only preceded consciousness by millions of years but continued to be interpolated and annexed in it. This is not the Freudian unconscious but a whole other kettle of fish.

From the chemicotropism of amoebas and neural pulsations of jellyfish through the nerve hierarchies of worms, mimeses of insects, and relations between synaptic ladders and emerging ganglia in mollusks, amphibians, and reptiles, subliminal autonomic functions have been incorporated and insinuated many times over in the neural system and the brain. Mammals no doubt installed their own deep syntaxes. These templates are not all (of course) in the human lineage, but they belong to the general hydrocarbon, polypeptide grid from which that lineage arose. The mechanism and contents of biopolymer networks built out of organic chemistry by DNA are unconscious by the nature of how they are constructed—by the system itself.

Such unconscious elements are also part of the greater realm of consciousness—nothing conscious exists even unconsciously without a base in *systemic* consciousness. Yet they cannot be *egoically* conscious; they cannot even be unconscious in a Freudian sense; they are not libidinal or id-like; they lie outside of cognitive psychology too. Their unconsciousness "exists because of the way our perceptual cognitive system is constituted and lies *in principle* outside our access."[1] This realm has been called "the new unconscious," though its installation in living systems probably preceded the "old" Freudian unconscious by millions of years—that is, if the distinction even means anything. Its modernity is based only on the relative lateness of its discovery; it has been around for a very long time.

There is "nothing confused or contradictory about the idea of a consciousness experience that one is not conscious of having."[2] The majority of effects of the billions of neurons in the human body are unconscious not only by a theory of instincts or in the manner of the Freudian unconscious or in any psychological or cognitive dispensation but by the hierarchy of their organization: their pre- and post-synaptic states, the inherent limitations of awareness and attention, and the manifold filtering, abstracting, and condensation that go into neural processing. The data-rate of the nervous system is very low (relatively speaking) and the conscious awareness of even that bit-flow is minuscule compared to the flow itself. Then only one part per hundred billion of the brain's processing power is represented in consciousness. The rest is washed away in selection and specification.

Mind is a multiply-folded-over mechanism that presumably can direct itself upon representational subsystems and stages, selectively and piecemeal. "The phenomenal representations that are constructed or activated in conscious experience are normally transparent to us. We know what they represent in virtue of our capacity to instantaneously and effortlessly connect these representations with other semantically related representations."[3] Their operations depend on "control windows and other elements of conative context"[4] eliciting "awareness of current states and activities of our environment and our body."[5] We experience the outcome not the process of perception.

Far from being a random or contingent feature, the burying of activity below prior levels and the folding over into prior folds (both of brain topology and mental representations) seems a longstanding evolutionary priority. Nonconscious states are not an accident or a countermanding or restriction of consciousness or even representative of a consciousness "dump"; they are a major systemic goal of consciousness, how it builds its stability, depth, range, and capacity for instantaneous selectivity.

This is a critical point. Whatever consciousness is in the universe, its biological expression evolved in nature not merely to generate conscious activity, but to form consciousness-like layers and then bury them in non-conscious modules for efficient application. The process has no moral or ontological implication: the modules could either be the ultimate "goal" of consciousness insofar as a "goal" has any meaning in a Darwinian system, or they could be foliated so as to handle the operation of lower-tier, mostly invertebrate aspects of consciousness and to free consciousness ultimately to create a phenomenal world with self-aware windows. In the latter instance consciousness is *meant* to be the tip of an iceberg, and a late-arriving one at that. Stated otherwise, even consciousness is structurally nonconscious.

> Action tendencies can be activated and put into motion with-out the need for the individual's conscious intervention: even complex social behavior can unfold without an act of will or awareness of its sources. Evidence from a wide variety of do-mains of psychological inquiry is consistent with this propo-sition. Behavioral evidence from patients with frontal lobe lesions, behavior and goal priming studies in social psychol-ogy, the dissociated behavior of deeply hypnotized subjects, findings from the study of human brain evolution, cognitive neuroscience studies of the structure and function of the fron-tal lobes as well as the separate actional and semantic visual pathways, cognitive psychological research on the components of working memory and on the degree of conscious access to motoric behavior—all of these converge on the conclusion that complex behavior and other higher mental processes can proceed independently of the conscious will. Indeed, the brain evolution and neurophysiological evidence suggests that the human brain *is designed for such independence.*[6] (My italics.)

Our actions are (again) not unconscious merely in the way that the repressed and universal contents of the psychological unconscious are; they are unconscious in the sense that the driver of a car does not have to know how to assemble an engine or generate internal combustion in order to op-erate his vehicle. The driving may be conscious, but its commutation of the mechanisms underlying it is not. The psycholinguistic structures of universal grammar, the computational mechanisms underlying vision, and the innate parameters of strategic planning, problem solving, locomotion, navigation, and stalking (vertebrate as well as invertebrate) are likewise systemic—part of consciousness ontologically and mindlike—yet absolutely and generically unminded.[7] "Much neural processing of visual representations [as well] operates without consciousness awareness."[8] In fact, we accomplish the preponderance of our learning and behavior without actual awareness.

Take a Martian's view of the action on a busy urban street: massive nonconscious neural output, most of it on autopilot, looks a lot like the tropisms of insects. We have more neurons and more concentrated ganglia than bugs do, but that's not the point. The subliminal system encroaches upon all behaviors, individualistic goals, and executions of free will, encompassing cultural acts and personal styles too:

> [C]onscious acts of will are not necessary determinants of social judgment and behavior; neither are conscious processes necessary for the selection of complex goals to pursue, or for the guidance of those goals to completion. Goals and motivations can be triggered by the environment without conscious choice or intention, then operate and run to completion entirely nonconsciously, guiding complex behavior in interactions with a changing and unpredictable environment, and producing outcomes identical to those that occur when the person is aware of having the goal. But this is not to say that consciousness does not exist or is merely an epiphenomenon. It just means that if all of these things can be accomplished without conscious choice or guidance, then the purpose of consciousness (i.e., why it evolved) probably lies elsewhere.[9]

●

One of the central premises of consciousness, in fact, is that it requires "the postulation of intentional but unconscious states."[10] A creature could not operate if the semantic and mental features of its repertoire had to be held in attention mode and their remainders appear on the central monitor. If thought and behavior were fully conscious, they would be too unwieldy and slow for this world, and their "experiential mismatch"[11] would necessarily paralyze action. We have neither perceptual nor computational capacity to notice every detail and log every anomaly. Instant nonconscious operation is how anything gets done or represented before its moment passes. Consciousness is not a primary output but, as noted earlier, a systemic remainder.

The nonconscious modalities of consciousness have no need to pay even lip service to aware consciousness. They run their systems beneath it as if that is what they were meant to do, and we go along for the ride. This is the "new unconscious" that neuroscience is proudly excavating, thereby eroding the long-held traditional view of how conscious and intentional human beings are: "After nearly three decades of research on automaticity and construct activation, it is increasingly clear that much of human mental life operates without awareness or intent.... [O]nce triggered, the execution of automatic processes proceeds rapidly, effortlessly, and incorrigibly to completion, leaving no traces accessible to conscious recognition."[12]

Even many voluntary acts are initiated unconsciously in the brain. "Unconscious volition is not an oxymoron,"[13] as layers of kinesthetic and unconscious codes support conscious performance, and "readiness potential precedes conscious intention or urge, which precedes muscle movement."[14]

It takes constant practice and renewed attention to keep even a habituated neuromuscular activity from dropping further and further below consciousness until it becomes inaccessible. Physical and mental skills, acquired by discipline and attention, likewise sink below the threshold of awareness into the system itself such that we carry them out without minded involvement. The moves of a trained athlete—a basketball guard, a football quarterback, or a hockey center—are learned consciously, then become autonomic.

Making such a "lost" pattern conscious before altering (or correcting) is common to all educational systems. Somaticist Moshe Feldenkrais proposed that, contrary to what Freud surmised, most neurotic habits do not have bound traumatic rationales at their base; it is merely that the body-mind has lost neuromuscular awareness of their innate patterns and melded them parasitically together. His "Feldenkrais method" provides counter-intuitive exercises that bring a subject to awareness of his uncon-scious movements and then get him to separate them from each other's performance—that is, to make them conscious and contrasting. Both psy-chospiritual and athletic trainings involve discriminating actions and then keeping them conscious *in order to be able to change them.* Freeing a single unconscious pattern may release other habitual patterns in parts of the body seemingly unrelated to the exercise because of the way that psycho-somatic syntaxes are constructed and held in the brain.[15]

Martial artists learn to read the unconscious readiness in opponents and anticipate their moves, as "even an above-threshold stimulus is pro-cessed unconsciously a fraction of a second before becoming conscious."[16] Awareness intervenes to throw or block a "subliminal" punch. Such "war-rior consciousness" was trained for millions of years by canines, felines, and raptors who didn't worry about distinctions in consciousness. No wonder action forms in martial sets are named for eagles, tigers, snakes, even rhinoceroses (e.g., "monkey steals the peach," "wild goose leaves the flock," "rhino gazes at the moon").

That subjective consciousness is not necessary for conscious activity seems to support the functionalists' position, one of whom asks rhetori-cally, "If the brain can initiate a voluntary act before the appearance of conscious intention, that is if the initiation of the specific performance of the act is by unconscious processes, is there any role for conscious func-tion?" He answers his own question, but not in a way that rescues phe-nomenalism, only in a way that transfers access nonconsciousness into access consciousness: "The volitional process, initiated unconsciously, can either be consciously permitted to proceed to consummation in the motor act or be consciously 'vetoed.'"[17]

Here comes that sucker punch again! Duck this time.

Not only is a person's perceptual field a small subset of what passes into the senses but, at that, an incomplete and inaccurate rendition. We con-sciously process a great deal of which we notice a tiny fraction. The rest is present to consciousness, recorded at a latent level, but "there are things in the world we miss at each and every moment."[18]

Perceived events fall out of working memory into a twilight zone after they register, though many of them might be retrieved with the proper stimulus: "To be a sensation or other state with phenomenal qualities does not entail that it is phenomenally conscious at every moment."[19] Short-term memory is by nature diffuse, uneven, and makeshift. "The nervous system continues to represent habituated stimuli even after they have faded from consciousness."[20]

Even "phenomenal states can be phenomenally unconscious.... [S]tates with phenomenal qualities need not be felt at all times by an individual who has them."[21] Many passages of incipient awareness are brief, confused, temporally removed, or obscure, or comprise "imageless unarticulated thoughts."[22] They are mixtures of nonconscious and ordinary consciousness. We are "informationally sensitive" to them without being "experientially sensitive" or phenomenally aware.[23]

We are not commonly aware of a refrigerator rumbling, the sound of traffic, and other background noise. Unless they are called to our attention we don't see every brick in a building. Though these things are present to vision, they have "no direct consequences for the streamlike quality of consciousness."[24] They are not selected as patterns out of the nonconscious flow.

William James put it this way: "A thing may be present to a man a hundred times, but if he persistently fails to notice it, it cannot be said to enter into his experience. We are all seeing flies, moths, and beetles by the thousand, but to whom, save an entomologist, do they say anything distinct?"[25] Of course even to an entomologist these countless insects go unnoticed. Conscious attention needs to be disturbed by a novel stimulus to break through the threshold.

James referred to the raw, unorganized material of thought as "an undistinguishable swarming *continuum,* devoid of distinction or emphasis [that] our senses make for us, by attending to ... the infinite chaos of movements, of which physics teaches us that the outer world consists...."[26] He added that attention "out of all the sensations yielded, picks out certain ones as worthy of its notice and suppresses all the rest. We notice only those sensations which are signs to us of *things* which happen practically or aesthetically to interest us, to which we therefore give substantive names."[27]

Where are the feeling of a chair after sitting in it, our orientation to gravity, the pattern of leaves caused by wind in the trees, ambient light and traffic, one's own heartbeats? They are not of ongoing consideration or concern to the "I." Mind continually misplaces concreteness, as all except the most relevant inputs drop into the void. In most cases they never existed. At the same time, there is no place in which these images and events could be contingently stored except consciousness or, more accurately, among the "nonconscious" systemic aspects of consciousness.

Representational reality is and always was a fiction, even for coyotes and bears. But that doesn't foil literal-minded prowling because you don't need accurate representations; you need only experiences with outcomes.

Even the line between so-called reality and our conscious reconstruction of that reality is circumstantial and provisional, as everything is reconfiguring and rearranging itself from the instant of perception, and then again the next instant. There is no fixed map or viewing point by which to anchor the frame through which semblances pass. At each viewing, "the phenomenal field is ... profoundly *changed by* the process of coming to attend to a sensation."[28] So the little bit that is 'worthy of notice' belies itself anyway.

The mind is neither a faithful replicator nor an honest broker; it constantly fills in and, in the course of presenting reality to itself, edits and reedits multiple drafts in running the scroll of consciousness. Nothing is ever captured complete and whole without selection, reconstruction, synopsis, and censorship. We regularly improvise and reinvent our narration to ourselves, especially when reality does not accord with our expectations. We fill in blanks, absences, lesions, blind spots, supplying what we expect. In keeping experience tractable and smooth, we supply a broken or disrupted pattern with itself as if a gap were not there. Our foot reaches for the last "step" on a staircase and is jarred if it hits a floor. Experiments with people viewing wallpaper patterns prove that subjects will look right past or "correct" an inserted variation in an otherwise cohesive and consistent pattern: a mouse in a field of roses, a few lines going the wrong way—Betty Boop in a Marilyn Monroe motif becomes Marilyn. Similarly, distorted rooms of joined trapezoidal surfaces are perceived as rectangular, and not just because rectangles project trapezoidally on the retina.

Experiments also show that the mind is physiologically unable to differentiate between artificially projected realities and actual realities, so we can be fooled by virtual projections, optical illusions, laser-created landscapes, and amusement-park funhouses. The apparent always wins out over the real. The Internet is filled with flat sidewalk paintings which appear like full anamorphic three-dimensional objects or declivities and holes, as people react with shock to a giant lobster, a pond with a water lily, or a seal poking up through ice floes when there is only chalk on stone. Check out www.julianbeever.net.

●

So much for the pictorial comprehensiveness of consciousness, but what about its historical reliability? How much of what we remember and come to believe really happened? How much else is inculcated by a combination of fantasy, fiction, and the propaganda of ourselves and others?

Memories are images forged by neurons firing together. For recall those neurons must fire again and, at least by scientific explanation, remake the same actual chemistry in which the events reside (hard to fathom all this happening and only this happening). In any case, there is ample opportunity for deviation and innocent alteration, e.g., for a person to remember something the way that he or she wanted it to have happened (or feared that it happened) rather than the way in which it did. "[A]n early memory takes on new meanings and changes as a person matures":[29] memento replaces "mneme."

Images borrowed from elsewhere, even outside one's actual experience—for instance, from books or movies or paintings—get grafted into flashbacks of one's own life identities, and world-views become fusions of selective focus, partial recalls, mythologies, revisionist histories, conflations, idealizations, synopses, and summaries.

As the mind regularly uses words to elicit pictures and pictures to elicit words, so associations of images, reminiscences, idylls, and real and imaginal *déjà vus* get conflated with one another, producing magical-realist landscapes and dreamlike retrospections held together by mnemonic devices. They don't seem magical-realist to the people sustaining them, but this is what the streaming of consciousness, when interrogated, comes out to.

"You invented a life and filled in the gaps": words spoken by his handler to Dr. Martin Harris, a character played by Liam Neeson in the movie *Unknown*—essentially a make-believe person, an alias created as a cover for an international assassin. After a near-fatal car accident, "Dr. Harris" awakes in the hospital believing his cover, as his mind fills in the blanks to make an entire lived life out of it, including turning his bogus marriage to another play-acting spy into a seemingly deep-seated memory of a fond and devoted union. Discarding his assignment and long-standing profession as a killer, he fabricates a "decent" man's autobiography based on his present spurious guise as a botanist and then tries to tide himself from moment to moment by using its sparse landmarks (a photograph, a few-minutes' memory of an interaction in a hotel room, a fake bio posted online).

But this is the manner in which any of us build and follow the presumably real landmarks of our "true" autobiographies. "Dr. Harris" ultimately turns into another, third person and chooses that alias's selfhood over his prior selves and guises.

Almost the identical words ("you invented a life and filled in the gaps") were spoken to Justin Fisher in *The Language of Secrets,* a novel by Dianne Dixon. As a kidnapped child, Justin turned a homemade song and a scrapbook into a full, happy, normal life he never lived while forgetting the unhappy, fractured one he did.

These are fictions, but they are based on substantive events. Consider how kidnap victims Steven Stayner (for seven years), Jaycee Lee Dugard (for eighteen years), Shawn Hornbeck (four years), and Elizabeth Smart (nine months) adopted bizarre apocryphal identities in order to survive their ordeals—they *became* someone else.

We all do a bit of it, reinventing and idealizing our own stories, taking away rough edges and imperfect acts, unconsciously adding a dash of heroism or glamour.

A woman wishing that she had refused her diploma at graduation remembers that in fact she did. But then how did she get the calligraphed scroll she comes upon occasionally in a neglected drawer? A friend who took an 8-mm film shows it to her decades later, and there she is, reaching out to receive the ribboned document.[30]

A middle-aged adult swears that her father forced her to drive on a frozen lake at age twelve and attributes her later difficulty in learning to operate a vehicle to the trauma of that event. In fact she was twenty-two and it was an empty snow-covered parking lot with protective banks around it. One virtual reality replaced another.

People remember attending weddings and funerals that they think they should have attended or missing parties at which they appear in photos. Millions recall being present at signature sporting events in stadiums that hold less than one hundred thousand.

For all these reasons (and others) consciousness suffers from verificational indeterminacy. We cannot validate perception and memory or the chronological relation between input and cognition. The days of our lives, even as they are happening and perceived seamlessly moment to moment, lose their discrete identities at once; their events altered, simplified, and smeared. Faithful portrayals deteriorate, erase themselves, fade instantaneously. Whitehead nailed it:

> We sleep; we are half-awake; we are aware of our perceptions, but are devoid of generalities in thought; we are vividly absorbed within a small region of abstract thought while oblivious to the world around; we are attending to our emotions—some torrent of passion—to them and to nothing else; we are morbidly discursive in the width of our attention; and finally we sink back into temporary obliviousness, sleeping or stunned.[31]

●

In a classic thought experiment, someone briefly glimpses a woman running without glasses yet remembers her *with* glasses. This could be a case of the immediate image being overwritten by a more vivid memory of *someone else* perhaps seen earlier with glasses (an accurate phenomenal record cognitively contaminated) or an instantaneous misperception. Was there a moment, however brief, in which the subject saw the woman without glasses or was the image replaced at perception? It is finally a question of whether the eyes are playing tricks or the memory is playing tricks.[32]

There is no way to tell and no true rendition because "the experience would 'feel the same' on either account. Perception turns imperceptibly into memory, and 'immediate' interpretation turns imperceptibly into rational reconstruction, so there is no single, all-context summit on which to direct one's probes."[33]

The philosophical literature holds countless discussions of "Orwellian versus Stalinesque" models. In the Orwellian interpretation, reality is constantly revised, while the Stalinesque view is that the mind instantly substitutes or imposes its own Potemkin reality.

"I'm not really sure," novelist Ariel Dorfman concludes similarly in *Mascara,* "if others fail to perceive me or if, one fraction of a second after my face interferes with their horizon, a millionth of a second after they have cast their gaze on me, they already begin to wash me from their memory: forgotten before arriving at the scant, sad archangel of a remembrance."[34]

●

If conditional awareness and tainted memories of past awarenesses shape identity, and identity is consciousness, and consciousness is all we know of reality, what does it mean that we routinely fudge, cheat, prevaricate, fabricate, and alibi? Science was undertaken as a collective human endeavor generations ago in order to check our imperfect or biased renditions of reality against one another (beyond the fallibilities of subjective minds) and to come up with a working model of what's actually going on. Yet here we are, back at the crossroads.

During the 1980s lost memories of sexual abuse sprang up like mushrooms after rain, as if our society had been harboring criminals and perverts in its neighborhood childcare centers and just now awoke to their presence. As children told psychotherapists and police tales of bizarre pedophiliac orgies and acts of genital violation, a few of the previously upstanding directors and employees of these centers went to prison. During roughly the same time period, the United States experienced an epidemic of alien kidnappings including medical experiments on abductees in spacecraft.

By the early 1990s most if not all the "memories" of sexual abuse were proven to be fantasies implanted in impressionable, compliant children by interrogators using combinations of hypnosis, drugs, persuasion, and intimidation (the UFO ones are still somewhat up for grabs). I don't mean to imply that sexual abuse of children didn't occur elsewhere at the same time at the hands of teachers, counselors, priests, and coaches (for instance, in church programs and numerous schools, camps, scout troops, and youth sports leagues), but for the most part it was *absent* from the daycare centers implicated during the epidemic. Both episodes seemed to end with the public pronouncement of False Memory Syndrome. Yet the "memories" were real while they lasted, the society at large believed in the accounts (and many people continue to believe in them); innocent people served real jail time for molesting children, children who remembered "events" that were totally real to them. Citizens banded to expose and confront the invasion of the Earth by alien Greys. Again, this is not to question legitimate instances of child molestation, which exponentially outnumber those of false memory syndrome; it is simply to point to contagious outbreaks of false memory.

Rare but striking exceptions to natural forgetfulness and memory attrition hint at the baseline capacity of the human brain. Savants and stage magicians use variations of eidetic memory, mnemonic devices, and number grids to store vast amounts of conscious information. Former pro basketball player Jerry Lucas entertained his teammates with the feckless feat of memorizing whole pages in the New York City phone book.

Even without pure photographic memory, the brains of ordinary people record picture-perfect memories too, though the mind instantly ceases to be aware of them. They remain outside ordinary access and cannot be easily fetched. However, a license-plate number of interest in the investigation of a crime may suddenly be recouped by a witness under forensic hypnosis—his passive storage memory having recorded what he didn't "see."

Dustin Hoffman as the idiot savant Raymond Babbit ("Rain Man" in the movie of that name) carried out extraordinary feats of mental calculation, "counting" playing cards at casinos and adding up splatters of scattered toothpicks instantaneously. But he could not begin to conceive how to do "two plus two" arithmetic. (The cinema representation is based on actual savants.)

When people with photographic or eidetic memory or savants (idiot and otherwise) are asked they how find the targets, counts, or answers in their recall and instantaneous calculations, they say they "just see it."

Other people have autobiographical or instant-access memory and can recall each calendar day of their lives, week by week, year after year, including an astonishing array of ordinary and routine events from 20, 30, and 40 years prior: exact weather conditions, plays in a pro football game, meals by menu and dining locale, endless picayune details. If you said, "February the eleventh, 1972," they would quickly reel off what happened to them on that day.

A *Sixty Minutes* segment (December 19, 2010) profiled a study of this talent by James McGaugh, a neuroscientist at the University of California at Irvine. Numerous claimants of autobiographical memory were tested for various calendar days decades earlier and recalled substantial numbers of events accurately (as confirmed by researchers' fact-checking). Like the savants, these people "just see" stuff (though, remarkably, with a calendar and whole-life datebook attached).

Shortstop Jimmy Rollins of the Philadelphia Phillies is said to have an instant recall of all his at-bats *vis à vis* the pitch sequences used against him, an ability he activates in clutch situations; he can sort through thousands of such sequences instantaneously and "know" which pitch is coming. Wolfgang Amadeus Mozart likewise "recorded" musical notes as he heard them and committed whole symphonies to memory.

There is no credible explanation for how any of this works or why, if it occurs for some, it doesn't happen all the time for everyone. But then memory is nonlocal in the brain, an organ of which we are using only one one-hundred-billionth of ostensible capacity anyway.

●

Dreams and trance-fugues provide their own nonlinear unconscious phases. Those in states of non-REM sleep, dreaming, and hypnosis may not be consciously awake or aware of the world, but they possess the attributes of full waking consciousness. They have consciousness as a property (because consciousness as an ontological state, as noted, is made up of unconscious, subliminal, and nonconscious contents), but they are not presently cognizant of it:

"The totally unconscious person does not lack knowledge and beliefs. Suppose him to be a historian of the mediaeval period. We will not deny him a great deal of knowledge of and beliefs about the Middle Ages just because he is sound asleep."[35] These may be "causally quiescent,"[36] but the "historian aspect" of the person's mind is only "like a computer that has been programmed in various ways" but is shut off.[37] It does not cease to exist. A computer does not lose its computational grid or memory when it is shut off. Consciousness likewise can be shut off (go into hibernation) without crashing.

Events in the outside world like loud noises, the kick of a bed partner, migraine auras, aches and pains, etc., often meld fluidly into oneiric landscapes, using a dream's narrative structure to give them new identities, contexts, and meanings. This shows peripheral apprehension and recognition, even of events that do not run through awareness and do not arrive as such in memory.

How does sublimation transform objects and episodes in one medium into those in another and then modulate a steady flow between them? How does bed movement, for instance, turn into a boat in a stormy sea? What is the mediating function, and how does it effect its transformation?

In somnambulism (sleepwalking), inner and outer worlds become totally fused. The environment is traversed by a "walker" who is sleeping. His subliminal perception and peripheral vision operate without his being aware of them. Here again, the line between conscious and unconscious is situational. What actual world does a sleepwalker negotiate in his fugue? These are not incidental questions; they strike at the core of the nature of consciousness, and the fact that scientists have no answer for them is indicative of the unresolved status of mindedness.

Some neuroscientists propose that dreams may actually not take place during REM sleep but transpire instantaneously upon awakening or in a later conceptualization of dormant modes.[38] I doubt it, but how would we know, especially since the act of waking immediately contaminates a dream anyway? Even the memory of a dream changes it; in fact in each recall an original dream turns into a new dream. Likewise each retelling not only changes the presentation of the dream but *what was dreamed*—in that sense there is no actual dream, so how could it matter if it is invented whole cloth upon waking? In fact, we cannot determine whether any remembered event actually took place because we cannot distinguish sensory flow from subsequent proprioception and cognition or either of them from later contaminations of memory and languaging—"Orwell" and "Stalin" ever at work.

We are left with a topology of holes, lesions, dislocations, approximations, artifices, sublimations, and adaptations; yet as with any internalized topology, our mind is not aware of the warping because it reads its own fabric and time-line as continuous and complete, and tracks distortedly right through its very distortions, obliviously around holes and hiatuses, as if the sheet of extant memory were an immaculate replica of its own experience. Here again the map is not the territory.

What follows is an informal catalogue of additional ambiguous phenomena along the consciousness/unconsciousness threshold.

- People undergoing *petit mal* seizures wander about aimlessly. They make "no record of a stream of consciousness," thus have no memory of it.[39] "[E]pileptics who have a seizure while walking, driving, or playing the piano … continue their activities in a routinized, mechanical way despite a lack of consciousness."[40] While they have no P-consciousness, they have A-consciousness; otherwise the car would crash.

- Long-distance truckers enter trances in which they operate their vehicle and navigate the highway perfectly, make the appropriate turns, use the brake and the clutch, respond to other vehicles and hazards in the highway. "After driving for long periods of time, particularly at night, it is possible to 'come to' and realize that for some time past one has been driving without being aware of what one has been doing.… There was mental activity, and as a part of that mental activity, there was perception."[41] The lost visual and narrative memory of the trip is not recovered even when the driver suddenly regains active consciousness. "There is transitive creature consciousness of both things (the roads, the stop signs) and facts (that the road curves left, that the stop sign is red, etc.), but no awareness of that consciousness."[42]

- Victims of amnesia routinely lose large chunks of their personal history and identity, yet they still know how to act as human beings, as their behaviors implicitly reflect their forgotten histories. The lost memories are present and decisive, governing but not consciously known. Again, where do they go, and why can't they be retrieved? It is as though a simple circumstantial block can replace a entire detail-rich, decades-long chronicle. So what is identity?

In the 1940s movie *Random Harvest* based on a James Hilton novel of the same title, Charles Ranier (played by Ronald Coleman) is an amnesiac who loses his prior history after becoming shell-shocked in a foxhole during World War I. He refers to "a wisp of memory that can't be caught before it fades away."[43] In a brilliantly entangled plot, he recovers his previous life after being bumped by a car but, at the same moment, surrenders his three years of "amnesiac life" following rescue from the foxhole (which he remembered in normal fashion as its events elapsed, giving him a three-year-long adult life without a childhood or pre-shellshock past, including a marriage to a woman played by Greer Garson).

Returning in confusion to his original residence after the street accident (because he knows where to go and doesn't recall his second identity or his address or wife any longer), he eventually ends up married twice to the same woman under two different identities, neither of which remembers the other until the final scene. In the double-amnesiac's second marriage, neither husband nor wife can fully commit to each other because they are each in love with another person, who happens to be their partner!

- Psychiatric or neurological patients who have undergone prefrontal lobotomies emerge from surgery as pretty much the same individuals. A portion of the ostensible physical basis of their identity (the brain) has been removed, yet self-identity and personal meaning are unaffected. So what was cut out? Was the "personality" simply replicated by other neurons? What state of "being," if any, was lost? What then is the "meaning" of the treatment?

- Alzheimer's victims forfeit their memories by degrees, but they remain conscious and, like sleeping-sickness victims, can have dramatic awakenings. They "know" who they are but have no access to it or its meanings. They show some capacity for self-reflection, but without the tags or links that hold their separate cognitive programs together and give them context and meaning. They know "everything" but not "anything." Each thing is absolute and timeless, without context. They smile or become animated when a favorite topic or beloved person is discussed, but they cannot represent to themselves what they are responding to. These people have identity, existential being, even purpose, but they do not understand what any of them are.* A friend sent out this record of the ups and downs of his wife:

> Dementia diagnosed 4 years ago, most probably AZ. It's a slow gradual slide, slips in little steps, stays steady, slips some more. Two months or so ago she slipped another noticeable notch to me. She lost some more contact with the world, some more of her spark. She did not kid around with the young guys in restaurants anymore, withdrawn more into her own world. Driving around, she looked vacant rather than smiling, did not really notice the fall color unless I pointed it out and even then her response was lukewarm; took much stronger stimulus to get a smile—language down to single words pretty much except for "I have to blow my nose." I felt I had lost another part of her.

> Then out of the blue she showed not only significant recall of a recent event but a sense of humor about it.

> This is not a one-night wonder. She continues to improve— she is smiling more when we are driving around, noticing folks in restaurants again. She is more active in looking at books; she engages when we watch movies on TV together. One night we almost had a true conversation back and forth in bed. This past Thursday when I asked her if she wanted to call our son she said, "Yes, I want to talk to him," whereas for the past few months she said no or said nothing, looked

---

*It could be argued that none of us recognize what any of these are, but most of us at least have provisional terms for interacting with them.

confused, did not know who he was—I had to tell her it was her son. So I call him and she gets on the phone and says: "Hi Andy, this is your mother. I love you. Tell me everything." 'Tell me everything!': her favorite phrase for years and years when getting on phone with family and friends. Had not heard her say it for at least a year!!!!! My turn to be totally utterly blown away. The clock is turned back at least six months at this point and who knows what will happen next?[44]

So what is memory, what is identity, what is self?
And what is love?

- People in comas are not aware of their body or environment but are incipiently conscious. Oliver Sacks reports: "Patients who suffered but survived an extremely severe somnolent/insomniac attack of [encephalitis lethargica] often failed to recover their original aliveness. They would be conscious and aware—yet not fully awake; they would sit motionless and speechless all day in their chairs, totally lacking energy, impetus, initiative, motive, appetite, affect or desire; they registered what went on about them with profound indifference.... [Yet they] retained the power to remember, to compare, to dissect, and to testify."[45] They kept their identity and beingness but without awareness of or interest in it.

- Patients under anesthetics exist in a twilight state such that they may be unconsciously aware of events during surgery. These suppressed "memories" can be traumatic because of the invasive, even horrific things being done to their body, or critical comments that doctors make about their innards or medical condition. An outsider literature documents the emotional and behavioral consequences of "remembered" events under surgical anesthesia.[46]

- Metabolic and hormonal chemicals, neurotransmitters, enzymes, and coenzymes, including dopamine, cortisol, and serotonin (both at base levels and spiking frequencies), play profound roles in the activation and texture of consciousness through neurotransmitters (as already discussed), yet their effects are unconscious.

Psychotropic drugs introduce exogenous chemistry through these cycles into the mind and personality of a patient, but in such a way that he does not thereby recognize his emotional shifts and thoughts as artificially altered chemical events (either before or after medication). He accepts and believes in his own moods as indivisible, existential events.

By the manner in which they inform identity and generate experience, the hormones (and drugs) become as real and valid as anything else.

Of course everyone operates all the time as if he or she were *not* chemicotropistic colloids, as if they owned their own experiences. They do not attribute their pristine states of consciousness, beliefs, or desires to their body's chemistry or or neurotransmitters, nor do they believe that drugs (of any sort, in fact) can change *who* they are. Even a person deep inside an LSD or ayahuasca trance regards it as arising from a baseline of selfhood.

To patients in psychiatric treatment, drugs and chemicals are mood adjusters, amplifiers and suppressors of existing states of being, not those states themselves or their innate meaning. Most people who take Prozac or Xanax do not think that *it* is reality, though they *do* think that it is influential, in fact the difference between suffering and happiness.

Again, where does mind start and the brain end? Where does the window of identity open out of an impressionistic flow of chemicals? Is there anything to us at bottom? Or is it all just hallucinations and mirages in chemicohormonal fog?

The fundamental unresolved issue of whether a feeling and the behavior engendered or instigated by it have an existential "meaning" (as well as their obvious biochemical/hormonal meaning) lies at the heart of psychological life as well as psychiatric treatment. If all that is requisite to antidote a painful or unwelcome emotion or to change dysfunctional or socially marginal behavior is a pill, and if there is no legitimate and authentic route to clarity via talking therapy, psychospiritual meditations, or manual (somatoemotional) medicine, then the independent basis of a mind, a self, a personal identity, and even a personal existence is medically delegitimized. Acts of mutual care, therapeutic intimacy, comradeship, basic rapport, and community aid going back to the Stone Age suddenly are downgraded and marginalized as if ingenuous and fatuous distractions from the sole "real" human movers: drugs, enzymes, hormones, and medical triggers.

If, as seems likely, pharmaceuticals (as well as shock treatment and lobotomies) merely shift, deflect, or modulate identity and its meaning or temporarily suppress some of its dysfunctional behaviors, then something else (e.g., rigged metaphors of self and of the primordial sensations of being and well-being) is occurring under their influence. Their chemical subtext functions only secondarily as a kind of psychosomatic propagandizing *vis à vis* the baseline of normality and health. Then mind, identity, and self retain their discrete basis, even though it may be camouflaged or embargoed from personal access.

This debate doesn't have a single solution or even a single trajectory, but it extends the basic mind-matter conundrum: How does an experienced interior self differ from a biological self, a pathologized self, a medical-object self, or a societal self (e.g., a mind from a brain under successive indoctrinations)? How do we get inside an inside, which is not just our body? How do we create, encompass, understand, and live a dual mind-body reality?

- The mind seemingly originates in the brain, but its content and messages may well come from differential layers of molecularization within cells and their interpolation in tissues. The intestines alone constitute fully decentralized ganglia—a so-called "gut brain" that functions similarly to the cerebral one at a noncognitive level, regulating not only digestion and elimination but their images and moods.[47] Other organs likewise have consciousness-like intelligence and even personality signatures and expressions that don't intrude consciously on the cognitive stream.

Early embryologists considered that various discrete tissue complexes might actually function and "think" independently like jellyfish, mollusks, worms, etc., for having evolved from them over eons; that is, conferring subliminally with one another like vestigial animals in symbiosis. If so, it is no surprise that such "animals" would each retain their rudimentary personae and mindedness.

116

Fragments of thoughts and behaviors stream psychosomatically from all our organs—liver, kidney, heart, spleen, etc.—into personality states, expressing their requirements and meanings in attitudes, bodily postures, personal concepts, and styles of behavior: facets of the serial flow that get bound as mind. One can actually feel heartfulness or lily-liveredness or exhibit both gall and spleen, and these may be more than figures of speech; they may be attributes of organs passing into partial consciousness.[48] Even individual cells and subcellular units (organelles) contribute their quanta to the overall throb of being.[49]

In numerous incidents (beginning with the transplant of the heart of twenty-four-year-old car-accident victim Denise Darvall into the chest of fifty-four-year-old Louis Washkansky in Cape Town, South Africa, 1967), heart recipients have reported developing emotions, personality elements, feelings, and even fragmentary memories foreign to them that turn out to be, upon investigation, almost certainly from the donor (also in cases of transplants of other organs to a less documented degree).[50] An explicit, detailed set of images of a high-school dance from a deceased woman that arise spontaneously in the thoughts of an extraneous man (who merely received the attendee's heart) is shocking in its implications. It is not disputable or vague in the manner of a mere mood or hormone like picking up a donor's anxiety or phobias. Transferred organ memories are discrete artifacts of something crucial about the nature of mind and being, and they speak to an unexplored scientific heresy.

Neuroscientists and psychiatrists file such testimonies (if anywhere) in the outermost alcove (or the trash receptacle), but tantalizing hints of their veracity open a whole new level of the issue of how and where organic consciousness establishes and then represents itself in body-minds. If some meanings and feelings can be transferred with an organ, especially one that's not a brain, then *what's that?*

- A wide range of identity disorders and neurological deficits lead to fissioning of consciousness such that some people experience their actions as those of another and refer to themselves in the third person. A child will tell a story about "Charles going to school" when he himself is Charles. These "evictions" and schisms are not game-playing; they are legitimate displacements of self.

Other people dissociate parts of their body tenably enough to think of their arms and hands as committing acts of someone else for which they are not responsible. Again, this isn't a game; it is perceived reality, an extrusion of part of the mind-body self: "a split develops between an 'I' who issues orders and the 'he' who carries them out."[51] There are now two consciousnesses, two selves, in a single mind.

One person in this divided state describes the circumstances whereby he goes out while staying at home: "I don't want to go. My brother takes me by the hand and leads me to the cab waiting outside. I see the driver there munching a carrot. But I don't want to go.... I stay behind in the house.... I see how 'he' stands by the window of my old room. He's not going anywhere."[52] Pronouns get scrambled, as "an imaginary double ... performs on the stage of memory."[53]

This also speaks to the controversial "multiple personality" or "dissociative identity" disorder as it has been reported in cases like those of Billy Milligan or the pseudonymous Eve White—controversial because some scientists believe either that the people are "faking" their add-on identities (like theatrical roles) for various self-serving purposes or that their psychotherapists are misreading manifestations of psychosis as if distinct, immaculate personalities.

Alter-personalities in these individuals, if authentic, differ radically from one another. A wild, fun-loving persona may be followed by a compulsive, moralistic one. A talented dancer becomes a bumbling oaf with a distinct foreign accent. One persona may be a macho man, another a coy woman. Again, these are totally distinct people who seem legitimately not to know each other or even, sometimes, of one another's existence.

Billy Milligan presented twenty-four distinct individuals: One had almost supernatural strength; the others lacked that capacity while some were actually frail. One was highly knowledgeable about technology; the others were tech dummies. Only one could play the harmonica and did it very well. "Billy" was alternately a thug with a Brooklyn accent, a woman named April with a Boston accent, an upper-class snob with a Cockney accent, a feminine cook, a class clown, a self-identified Jew with a Jewish accent, and so on. When one of these subpersonalities was charged with rape and murder, it brought the whole complex to light (while raising doubts as to whether it was real or Billy's strategic defense).[54]

Do legitimate multiple personalities in one person comprise different phases of an individual consciousness or different conscious identities altogether? What is the compass of private aware "beingness" or ego self if identities can split and generate phases that cannot or do not access one another?

- Embryos *in utero* are conscious but have no symbols or social concepts. Are they cognizant (though without language or an external world to project into), or do they gestate in a state of dormant percipience, or are they completely unconscious and unaware? When does each life (identity) begin?

A related issue: do newborns feel pain (for instance, during circumcision), meaning not 'do their neurons work?' but 'do they conceptualize their input?' The same argument has been applied to circumcision of infants as to animals being butchered: the knives are felt but the pain is not grasped or represented so does not really hurt. (But is not all pain adventitious, the result of a synapse tripping a signal into the central nervous system, which gets interpreted as either irrelevant or ouch! in the brain? See immediately below.)

The general obstetrical position is that babies in the womb (and immediately after birth) are undeveloped personalities without organized knowledge, desires, or concepts. They are raw genetic templates, ciphers and blank slates upon which lives will be written. However, a large body of secessionist research regards pre- and peri-natal infants as possessing a sophisticated spectrum of knowledge and a range of intentions and agendas as well as the telepathic capacity to read and interpret events occurring in both the mother's consciousness (and unconscious mind) and in the world around her. Not only can the unborn baby read these in a novelistic sense; he or she can discern latent meanings beyond what the parents and other "born" people know. In fact, the unborn baby seems to pick up things well beyond its cognitive capacity, so some of this capacity would have to be paraphysical—not associated with the brain at all, as if the fetus were not only human but a transpersonal telepath.[55] Babies almost always lose their prenatal clairvoyance after leaving the womb.

If hearts can cache and transfer discrete experiences and the feelings associated with them, where are the system's boundaries? The sky's the limit. See Volume Three, Chapter Seven, for a full discussion of this phenomenon and its implications.

●

As just noted, pain is not a direct experience with a fixed meaning but a stream of signals sent to the brain for valuation, a fusion of neural synapses with sheaths of cognitive, cultural, and psychological concepts making the input "hurt." *Pain is an interpretation.* Mostly, of course, it "ouches," but that is still a judgment, a reaction to a chemico-electrical event. Under some circumstances "pain" hurts a lot. Under others it hurts, but it doesn't matter. Under still others it doesn't even "hurt." I return to this deceptively complex topic in Volume Three too.

People with various neural deficits or after lobotomies report feeling pain but not minding it. That is the whole point of drugs like morphine, Demerol, and Vicodin.

Acts of reckless bravery and martyrdom in combat require separating phenomenal awareness from the pain feedback loop. Some people stoically undergo extremes of sensation that make others crack, while military torturers connive horrific new techniques to get "intelligence": waterboarding, hanging in midair, genital shocks, decibels of noise, sleep deprivation, dismemberment. They are vicariously assessing the degree to which pain is situational or existential by varying the input of agony according to their own fantasies. Torture becomes a furtive mixture of science fiction, performance art, sadistic theater, and snuff pornography.

This is a plea for Dick Cheney, Donald Rumsfeld, John Yoo, and their fellow advocates of "waterboarding is not torture" to cop—not that they deny the reality of pain but that they define torture relativistically so that tolerable (or at least legal) extenuation is distinguished from intolerable (or illegal and unethical) pain. A victim of course doesn't have similar leisure.

Ethics, international law, and our social compact rest upon such elusive standards. The choices we make about the link between identity and consciousness (neuronal flow and pain) inform and become choices we make about self and other, reality and illusion, and "fake" and "real" suffering, whether of torture victims, underpaid workers, partners in forced marriages, or bullied and hazed children.

If a particular pain is brief and instantly fatal, it may have minimal existential consequences—a bottomless *koan* all by itself. A surrogate argument made by jihadist leaders to suicide-bomber candidates while training them is that death so quickly follows device ignition that the explosion tearing oneself apart is felt as "a mere pinprick." A pin is then briefly inserted in the trainee's finger to show how meager the sensation.

Fly a plane into a skyscraper; create and then enter a towering inferno. Yet you and the passengers feel nothing because the sensation has no time to register before experience ends. This recalls the quip: "I don't mind dying; I just don't want to be there when it happens."[56]

The dying jihadist is also taught to categorize his or her experience not as pain but euphoria: *"Allahu Akbar!"*

Underneath all of this too hovers the ubiquitous question, "Who am I?" and its corollary, "When or how do I divorce myself from things happening to 'me' and their meanings? And then am I still me?" This is likewise (as discussed earlier in this chapter) the crux of psychotropic or psychedelic phases of consciousness, just in a different format. We already (before any mind-altering substance) hover above an irreconcilable, often terrifying *reductio ad absurdum:* If pain is contingent, identity is too. Once again, what is consciousness, and what is a first-person self?

- Sensations and pains "do not essentially involve relations between persons and parts of their bodies,"[57] as limbs that have been amputated continue to be perceived as phantom appendages. Likewise, sufferers of severe neurological disorders are often unaware of them, as they are "influenced by contents that find no access to conscious representation."[58]

  Most of the population, conversely, experiences some measure of so-called psychosomatic pains that are "only in their minds."

- Each of the cells in multicellular bionts is potentially an independent life form. This suggests a bottomless conundrum like imagining alternative universes. Conjure up each of the separate potential consciousnesses in *every one* of your cells. Are they also *you?* (See Duncan Idaho's gholas, pp. 74–75.)

Further up the neuronal ladder the brain has twin hemispheres as well as quarantined levels within these, so consciousness is routinely split into activities that "coexist but mutually ignore each other."[59] There are at least two independent consciousnesses, left and right, in all people, each a fairly self-contained processing system: "[E]veryone has two minds, but ... we don't notice it except in these odd cases because most pairs of minds in a single body run in perfect parallel due to the direct communication between the hemispheres...."[60] Each hemisphere is "said to be dominant for different functions."[61] They communicate with each other through four neuronal cables, the thickest of which is the corpus callosum.

Is our internal cognitive stream a monologue or dialogue with ourselves or multiple dialogues at different levels involving separate personae? Do Billy Milligan and his "friends" help clarify this dilemma or simply blast it into a whole other dimension? What about the multiple layers and "personalities" of all unconscious or interconscious exchanges?

"[F]or persons whose corpus callosum ... has been cut, the left-hand side of the brain (for right-handed people) is not aware of the activity in the visual system taking place on the right side, whereas in a normal person it is."[62] Is this a matter of actual awareness or simply a loss of capacity for representation? "[T]he right hemisphere of a split-brain patient might have awareness although its disconnection from the verbal left hemisphere precludes verbal report."[63]

Can each hemisphere of a split brain potentially cultivate its own individual first-person identity? Or is there a root self that transcends brain anatomy? Neuroscience explorer Siri Hustvedt elucidates:

> In the 1960s, Roger Sperry began his experiments on patients who, due to intractable epileptic seizures, had undergone an operation called a commissurotomy: their corpus callosums had been severed. "Each disconnected hemisphere," Sperry said..., "behaved as if it were not conscious of the cognitive events in the partner hemisphere.... Each brain half, in other words, appeared to have its own, largely separate cognitive domain with its own private *perceptual* learning and memory experiences." Are we two or one?

Among the stranger stories in neurology are those of people who appear to be torn in half. Their doubles take up residence in their own bodies, and their right and left sides do battle. [See above.] When the right hand buttons up a shirt, the left one unbuttons it. When the right hand opens a drawer, the left slams it shut.[64]

Hustvedt is asking, "Who owns the self? Is it the 'I'? What does it mean to be integrated and not in pieces? What is subjectivity? Is it a singular property or a plural one?"[65]

- We also pick up our familiar "selves" and identities subliminally and nonconsciously from the behavior of others and the culture itself: "The nonconscious mimicry of others is a continuous part of everyday existence and has become seamlessly woven into our daily lives. Because it is such a fundamental feature of our interactions with other people, we are not usually aware of the large role that mimicry plays. In our evolutionary past, such automatic mimicry was probably necessary for physical survival. If this was the case, then those who developed this capacity to automatically 'do as they see'—acting first and thinking later—were the ones to survive. In this way, they may have passed on the tendency to automatically mimic to their descendants, and it became the default tendency."[66]

Whether or not the Darwinian explanation by itself is sufficient or even valid, we certainly resonate toward observed and modeled behavior patterns. Psychosomaticized empathically to a sublime degree, *Homo sapiens* enacts mimicry in nonconscious waves across whole societies, creating fashionable and hip personae and other strategic and "hoodlum" guises, personalities that become aware only secondarily (if at all) of their own exogenous sources. To themselves the behavior seems personal, intentional, free-will-motivated. In that sense we are all actors in biocultural dramas, unaware that we and everyone else are matching each other's costumes and masks, performing others' gestures and countenances—some ancient, some recently in vogue.

- People with simultanagnosia have an inability to view multiple things simultaneously, while those with anosognosia are unaware of their own deficits such as blind spots, garbled speech, paralysis, etc. Prosopagnosiacs have impairment of face-recognition capacity following brain damage. They see the faces, and they recognize the names of people, but they can't make a link between the two, failing even to recognize pictures of their closest relatives or of themselves. "[I]ndividuals with speech impairments such as global aphasia may be unable to report their inner states." Yet there is a covert recognition factor in all the above conditions: "unconscious perception of neglected stimuli."[67]

Blindsight is the capacity of individuals with damage to their primary visual cortex to receive semi-subliminal information in those optical areas regardless of the debilitating condition. For instance, when figures of two houses, one normal and another on fire, are shown to subjects in such a way as to require them to use their injured pathways, even though they ostensibly can't see either figure they choose the normal house over the burning one at a rate well above chance. Likewise, they evince consistent preference for a completed image of a cocktail glass over an interrupted one. They also prefer a figure of a vase with flowers to an empty one. When asked what they see, they say, "Nothing"; they claim they are merely guessing. They give as their reason for choosing one unseen house or piece of pottery over another the fact that it "felt smoother" or was otherwise more comforting.[68]

Some researchers interpret blindsight as "residual functioning of primary visual cortex," and "small island[s] of functional visual cortex."[69] Others implicate "light from the scotomatous region of the visual field reflecting off other surfaces into regions of the visual field represented by intact primary visual cortex, or directly by residual functioning of lesioned areas of primary visual cortex."[70] The scattering also could come from the retina itself or objects outside the eye, but it is unclear how that would contain sufficient information for accurate guesses.

Additional suggestions include the use of the lateral geniculate nucleus as a pathway between the retina and extrastriate visual cortex[71] or a "subcortical visual system, which consists of projections from the retina to the superior collicus, and onto the puvilnar and cortical visual areas."[72]

Prosopagnosiacs, when shown pictures of familiar faces or pictures in sequences in which a famous semantically related face is presented first (Prince Charles before Princess Diana), select the correct name (Diana Spencer) more often than when the pictures are in random order. Because "there is a breakdown in the flow of information between conscious and nonconscious mental systems,"[73] it would seem that "the face recognition system is intact in these patients but prevented from conveying information to other brain mechanisms necessary for conscious awareness."[74] Once again, lesions as well as links form between domains of recognition and knowledge—between the material basis of consciousness and consciousness itself.

The means by which organisms receive information and experience are never limited to explicit neural and sensory channels but also come from either unknown acuities or unrecognized alliances of known capacities, in either case operating at subliminal levels. Everyone picks up ulterior, latent, or other buried and suppressed stimuli and receives information subliminally without a clear sense of how it is arriving in their awareness.

These capacities and compensations ironically open the door a crack to telepathy and other psi phenomena while showing the extremely complex, diverse, diffuse, and subtle ways in which organisms collect and process information: "Consciousness [is] a graded property."[75] There is a great deal of subliminal perception in both normal subjects and those with neural deficits (and of course paranormal subjects in psi-testing labs).

As with blindsight and autobiographical memory, psychic students are encouraged to "see" auras, images of disembodied entities, and past-life pictures by "looking with feeling" rather than through conventional vision. A seemingly impossible object or event is then spontaneously viewed. The seeing is automatic, instantaneous, and mysterious without a physiological mitigation. Aside from the obvious issue of the reality or validity of the "vision," corollary questions arise: Is a live image actually stimulated on the optical cortex? Is it subliminally intuited? Does it come through extrasensory senses and/or unflagged neural channels? Does it exist in consciousness but not the senses? Or is it solely imaginal and contrived like a figment evoked in a daydream?

There will always be "subliminal cues in problem solving and semantic disambiguation."[76] How do we "know" what we know and organize the inputs we receive? We cannot even disentangle the conscious from the unconscious or either state of mind from potentially paranormal sensory sources because there is no paradigm or experiment that can discriminate or isolate modes of information—how either get into the neural flow and how the ganglia use their own "syntax" to sort and interpret them (or in fact what is supplying that syntax, and where and how *it* arises).[77]

To draw a firm line between normal and paranormal perception disregards the logical verity that there is only one universe, so any capacity, if it exists in that universe, is both normal and paranormal—the distinction between them a semantic (or ideological) quibble. It is "paranormal" that we have sensory pathways at all; likewise it is "normal" to use whatever operating channel presents itself, sanctioned or not, explicit or borderline, arrant or subliminal. They all go into and come out of the same wash.

As long as consciousness is up for grabs, the conditions and sources of particular consciousness are also up for grabs, and we can never be sure of where information is arising and from what nexus or position (self-identity) it is being actively organized into images, thoughts, contexts, meanings, and actions. Experiments alternately expose and exclude neural and cerebral mechanisms and discriminate them from one another, but they do not get at the source illumination of consciousness or the existential cohesion and sense-making penchant of its beam of awareness. Biological consciousness is not only a scientific hodgepodge; it is a fugitive state resting upon a hub that is forgeable and frangible.

CHAPTER SIX

Qualia or Zombies?

"I started a joke, which started the whole world crying,/
But I didn't see that the joke was on me, oh no."

—BARRY GIBB, MAURICE ERNEST GIBB, AND ROBIN
HUGH GIBB, THE BEE GEES

Philosophers use the term "quale" to express "a unit of interior reflectivity or of qualitative character of experience";[1] its more commonly used plural is "qualia." Qualia are phenomenal properties that define how consciousness *seems* to its possessors—its "'raw feels' ... [its] unique subjective infusions in the mind."[2] The introduction of qualia to the philosophical lexicon acknowledges that "there is something it is like [for an organism] to *be* that organism";[3] furthermore it is a way to characterize a subjective basis of experience that is, in principle, applicable to a discussion of all life forms that act in relation to their environments, including organisms on countless "other planets in other solar systems through the universe."[4] The way in which a quale is a unit of consciousness is intended to be comparable to the way in which a joule is a unit of heat, though it is hard to validate it as anything more than a metaphor. Yet it expresses a belief system: if it exists, it can be measured and quantified. Although functionalists do not believe in qualia, the mere use of a term applicable, like hydrogen or carbon, to other planets in other solar systems legitimizes consciousness's universal *material* basis.

There are three schools of thought on this: the idealists for whom qualia are absolute (and who do not even require such a neologism), the medium functionalists for whom qualia are a placeholder for a universally applicable (if undiscovered) link between matter and consciousness, and the extreme functionalists who deny that qualia exist.

Even in the most transparent observational circumstances qualia are both obvious and elusive, simple and complex, immediate and unanalyzable. They have no structure or location, yet are conceptually and phenomenally irreducible. They are at once atomistic, homogenous, "ineffable ... private ... directly or immediately apprehensible in consciousness."[5] It is their spontaneous "intrinsic, introspectively accessible properties that are wholly nonintentional and that are *solely* responsible for their phenomenal character."[6]

Nonintentional? For sure! Who asked for any of this?

Baseline qualia include blue, green, red, yellow (e.g., any hue or blend—supposedly humans can distinguish over a hundred thousand of them): brightness, saturation; loudness, softness, shrillness, or droning (the ear has an auditory range across almost ten octaves); sweetness, sourness, bitterness, stringency; muskiness, pungency, putridness, peppermint, cinnamon; pressure, heat, cold, pain, contrast (the fingertips can discriminate differences down to 1/10,000 of an inch as per the raised dots of Braille). Compound qualia characterize moods, emotions, states of being, and so on, as long as awareness of them is subjective. Since qualia are immaterial though indivisible, they have no true boundaries from one another, and they exist collectively in (and as) our sensory awareness of aspects of our own existence.

Materialists dispute that mental phenomena require a subjective designant, as this would mean gratuitously extending thermodynamic laws to intangible entities, and inviting unmeasurable properties into nature. Qualia to them are a frill; reality can be conducted satisfactorily without them, as creatures are operationally complete prior to any subjective experiencing of the world. Neo-behaviorism substitutes an objective (or solely behavioral) parameter for an existential one, "in order to have nothing left over which cannot be reduced"[7]—"no room for the possibility of the presence of consciousness in the absence of external behavior."[8] Consciousness *is* what consciousness *does*. End of story.

In a piece entitled "Quining Qualia" in which he intends to demolish the vermin—that is, facts with "special properties, in some hard-to-define ... or special ... way"—functionalist Daniel C. Dennett adopts a disingenuously empathic tone:

"Look at a glass of milk at sunset; *the way it looks to you*—the particular, personal subjective visual quality of the glass of milk is the *quale* of your visual experience at the moment. The *way the milk tastes to you then* is another, gustatory *quale,* and how it sounds to you as you swallow is an auditory *quale.* These various 'properties of conscious experience' are prime examples of *qualia.* Nothing, it seems, could you know more intimately than your own *qualia;* let the entire universe be some vast illusion, some mere figment of Descartes's evil demon, and yet what the figment is *made of* (for you) will be the *qualia* of your hallucinatory experiences."[9]

To Dennett, qualia are akin to unicorns or Santa Claus—an excrescence: "They *do* nothing, they *explain* nothing, they serve merely to soothe the intuitions of dualists, and it is left a total mystery how they fit into the world view of science.... Epiphenomenal qualia are totally irrelevant to survival."[10] The "real world" needs qualia as much as it needs Bugs Bunny or the Abominable Snowman. They are a superfluous overlay of *something else,* concocted in order to privilege personal reality and valorize our existence. That "something else" (to Dennett and crew) is the nebulous data-processing activities of organisms. Just how quixotically nebulous (and shifty) these are—even after the last four chapters—you will soon see.

But Dennett's central argument is that not only do qualia not exist and do not *need* to exist; they *could not* exist.

Epiphenomenalism is materialism's Occam's-razor stab at ridding science of qualia for good. It posits that, while phenomenal consciousness may "exist" (epiphenomenally in fact), it has no "bump" or thermodynamic effects, hence exhibits no impact anywhere in nature. Thought is a mirage without mass or gravity—a ubiquitous and resplendent mirage but a phantom all the same, an equant invoked by idealists to "explain" presences and activities that have a far simpler explanation. "If consciousness is an epiphenomenon, that is, not *essentially* linked to causal processes, or is only a recipient of but not a contributor to effects in a causal network, then there exists the possibility that the same organism that is taken to possess consciousness could be going through the very same mentations and behaviors even if it had no phenomenal consciousness at all."[11] We've been there back in Chapter Two. P-consciousness is superfluous, so who needs a discrete category for every mirage? What appear to be concrete phenomena of consciousness are merely epiphenomena of functions, the spin-off of what organisms *do*. Pigs and puppies are fancy machines holding varying illusions of their higher status.

Whether, by granting "mind" its own rubric, even though an immaterial one, epiphenomenalism is more permissive than hardcore materialism is moot. In materialism "mind" has to be material in some fashion because it exists at all, so it is considered a secondary attribute or by-product of material systems. In epiphenomenalism "mind" does not have to be even a secondary effect of matter; it can be a hallucination or phantasm of some unknown sort. The two definitions effectively converge.

But either is also saying that mind apparently exists but *does not really exist;* i.e., exists as a by-product of real activities but is not itself substantial because the same activities could be carried out in its absence, with the same meanings and the same entropic displacement. It is not even as bump-like as electricity or puff-like as air.

So what's consciousness doing there, and why weren't "its" activities carried out otherwise? Why do we have, so far as we know, only consciousness with qualia? Why did the universe create "Daniel Dennetts" with convictions that qualia do not exist? What is *their* thermodynamic basis, complete with biases? "Epiphenomenalism ... can't explain how its putatively causally impotent phenomena could get one to worry about the mystery of their existence in the first instance."[12] After all, "beings without qualia" (or who declare themselves "without qualia") still have enough of a sense of what kind of excrescence they are to employ language "accurately." They depict their own situation and deny qualia, as they discredit the very basis of their experience. It's a paradox inside a paradox. What mirage could they be experiencing so powerfully that they would dream up a concept like qualia only to misapply it to something else? How and when does a humanoid machine decide to prevaricate or deceive?

Subjective being disappears behind its own grin but still grinning. And materialists never did like the confident mien of that grin.

The value of epiphenomenalism at its origin was that it functioned as a stand-in for the thing that was presumed to exist but didn't, so it set new terms for the discussion about "whether consciousness was part and parcel of the causal network that was responsible for the decisions we make, actions we take, and so forth or whether it was just an idle spectator, riding along causal processes, perhaps being caused by them, but without exerting any causal effect on those processes itself."[13] If the latter (as implied), then consciousness is an illusion. Still that grin....

Nineteenth-century Darwinian biologist Thomas Huxley, an early advocate of the epiphenomenal position, reasoned: "The consciousness of brutes would appear to be related to the mechanism of their body simply as a collateral product of its working, and to be as completely without any power of modifying that working as the steam whistle which accompanies the work of a locomotive engine is without influence upon its machinery."[14] He went on to say that "in men, as in brutes, there is no proof that any state of consciousness is the cause of change in the motion of the matter of the organism."[15]

He was proposing that, just as you can turn water into a whistle without any subjective intermediary (like consciousness), you can turn biochemistry into behavior. Concentrating on access consciousness while ignoring its phenomenal stepsister, functionalists presume (as noted in Chapter Three) that once we become sophisticated enough in our tools of physics and neuroscience, we will get to the real source of these exotic hallucinations and do away altogether with qualia (and any other props that falsely distinguish subjective from behavioral actions).

By making consciousness into a hallucination, these folks have turned a mystery into an oxymoron.

To dualists (who support the independent existence of qualia and phenomenal consciousness as elements of a radically different domain from matter, yet an equally autonomous one), there *is* something extraordinary about consciousness, a quality *at least* as basic as mass, gravity, or momentum, without which a full description of nature and the universe is impossible. Consciousness is the universal medium in which everything (and everything else) is suspended or subsumed. More than that, though, it is "a fundamental process in its own right, as widespread and deeply embedded in nature as light or electricity.... Along with the more familiar elementary particles and forces that science has identified as building blocks of the physical world, mind must be considered an equally basic constituent of the natural world. Mind is, in a word, elemental, and it interacts with matter at an equally elemental level, at the level of the emergence into actuality of individual quantum events."[16]

That couldn't be a more opposite diagnosis. What to the epiphenomenalist is nothing, to the dualist crowd is at least half of everything and probably a lot more: electrons, photons, and gravity are pretty fast company, and states of mind leave even that elite club in the dust. It is not an unwarranted status given that, after epiphenomenalism's fancy kinetic shortcut and the dualist and phenomenalist critique of it, we still do not begin to know what "thought" and "behavior" are or how we experience their existence, or in what substantial medium their illusion emanates. Functionalists do not recognize any of these as problems, for they are, after all, only epiphenomena.

To a phenomenalist, the perception of the color "red" is fundamentally and ontologically different from a physical description of "red" and how redness is produced in nature and then transmitted and made neurally red in the brain. This quale is red's *experiential redness*. It is different from oscillations of electrons or the physical properties of light that produce a vibration and the synapse through which that vibration enters the mind.

To functionalists, though, there is no separate phenomenal red, for "two experiences with the same intentional content must be the same in all mental respects."[17] "Red" itself is an independent fact that does not have to be experienced to be totally and fully, irreducibly whole-enchilada red. Furthermore, red's so-called quale adds nothing to real red except an abstract and indefinable leftover (that we happen to call "red")—"red" is red enough without it.

The phenomenalist rebuttal is commonsensical: "The intentional object of experience should not be confused with an intrinsic quality of the experience itself."[18] If we grok something in our mind, the grokking is a reality too, a different reality from the thing: "[A]n important aspect of a person's mental life cannot be explicated in purely functional terms."[19] The thought or sensation of an object is not the object itself or an incidental adjunct to it; it is its own thing. There is "a gap between intentional and qualitative contents of experience"[20]—and what fills that gap is more than just an empty set sailing off the edge of reality: qualia are *real and additional.*

Even if you "know exactly what the chemical make-up of ammonia must be," you would still be unable to predict the effect that a molecule of ammonia with its molecular structure must have "when it gets into the human nose. The utmost that [you] could predict ... would be that certain changes take place in the mucous membrane, the olfactory nerves, and so on...." You cannot know the peculiar smell of ammonia; that smell is new information.[21] That acrid jinni enveloping your being as you yank the stopper from a bottle *is something.*

Likewise there is no way to predict from the chemical formula of a skunk's effusion what it is like to smell "skunk" from a passing car. The aroma of bread is different ontologically from the chemistry of the loaf. Fear has a profound interior resonance above and beyond its "fight or flight" trigger. "This [intrinsic] quality of experience cannot be captured in a functional definition, since such a definition is concerned entirely with relations, relations between mental states and perceptual input, relations among mental states, and relations between mental states and behavioral output."[22] That quality is necessary, in effect, to complete the universe because *this* universe is also experienced, known.

No robot lion comes out of a lioness's brood; a flesh-and-blood cub not only behaves differently from a robot; it *is* different. It has secondary qualities that require primes (qualia) in order to run. No network inventory, however thorough, can explain either the source or properties of those primes in a living creature, either an animal or a human. It is just different—dead-reckoned and "smell test" different, both *a priori* and *a posteriori* different—from the same occurrence without them: "[S]omeone who had a complete knowledge of the neurophysiology of a mental phenomenon such as pain would still not know what a pain was if he or she did not know what it felt like."[23] There would always be something left over: a subjective experience of "pain" (or "red") in the flesh—a quale that is part of the universe and without which reality is a ghost of itself.

The Christian slave on whom the Roman character Androcles was based would have achieved a far less satisfactory result had he removed a thorn from a robot's foot than he got by removing one from the paw of a live lion, even if the programmable aspects of the animal's response— a voice track and set of wind-up-like gestures indicating gratefulness— would have been represented the same way. The fact that both lions are only hypothetically and fictively real notwithstanding, there would have been no fable and no moral with a qualia-less lion. Try retelling the story with a robot Simba and you'll see what I mean.

When qualia become only illusions thought up by other illusions, pretenses conceived pretentiously, the critique that delegates them to their phantomhood does not so much make subjective experience delusional as make *all* of reality suppositional or apparitional. Qualia may be phantasms trancing out on superfluities, but they are likewise the forms that arise whenever creatures and objects emanate at the same frequency as each other. Everything at core is only energy and space plus the radiation and curvature arising therefrom, so why should qualia need to be any more palpable and hardcore? Couldn't they just be a highly gossamer rendition of space, curvature, and energy; subtler in their way than quarks or strings?

The problem is not that the qualia are uniquely phantasmagoric; it is that they are being held to a standard of reality that nothing else in nature can match either.

●

In a thought experiment a vision scientist, Mary, raised in an achromatic black-and-white environment, studies colors over a black-and-white monitor and learns everything there is to know about them—what they are, how to produce them. Upon being released from her room and seeing a ripe tomato for the first time in her life, she apparently is startled to learn something new, red's nonphysical phenomenal properties; she gains a whole epistemic state. The appearance of a ripe tomato on her windowsill, then a fire engine racing by the building, are absolutely novel and transformative. Shocked by the quale, Mary becomes a different and wiser person. She evolves through her experience of actual red.

On the surface this seems indisputable, but functionalists dissent: "Mary gains new knowledge only in the sense that she comes to know in a new way facts and propositions that she already knew."[24] She gains nothing more than a different form of what she already possesses in her intellect: red. She "acquires new skills, not new information."[25] The rest is steam. She does not learn anything really novel; she already knew absolutely everything there was to know about red by knowing everything physical. "[B]lack-and-white Mary knows all about the intrinsic characteristics of color-experience," unlike most uneducated people who are "totally ignorant about its intrinsic character."[26]

It is an odd universe indeed in which scientists have deeper realities (because they know certain facts) than, say, farmers or polar bears, merely by dint of their education regarding the physics of color. Provincial scientism of this sort defies even common sense. (Sorry, this is an oversimplification of the functionalist position on qualia, which is based on objective facts rather than the subjective knowledge of facts, but it speaks to a deeper reality: No one denies the existence of qualia when the shit hits the fan, tenure is denied, the Mongols invade, and everyone heads for the hills. In fact, this becomes a qualia-wild universe.)

From a functionalist standpoint, the phenomenal experience of red is at best *epi*phenomenal (whatever that really means), at worst delusional and, in any case, of no significance or use. A so-called quale like pain "does not pick out its referent via a contingent mode of presentation; it conceives pain directly and essentially"[27]—no *thing*, even no phenomenology, gets in its way. We know that a person has its physical property by his behavioral "ouch!" without needing to suspect that he also has an experiential "ouch!" but, more salient than that, even our own "ouch!" does not need the neural concomitance of pain, only a synapse confirming a referent. We "ouch" because of abstract data-flow, not because of a separate concrete entity in nature like joules called pain. Subjective pain itself doesn't exist thermodynamically. The fact that pain hurts is information, not experience (qualia)—or its so-called quale does not add to its information.

The rebuttal from the phenomenal camp is, yes it does hurt, and the pain means something independent, often important and reality-changing (Christ at Calvary, for starters). Mary acquires an absolutely new type of knowledge upon seeing colors (or feeling pain) for the first time—she acquires qualitative as well as functional information about the universe (skeptics of course consider that she acquires only a different view of the latter[28]). "The contention about Mary is not that, despite her superb grasp of neurophysiology and everything else physical [one type of knowledge], she *could not imagine* what it is like to sense red; it is that, as a matter of fact, she *would not* know [have that second type of knowledge]. But if physicalism is true, she would know; and no great powers of imagination would be called for."[29] She would lack nothing and would not have to use the imagination that she ostensibly doesn't have—and nature would be happy as a clam and not bat an eyelash (forgive the mixed metaphor, but this whole topic is metaphors in wrong lanes).

"Happy as a clam" is not the case phenomenologically. Try running a universe without qualia. "What [Mary] lacked and then acquired, rather, was knowledge of certain such properties couched in experiential terms,"[30] and this is genuine knowledge not an epiphenomenal excrescence or mere remainder.[31] Phenomenal pain is different from neuropathic or inflammatory C-fiber firing. It not only elicits functional behavior but opens its own vast realm of self-arising designations and meanings. Again, subjective awareness is needed to complete reality, not because reality has to be that way but because it *is* that way.

●

As you can see, the underlying debate always toggles back to the same pivot: whether the qualia amount to something ontologically real or are merely epiphenomenal artifacts (collateral signals) of a different convention of representing (or acting in response to) the same thing—yellow, pain, whatever. Underneath lurks the issue of whether life itself "means" anything.

Physicalists believe that they do not have to grok anything other than physical red because red's so-called qualia are not unique properties, nor do they—nor can anything—"hide forever from objective science in the subjective inner sancta of our minds."[32] Reddish vapors need not apply!

This (by the way) is the real rub for behaviorists, not the poignant redness of red (which admittedly doesn't thrill them either) but its imperiously exclusive range of jurisdiction. To them physical red is sufficient for all occasions, so red's so-called qualia are gratuitous, already covered by (if one is going to get picky about it) heat, water, chemico-electric synapses, and molecular motion. As long as there is no additional irreducible red apparel, no special redness lurking in the mind (or anywhere in nature), red's physical expression as a wavelength within a spectrum and a light-source specification based on objects' physical properties such as light absorption, reflection, or emission spectra should be adequate and inclusive—all over but the shouting—*pre-qualia*. Then "if our idea of the physical ever expands to include mental phenomena, it will have to assign them an objective character—whether or not this is done by analyzing them in terms of other phenomena already regarded as physical."[33] Registry under "physical causes of purple glows": case closed.

There is no escape for qualia; the materialist weir covers the entire outlet from their cove. Even if they turn out to exist, they will be caught and proven mere electrical impulses (or whatever) by some super-subtle scanner in the future. They won't get to be real fish, ever.

The behaviorist position is not as absurd as it appears at first glance, for its supporters are trying to assert not that the experience of red's redness is no different from the nonexperiential knowledge of red's redness but that the difference doesn't amount to anything—anything ontological and game-changing, that is. Stated otherwise, it isn't a necessary or useful component of a unified gravitational-thermodynamic model of the universe nor is it needed for an inventory of reality. Where in physics or astrophysics is mind? In biology where is mind except posing as a carryover of metabolic activity and system information? It does not open a conscious or intelligent universe; it is just another state of entropy.

Both dualists and functionalists face the same original dilemma of our being unable to conceive "how a single event or thing could have both physical and phenomenological aspects or how if it did they might be related."[34] It doesn't matter if qualia get reduced to molecular effects someday or remain epiphenomenal and figmental forever; they still have to be tied to the system somehow and they can't. This is a unique dilemma in nature.

The map is once again not the territory, and the two might not even go together. In fact, the relationship between the two (physical reality and phenomenological reality) is especially baffling (as noted in Chapters Two and Three) because nothing in the brain or nervous system suggests a coupling device. Nothing in the brain concocts the *meaning* of red or looks like an artifact or process that might do the trick. Yellow paint or the neuromolecular equivalent of paint doesn't flow across the cortex.

Not only is the brain not a container of redness or clouds or anything remotely like them, it has no evident apparatus for reaching across the dualists' famous matter-mind barrier into the mind part; it "offers no workable idea how, why, or where mind and matter stage their dubious liaisons; nor how a dualistic ontology could be reconciled with the principle of evolution by natural selection and the whole neo-Darwinian synthesis."[35]

How could mind and matter have been subject to a single regime of natural selection when neither is represented in the other's domain? Physical and phenomenal facts are separated by a chasm: "[t]he links between qualia and their neural bases are [arbitrary], unintelligible, and present us with an unfillable explanatory gap."[36] It is as though the two phenomena are stuck together randomly and can be connected only *a posteriori,* with no amount of *a priori* tinkering able to join them. And that is insane for any sort of universe.

The qualia enigma is finally a debate without a referee or a clue as to where one might come from or, for that matter, situate himself in order to arbitrate fairly.

●

Thousands of pages of philosophy have gone into another thought experiment that tries to trap qualia. Its narrative goes: What would it mean if someone consistently saw green as red; that is, their red and green receptors were reversed and connected as such to the rest of their visual and cognitive apparatus?[37] "[T]wo machines with different hardware might have different qualia despite computational identity, having the same computational structure and running all the same programs."[38] Meanwhile their "brains [would be] molecular duplicates."[39]

How would we know if qualia were inverted in two individuals, since both people would report the same experiences on cue? The inverted subject would still use the "correct" term ("red" or its equivalent) to express his own green quale. True, he would "experience" red when he reported green, though so far as anyone knew, grass would be just as "green" for him as for us. (Color is relative to adjacent colors anyway.)

Phenomenal red and green (by this argument) are epiphenomenal anyway, nonexistent hallucinations that have no bearing on reality. Who is to say that green is not really red anyway in some individuals, since these are contingent pulsations, interchangeable neural signals, deciduous concepts, and finally mere abstract phonemes representing something else? There might be two reds, a red red and a green red, and likewise two greens, at least from the standpoint of any individual's visual and cognitive apparatus inverted from another's.[40]

This verdict, of course, requires that green and red not have separate meanings with any of the aesthetic, emotional, psychosomatic, or psychic qualities that are reported by various philosophers, artists, and mystics. Green and red have to be physicochemical synapses generating signals for behavior and functions (eat that plant that emits green light, stay away from that berry flashing poisonous red).

Forget fashion shows, Cézanne's blue, or healing with color. Forget the innate luminosity of the chakras or any notion that "color acts on us preconsciously [preter-consciously] before we're able even to name the color we're seeing."[41] To epiphenomenalists, phenomenal color has *no* meaning, no mood, no mystic effect or aftertaste, no actual existence; it is a convention, a wavelength, a message-unit only.

A correlative line of reasoning is: are we speaking of phenomenal red or access red? If the latter, functionalists ask, why should its quale even matter as long as an "inverted" person behaves and communicates consistently in a "red" fashion, even in the context of green as red? After all, if "experiences" don't exist apart from a baseline of electrical impulses in grids conveying commands to centralized ganglia through control hierarchies, if creatures react consistently and appropriately to green as red (or red as green), then "red" synapses can effectively be "green": alternate wires flowing through channels leading to identical appropriate behavior.

The functional states of green "red" are "no problem, boss!" because they "interact causally in the required ways with inputs, outputs, and other internal states of the creature."[42] Again, presuming no phenomenological complications or other introspective exotica.

Versions of this inverted-qualia thought experiment include replacing intersubjective parameters with intrasubjective ones—a *noir* thriller in which a person's qualia are tampered with by an "evil neurophysiologist." Either early pathways are left intact but certain memory links are inverted or qualia-producing channels in the optic nerve are inverted farther downstream. The "mark" waking up does not know whether his qualia themselves or his memory-linked qualia reactions have been inverted. This puts a Stalin-Orwell twist on an already multi-tiered crisis of meaning. I won't discuss the host of both epistemological and ethical issues that these experiments raise, but suffice it to say that the inverted-qualia bibliography is voluminous.

John Locke foresaw this precise conundrum in the seventeenth century when he first proposed the concept of spectrum inversion by writing: *"[T]he same Object should produce in several Men's Minds different* Ideas at the same time; for example, the *Idea* that a *Violet* produces in one Man's Mind by his Eyes, were the same that a *Marigold* produced in another Man's, and *vice versa.*"[43]

This minor and incidental bugaboo did not go away, for it opened the floodgates to the relation of private experience to consensus reality. In proposing it as a major *reductio ad absurdum* twentieth-century philosopher Ludwig Wittgenstein commented: "The essential thing about private experience is really not that each person possesses his own exemplar, but that nobody knows whether other people also have *this* or something else. The assumption would thus be possible—though unverifiable—that one section of mankind has one sensation of red and another section another.... [I]f this can be so at all, why should it not always be the case? It seems, if once we have admitted that it can happen under peculiar circumstances, that it may always happen...."[44]

The exclusivity of sensory experience hits this puddle with one gigantic splash, but the behaviorists do not immediately notice it. When there is no objectively verifiable basis for any form of knowledge or information, the ontological question, once again, becomes the epistemological one (and *vice versa*). But does it even matter? If there is no phenomenal component to experience—only behavioral transcriptions and stimulus-response sets—then it is of zero concern whether a person has his or her cables (or nonexistent qualia) inverted. If nature has no inside, then interior inversion is inconsequential, and so-called privileged knowledge is privileged knowledge of zippo. In fact, there is no such thing as epistemology.

And physicalists believe, of course, that this inverted-qualia riddle will ultimately be solved by finding "the exact neural correlate in a human brain [such-and-such firing pattern, such-and-such fixed designator] of seeing red"[45] and then using it as a yardstick to compare your 'red' to my 'red.' If red and green receptors are reversed, then this would be a trivial neurophysiological anomaly, to be dealt with on a physical basis by physiological investigations.[46] The map *becomes* the territory. Mere tangled wires—no big deal!

In this debate's persistence over centuries, one senses more than just semantic quibbles and recreational jousts inside sanctuaries of polemics (while people elsewhere are killing each other or starving). It is not just that both behaviorists and idealists would drop the discussion in a heartbeat and rush to the feeding trough (or more in keeping with their endowed situation, the dessert tray) if scarcity were about to hit town big-time. It is that there are, beyond the cavils, outliers, and eavesdroppers, footsteps of sober consideration of the universe by our species mind, as we are roused to take ourselves and our words and conceptualizations—our enigmatic existence—seriously and to weigh each proclamation and demand that we make upon reality. We are not merely squabbling over qualia and nonqualia, essences and ephemera. We are trying to dead-reckon *our actual place in the universe*.

So while the qualia argument means something very different to the partisanships on either side of it, it also has a subtle resonance that transcends its terms. It is a meditation on the nature of being and not-being. In that sense I support both camps as well as their clash because, through the opposition, a primordial dialectic glimmers, a wick with a lumination of something eternal. It comes with being human and cannot be renounced or given back to the universe, even by behaviorists, any more than it could by Mesolithic sorcerers with antler crowns. Neuroscientists own it as much as shamans and idealists. They proclaim their ownership by negating qualia and subjective existence, which is a magical, sacred, and phenomenological incantation too.

As long as one faction doesn't obliterate the shared paradox, they collaborate on a deeper truth—one that neither party would amiably admit. What that is, I can't say, except to suggest that it arises when we go back and forth between polar positions. It has something to do with the fact that the universe, though subjectively conscious in its discrete entities, has stipulated, for some essential reason, that it must maintain a crucial separation between phenomena and phenomenology; that experience must operate *as if* it were only physical and thermodynamic (a point for functionalism) and in terms that make it seem as though *it* cannot have consciousness as an innate property (another point). Only by that enigma can it reveal its deeper truth—we are real (checkmate, idealism) but not real in the way we strive to be (checkmate, functionalism), so each has to pay the other piper, dearly in fact, in order to be here.

To demand the pure and separate reality of qualia is to petition a sort of immortality we can't have, while to destroy qualia altogether is to deny the poignant and precious existence, albeit contingent, that we do.

Perhaps qualia are reflections of intransitive, even disembodied aspects of consciousness, whereas the coercive extermination of qualia is a prerequisite for carrying out objective scientific experiments and medical procedures inside embodiment—e.g., for coring reality in a transparent, empirical way. In that case both are necessary for happy lives and good healthcare. Where they collide is the rift along which the universe is emanating. But collide they must.

For now, insofar as philosophy aspires to mere secular grandeur and precision, it needs consciousness yet cannot accept it *carte blanche*. That is why the argument spills ultimately into linguistic theory, quantum mechanics, cognitive psychology, and deconstruction of pretty much all of literature and social science, clear across the remainder of the twentieth century into this next. That is why immoderate geopolitical arguments go on for generations, while a yardstick on which all of us, or any two disaffected parties, can agree eludes us, and the planet.

Wittgenstein again: "The thing in the box has no place in the language-game at all; not even as a *something;* for the box might even be empty. —No, one can 'divide through' by the thing in the box; it cancels out, whatever it is.... It is not a *something,* but not a *nothing* either! The conclusion [is] only that a nothing would serve just as well as a something about which nothing could be said."[47]

That's eating your cake and having it too, qualia-wise—it's the intermediate position.

If you can't ever get at it, it might as well be a nothing as a something. Derrida certainly thought so. From a Freudian standpoint, it is systemically and eternally unconscious. From an anthropological standpoint, it is a series of tropes masquerading as laws. In the context of object relations and totemism considered in tandem (along with Derrida's erasure), the extreme scientific and materialist position suddenly bellies up as an inflated *reductio* under which (or over which) a bottomless reservoir of unsignified and unconscious meanings continues to overflow its banks and erode its premise. After all, there is nothing really in its box either.

In other words, while functionalism may relegate qualia to epiphenomena, deconstructionist anthropology and psychology relegate science (as well as functionalism) to tribal myths. Despite their Nobel pedigree, scientific narratives are still Stone Age etiologies that use atoms, cells, quarks, and neurons rather than tigers, spiders, armadillos, and farts as their signifiers. Each system of designation (functionalism and totemism) is a self-referential myth cycle, a matter I will discuss at greater length in the next chapter. For now let us slip back into more genteel accouterments.

●

A key physicalist objection to qualia is the infinite regress implicit in the empirical location of a subjective, immaterial property that is also unlocalized. In any abstraction placeheld by the mind, there is no limit to the degree of shagginess of the shaggiest, shaggiest, shaggiest dog in the world or the number of barrels of beer on the wall as long as "we are still appealing to the consciousness of a higher-order state to confer consciousness on a first-order state, which leaves the circularity unremedied. It also generates an infinite regress, since each nth-level state must be rendered conscious by an $N + 1$st-level state…. [R]eal human beings have nonconscious representational states that are monitored by other nonconscious states."[48] Where do you stop, and how does anything get to the end of the line?

The states are either qualia (to which functionalists say: boo!) or nebulous and shifty "something elses" (to which phenomenalists say: "made you look, made you buy a pocketbook"). Yet, if somewhere in that regress phenomenal consciousness is generated, even if only epiphenomenally, phenomenalists accept it as the sacrosanct metaphysical wafer inserted into the cosmic slot, while physicalists equally stubbornly aver that it can be a wafer but not a real one, for nothing can run its own program or get to monitor and affirm its own infinite states or be solely responsible for their verification. Nothing in the Newtonian universe has earned that privilege. No one can rig the game and scam the house indefinitely. Yet consciousness has been getting away with that and a lot more ever since its ghost arrived in town. One physicalist speaks for all:

"My visual experience of a barn is conscious, not because I am introspectively aware of it (or introspectively aware that I am having it), but because it (when brought about in the right away) makes me aware of the barn…."[49] That's consciousness as an elusive medium in which only *bona fide* barns and barnlike objects are embedded. The victory goes to the mundane object; e.g., the barn is real, reality is real, and there are, in principle, only ninety-nine (or however many) barrels of beer on the wall and will be ninety-nine forever—no n^{th}-level metaphysical complications to get us either around or over. The wall is a real wall too.

Testimony from the functionalist camp then leads the jury through a quite elegant lawyerly argument: "Perhaps we are equipped with an introspective faculty, some special internal scanner, that takes as its objects (the *x*s it is an awareness of) one's experiences of barns and people…. Perhaps introspection is a form of meta-spectation—a sensing of one's own sensing of the world. I doubt this. I think introspection is best understood, not as thing-awareness, but as fact-awareness—an awareness that one has certain beliefs, thoughts, desires and experiences *without* a corresponding awareness of the things themselves….

"What makes [beliefs and experiences] conscious is the way they make us conscious of something else—the world we live in...."[50]

Back to the mundane object that is the sole ship afloat in the wine of consciousness—a very red barn against a very blue sky. The object and its embedding are what count, not what it is embedded in. Reality rules, *sans* qualia or qualifications, *sans* Socrates, *sans* Proust, *sans* rock 'n' roll. Each item is flat, flat, flat; it elicits functional responses only, and its mental qualities are flounces and frills, by-products of the real McCoy.

Consciousness is merely a red-alert signal, a valet who brings us the world—sweet, juicy, and bright—on an unadorned platter. Awareness is a check digit not a reality window, and it is certainly not a wondrous appearance with metaphysical implications. Barn. Stone. Snake. Real, real, real. *"And if one of those barrels should happen to fall,/ninety-eight barrels of beer on the wall, ninety-eight barrels of beer...."*

Furthermore, a functional declamation of consciousness does not need mystic infusions or qualia, only enough referents (barrels, ping-pong balls, or internal states in proper relationship to one another). If qualia emerged somehow out of the ping-pong balls (or other relational units), functionalists still assert that *they are without metaphysical implication and do not require a separate ontological basis;* they could have arisen to sort semantic properties of networks, or relational interfaces (and to transfer data between modules). In fact, that is exactly what has happened; ping-pong balls (molecules) have combined to replicate the presence of a universe in monitors, but only in order to allow successful Darwinian predators. The steam and its whistle are just decorations.

But why should the Darwinian universe have tolerated such a huge expenditure of energy for such a luxurious function? Why the song and dance? "What is the evolutionary advantage of the inner life of humans and other conscious beings? ... [F]or an animal seeking to make a living in a competitive environment, what good does it do to have a mind? ... Some have suggested that consciousness helps build an inner map of the outer world, or is useful in planning complex tasks or learning new pones, but at this point in our knowledge of mind, it is difficult to see why jobs of this sort could not equally well be carried out by unconscious machines."[51]

Kind of a vapid universe, but why not? Entropy doesn't seek entertainment value. On the other hand, the operating system of a qualia-less tiger robot might have gotten confused by an overload of competing signals; it would have had to perilously oversimplify its data-streams or it might have attempted to process mixed or incomplete instructions and then malfunctioned and gotten chomped or starved. Its cousins with value-and-belief software had more options, solved more problems, and prowled long enough to breed. The ping-pong balls, barrels, and N+1 states somehow found each other, stopped their regress, and landed on a representation stable enough for them to hiss and growl.

Yet qualia did not evolve to create existential experiences; they inserted themselves as links, macros, system memory, and data-processing software for operational feedback, integration of parallel processes, and computational speed: "[B]ecause of their ability to store memories in spacetime, rather than in space, conscious computers can perform the same job as unconscious computers and require fewer parts to do so. An unconscious computer's memory is limited by the number of its present storage spaces; a conscious computer can store memories in the present too, but in addition it can access events that have happened long ago, events that lie 'outside' the computer's present state. A computer with inner experiences of the SRM [Spacetime Reductive Materialism] variety possesses in effect an extra storage medium in the past—a kind of invisible spacetime 'hard drive'—that could give consciousness a competitive edge in the Darwinian struggle for existence."[52]

Gradually, incrementally, conscious evolving systems sway toward a rudiment of something nimbler, a bit more adroit: "[C]onsciousness would not be the matter of arrival at a point but rather a matter of a representation exceeding some threshold of activation over the whole cortex or large parts thereof,"[53] somehow creating "belief in the content of the experience."[54] This content and the belief in it would make for a more serviceable, prosperous robot. But it wouldn't make the hallucination itself real.

In fact, this is the only way that functionalists permit a bit of conscious spume into their robots: as software supplements, a way to process a cacophony of signals bubbling over its own pot.

"I" is not an anointed witness; it is simply an add-on module with circuits to supervise those of other modules, a module which incidentally contains a working alias for its entirety, called "I" in the way that Joe is called "Joe."[55] You merely need to tell the machine to represent its whole as itself in one of its components and then to route any rumors of its existence (or nonexistence) into that component. This makes for a supremely functional (and happy) antelope or monkey (or sometimes a melancholy antelope or monkey)—in either case a qualia monkey in lieu of a null monkey that swings from the tree that no one, not even the monkey, hears as it falls. I will leave "Machine Joe" and his functional "I" for later in this chapter.

From a phenomenalist standpoint, something is of course unaccounted for: "[H]aving a functional inner 'I' does not suffice for being able to think of oneself as oneself; nor does mere consciousness as opposed to self-consciousness confer personhood or any moral status."[56] A module may process "red," but it doesn't have red experiences or red moods and is not able (like Cézanne) to paint red and blue abstractions or sing the blues like Leroy Carr or dance red-hot sambas like João Parahyba. So where do these extras come from? They are certainly not the ephemeral residue of wiring (stuff that can be spliced out of the system without causing fatal damage). And they appear to introduce their independent series of epistemologies, which have to be accounted for in any system with a claim of completeness. Not only that, but they make the antelope or python into what *it essentially is,* and even what it does. And, once again, on this planet anyway, we don't have complex levels of function (beyond "taxis" and "tropisms") without full-fledged qualia and consciousness.

Dennett attempts to sweep away this whole infestation with a master stroke: "Today, no biologist would dream of supposing that it was quite all right to appeal to some innocent concept of *élan vital* as one's name for DNA.... I want to make it just as uncomfortable for anyone to talk of qualia—or 'raw feels' or 'phenomenal properties' or subjective and intrinsic properties or the 'qualitative character of experience'—with the standard presumption that they, and everyone else, knows what on earth they are talking of...."[57] They don't, *but that's exactly the point.*

In Dennett's world mental processes remain ordinary physical processes in higher-order states, forever and under any circumstances. Their so-called "feels," their pseudo-qualia, are less than fuzz on the peach, a kind of afterglow or empty rattle of no purport. What matters is that the snake strikes for real and even fatally, not that he (or we) experience the oscillation of his keratin segments. The actual world is not only largely physical, *it is entirely physical.* And so is the philosophical world. It is made of widgets and gears, molecules and membranes, mass and heat, so qualia have to be one of those or nothing at all; the game is now "Animal, Vegetable, or Mineral"—no extraneous options. "Feels" are system noise: "This is why physicalists must hold that complete physical knowledge is complete knowledge simpliciter [universally true regardless of circumstances]."[58] There is no other, nonphysical universe or nonphysical operating system in this universe. No "second" rattle to tie the mouse to the snake.

Consciousness has to hit the ground in matter somewhere and, where it does, its prior free romp must turn into a burst of system steam. This is crucial to both camps in different ways. For dualists and phenomenalists, the steam (or whatever) exists but never bottoms out physically. It can't or it wouldn't have real existence. It is like a homeopathic medicine that has no actual substance but (ostensibly) performs as if it not only did but were more potent and curative for its absence.

Functionalists and behaviorists agree here: mind must return to and bottom out in matter. Otherwise, *the entire universe* falls into a rabbit hole of mock turtles, queens materializing off playing cards, lobsters dancing quadrilles. The phenomenalist replies, "So what and what else is new? You haven't seen the least of it." And there she rests.

In essence, phenomenalists turn the materialist trope around such that the actual world not only is largely phenomenal, it is only phenomenal, for without phenomenal representation it would not exist at all. Qualia are not only real but the only "real" thing, while matter is simply a mirror for their reflection as well as the reflection of minds in their mirror.

Dennett concludes his onslaught on idealist metaphysics by proposing that qualia are their own inevitable logical dead ends, representing the same sorts of fictions that literary characters engender in relation to the nonfictional "reality" out of which they are engendered: their creator's experiences. They "exist only by agreed-upon conventions and suspension of disbelief." Well, we already endow and personify the fictive substantiality of our lives. Do we ever! Dennett concedes, "Perhaps people just want to reaffirm their sense of proprietorship over their own conscious states...." After this brief sigh, he corrects the misplacement:

"The infallibilist line on qualia treats them as properties of one's experience one cannot in principle misdiscover, and this is a mysterious doctrine (at least as mysterious as papal infallibility) unless we shift the emphasis a little and treat qualia as *logical constructs* out of subjects' qualia judgments.... We can then treat such judgings as constitutive acts, in effect, bringing the quale into existence by the same sort of license as novelists have to determine the hair color of their characters by fiat."[59] Later he adds, "The properties of the 'thing experienced' are not to be confused with the properties of the event that realizes the experiencing. To put the matter vividly, the physical difference between someone's imagining a purple cow and imagining a green cow *might* be nothing more than the presence or absence of a particular zero or one in one of the brain's 'registers.'"[60]

Complete knowledge simpliciter! One system validates consciousness as what it seems to be, giving it the temporary benefit of the doubt, especially in the absence of either the territory or the map. The other system says, 'Modules, monitors, and wires into fuseboxes, zeroes and ones. Nothing mysterious or ineffable. Just the everyday fiction by which a person composes his or her perceptual life.' And, I would add, it hardly matters whether a purple cow is in the mind of an actual person or a fictional character because, to the functionalist, they are *both* epiphenomenal, both fictive—at first and second orders of regress, respectively. Qualia are qualia are qualia. You can even have a purple cow in the fictive mind of a fictional character in the mind of another fictional character, and so on: as many bottles of beer as your larynx will hold out. No need to petition the papacy for a "holy ghost" decree.

●

I feel the presence of a bully here. The ideologically dominant paradigm tries to make law out of dogma so that it always comes out on top. It ignores its own fathomless subrogations for which there is no redeeming true version, only a permutation group—an elaborate neo-Darwinian totemism. Just because scientists are using mechanical and digital referents rather than bullroarers, prayersticks, and totem poles doesn't mean that they get to peremptorily disqualify their own and everyone else's subjectivity and declare a sole physicalistic status for the entire universe, its habitants and species. They can't just decertify them and everything else as illusional in advance, because E.T.s may well have other ways of looking at things that change the equation, the question as well as the answer. You don't have to go to Alpha Centauri or the Pleiades to shop qualia; you can take the vantage of an indigenous Mexican who not only believes he can put his *nagual* into a crow* but effectively does so.

A shaman, whether Maya or Zulu or Aranda, doesn't intend self-teleportation as a fantasy of being a crow (or cheetah or snake); he means being a crow *as a crow,* a snake as a snake, an eagle as an eagle. Likewise the realm of nature from which these snakes and crows arise is not a function or property of something else, nor is it a mere subjective collocation of random physical states and objects; it is a magical operation for those who become conscious in its zone. And nobody knows what *anything finally is,* even in the dominant paradigm, so nobody can exclude *naguals* out-of-hand. It is not a case of epiphenomena versus phenomena; it is a case of the undiagnosed status, hence the boundary (human or other) of consciousness.

The indigenous agenda is one of leading with meaning: Does your mind move into the bird? If so, okay and what use can be made of it?

The modern agenda is one of reduction to *a priori* thermodynamics, a priori thermodynamics—no one in a scientific tradition would take the teleportation of consciousness seriously. The "experience" is not interrogated; it is merely assigned to primitivism or collective delusion.

The independent existence and possible extrabodily range of personal mind has been prematurely discarded by a civilization more interested in "commoditizing" and "proving" than "being"; yet "being" is far more profound than any extant marks or reality programs—and also probably

---

*Among Mesoamericans, the aspect of oneself that can inhabit another creature, usually an animal, is the *nagual.* This is battlefield conduct; it is not (and not intended as) theater. See also *phowa,* Volume Three, Chapter Four.

more faithful to the intrinsic mechanics of the cosmos—because that consciousness wafer can't just be pulled out of its slot without the physical world collapsing too. The universe was created and ourselves inserted precisely into it in the same fashion. Science counts, yes, but primarily within its own cultural and socioeconomic frame, and that could crumble at any time back into the "black ages" of evil eyes and totem crows, or anything else for that matter. If civilization fails, everything drops into the prior "meaning" phase anyway.

Existence rests not on the objects one controls or the propositions one proves: what it rests on is the utter drop cloth of being here at all. That's what's unique and special about the universe: its subjective depth. Otherwise, you might as well run a giant meteor gallery and galactic pinball machine in the void forever.

As Yaqui sorcerer Don Juan Matus reveals his shamanic secrets to anthropologist Carlos Castaneda, he extends a longing and a sense of loss and regret for a wondrous and fading emanation all around us, not because it has vitiated itself but because we have used our own "magic" to suffocate it under Potemkin villages and mall-enclosed realities. It doesn't matter if Don Juan is a hoax because we'll never know. At this level it's all just testimony and slogans anyway:

"For you the world is weird because if you're not bored with it you're at odds with it. For me the world is weird because it is stupendous, awesome, mysterious, unfathomable; my interest has been to convince you that you must learn to make every act count, since you are going to be here only for a short time, in fact too short for witnessing all the marvels of it."[61]

Both physical properties and qualia pale before the actual White Whale.

●

As regards "close encounters" of the third and fourth—or *nagual*—kind, one physicalist takes a creatively duplicitous tack by arguing that if astral bodies, psi energy, and auras exist as real items in this universe—though of course he very much doubts it—the second-class status of qualia still persists, for qualia are *even more metaphysical* than telepathy or psychokinesis. "If there is such a thing as phenomenal information," he declares, "it isn't just independent of physical information. It's independent of every sort of information that could be served up…. Therefore phenomenal information is not just parapsychological information, if such there be. It's something very much stranger."[62] And worse.

Right on! It is information in the absence of any originating substratum, physical or paraphysical, kinetic or telekinetic. It might as well be the Abominable Snowman or *élan vital.*

By existing at all, consciousness makes *everything* in the universe potentially paraphysical and potentially parapsychological, even the most ordinary datum and deed, because after all, what's that? My original question: What's anything, and how the fuck did it get here?

Mind just pops up out of the middle of nowhere like the "weasel," and demands everyone's attention: a puppet in a music box. Telepathy and the like are mere party tricks after the weasel's unprecedented feat of slicing open reality and stealing the show.

The basic emanation and self-awareness of substance are far spookier than telepathy ("mere" thought transfer between individuals) or telekinesis (moving physical objects by thought), spookier even than *naguals* or remote viewing. Consciousness is the ultimate interloper, the premier apparition and poltergeist. After all, the very fact that there are ghosts (thoughts) cluttering the world means that there can be many different kinds and statuses of ghosts. Every one of us is a ghost, a specter—that is, if the physicalists are right.

But what if we are *not* a ghost—or are a *real* ghost. Then that marks the end of physicalism, an outcome that physicists are sure will never occur. They are so sure of it that they are willing to give telepathy and the like a free pass. Here is why: *neither* Newtonian physics *nor* telekinesis (if it exists) *is rooted in a demonstrable or stable first cause.* They are equally up for grabs and for the same reason! In a universe that permits true qualia, parapsychological indeterminacy takes center stage. In a universe from which qualia are excluded, it is at most a minor nuisance.

No wonder physicalists want to keep consciousness from sneaking into the nightclub before telepathy. If telepathy gets in, so what! No one will notice—and if they do, they can pretend it's someone else. If consciousness gets in … ah, but it already has! The only way to get it out is to attach it to telepathy and then kick it out as a party-crasher or otherwise, presto, make it disappear.

If, however, we were to accept, in some fashion that *I* can't imagine, that consciousness is *only* chemico-electric and epiphenomenal, then our very existence, including science, is delusional smog—nothing more (or less) than another passing phantasm. The whole debate is moot. Make your choice: weasels or wraiths!

When physicalists aver that qualia are *even more metaphysical* than ghosts, they are returning under cover to their old Maginot line: that consciousness (to exist) must have its *sine qua non* in the brain, even if we can't presently find it. By parallel reasoning, even telepathy and telekinesis could gain admittance to a properly thermodynamic universe but only after they were satisfactorily frisked for qualia too. They would then represent mechanical activities of an order that has not yet been identified but will be—here or somewhere else in the cosmos or sometime in the future. They would work because of a thus far unidentified but wholly physical and natural mechanism translating neural into kinetic energy, mind into tractor beams. It would all be kosher.

In physicalist dogma the "epiphenomenal part of the phenomenal aspect of the world" may be "the residue left behind when we remove the parapsychological part,"[63] but consciousness is still, to that same dogmatist, a parapsychological phantom or, more properly, a paraphysical fiction. Once we quarantine run-of-the-mill parapsychology (as per his suggestion), we are left with the metaphysical elephant in the parlor: our privileging of our own first-person perspective: "If the causal basis for those [normal or paranormal—in the end it doesn't matter] abilities turns out also to be a special representation of some sort of information," the functional physicalist offers, "so be it. We need only to deny that it represents a special kind of information about a special subject matter. Apart from that it's up for grabs...."[64] This is like making mind get in line behind its own psi phenomena—a logical incongruity, unless the queue is to get a ticket to represent itself in physical matter; then the waiting order is irrelevant.

By physicalist proclamation, if you are a phenomenalist you are by definition an irredeemable parapsychologist too—a dirty word in polite academic circles. But, again from that same quaint functionalist viewpoint, if you're going to coronate potatoes as belly dancers, you might as well be claiming Vulcanic mind meld for them too. You might as well go for all the roses while you're running amok. In that sense we stand a far better chance of demonstrating telepathy than qualia. At least there is a series of experiments and hypothetical future experiments for psi phenomena, but where are you going to place the Geiger counter (or computer algorithm) for consciousness? We've covered that territory many times.

But likewise (still from the physicalist viewpoint) if we do not remove the parapsychological part, the residue might even—somehow, on some planet somewhere—move matter with thoughts. So what! Telekinesis (because we scientists don't really believe in it anyway) is forever less radical than an entire separate ontological status for consciousness and its qualia.

●

Every possible act of independent consciousness in beings possessing its odd plug-in is independently and *sui generis* presumptive of its own existence and (even worse) its free will separate of the physical system of which it is supposedly an epiphenomenon. It is flipping off its own programmed synaptic outcome. It is flaunting qualitative existence, immune to mechanistic probe or interference or even the hypothetical quantum states of its particles.

Consciousness, by lounging about and conveying information, has already claimed the parapsychological and telekinetic territory in a way that present rubrics of parapsychology cannot touch—but physics and functionalism can't touch it either. Subjective existence *is* the Holy Ghost; it cannot be made any more metaphysical or impart anything that is stranger than *its very existence,* so healing by prayer and remote viewing are well down on the overall strangeness list.

Mere telepathy palls before the appearance of stars and worlds, trees and clouds, creatures and desires inside a self-aware liqueur, an opaque, diaphanous effusion of being, because telepathy (again if it exists) must arise subsidiary and ancillary to matter and to consciousness as an outlying corollary.

The magic of producing phenomena in a universe dwarfs any regional feat of sending and receiving messages to and from remote colleagues wirelessly with the mind (whether by "outback hunters" or "migrating geese" or mediums on any world). A universe with consciousness is more different from a universe without consciousness than a universe with telekinesis and telepathy is from the universe we have—and by a long shot. Consciousness is the glow that illuminates everything (dreams included)—even the physical world, even what it is not supposed to illuminate, and that is a big one, the biggest one of all. It reduces every other coup or experiment to sleight of hand (or sleight of mind if you prefer). Turning the light *on* is the Big Kahuna. The physical world not only bows but dances to that light.

Once you get something subjectively aware, you have crossed an absolute divide in nature—from the abyssal void without form (or meaning or any kind of shit) to the lit room, a tavern filled with beer-drinking patrons, a juke box blaring, darts being thrown at a corkboard. If those customers are also covert Martians, or mind-readers, or shape-changers, it hardly registers on the meter. It doesn't make their existence a bigger deal than the fact that they are bustling about, and (on top of that) know that they exist and hardly care—as that guy orders another beer even though he is drunk.

Phenomenal existence is already a link between physical and the metaphysical universes as well as between self and being, self and other—any self and any separate (or not-separate) other. It is also a link between absolute self and absolute meaning. Trouble is, trying to explain that to a physicalist is like selling "red" to bloody functionalist Mary in her black-and-white prison.

Either way, the physicalist sticks the phenomenalist with being a metaphysician. Either way, if mind moves matter or transmits itself to other minds *sans* particles or waves (or even if it doesn't), it must finally have a physical, mechanical, molecular basis simpliciter or forfeit its right to participate in modernity's unified theory of all reality. That is why the physicalist concedes the realm of metaphysics and parapsychology to the unsuspecting phenomenalist (if he wants). It is because he himself doesn't believe in it anyway; then he gets to nail his adversary on qualia being *even more metaphysical.* It is a game of phantoms and wizards anyway (choose your sword).

The phenomenalist's most compelling argument is the insoluble paradox of first-person as opposed to third-person experience: Qualia are not merely barrels of beer on the wall, second-order monitoring states installed on a take-it-or-leave-it basis; they are properties of first-person experience, thus, have a unique status in the universe. They can't be summarily kicked out. Even physicalists grudgingly admit the same (though never letting go of their neo-Darwinian crutch): "Introspective consciousness is consciousness of self…. Inner perception makes the sophistication of our mental processes possible in the following way. If we have a faculty that can make us aware of current mental states and activities, then it will be much easier to achieve *integration* of the states and activities, to get them working together in the complex and sophisticated ways necessary to achieve complex and sophisticated ends."[65]

After all, how could you even contemplate conducting science without your first-person thoughts, your introspective speculations, without consulting your interior witness and homeboy devil's advocate? How could a hawk hunt rabbits and possums? "Without introspective consciousness, we would not be aware that we existed—our self would not be self to itself. Nor would we be aware of what the particular history of that self had been, even its very recent history."[66] You can't do science or even archery that way.

Physics is dead in the water without qualia; yet it's a Catch-22 because qualia cut physicists off from their privileged third-person vantage, which behaviorism then pretends to solve by shifting "the locus of scientific responsibility from an observing subject to the experimenter who becomes the observer of the subject."[67] This switcheroo is not up to the task. You couldn't both exist at the same ontological level and not get in each other's way. Check out this ledger of first-person affidavits:

• Phenomenal qualities do not themselves create or guarantee phenomenal consciousness; they "must be sensed or felt by the individual subject."[68] "[E]very mental state has to be *somebody's* mental state. Mental states only exist as subjective, first-person phenomena."[69] This is the cognitive impenetrability of self.

- "No one else can die my death or laugh my laugh, although, of course, someone else can certainly undergo qualitatively very similar deaths or laughs."[70] We can empathize with someone else and vicariously put ourselves in his or her place; we can internally approximate his or her feelings and experiences (the whole *raison* of pornography)—but we can't go there ourselves,[71] or know how much we are missing, qualitatively as well as quantitatively.

   In fact, the main purpose of art and literature is to try to convey an intricate or elusive experience to another person by proxy, since the experience itself cannot be transported between first persons. The more widely one inspires the *aha!* of recognition in others, whether by realistic pastiche or intuitive abstraction, the more powerful and successful a work of art is deemed.

- "When one imagines another's conscious state, there is no conclusive way of checking up whether one has done so correctly or not."[72]

- We cannot imagine "the battle of Cannae as it might have been experienced by Hannibal, though we can imagine *something,* and that is different from both our and Hannibal's experience."[73]

- "The ontologically rather ordinary fact that phenomenal properties of an experience exist only insofar as they belong to someone's experience..., when combined with the epistemologically rather extraordinary fact that experiences cannot epistemically be shared, and hence everyone can have 'direct access' to only his or her qualia, seem to make it uniquely, even surprisingly difficult to investigate the ontological nature of qualia."[74]

   This extraordinary fact is super extraordinary. Who ever heard of objects that exist only with access from one view? Yet that's the case with all thought objects. Each thing has phenomenological uncertainty states greater even than those of subatomic particles because it is always a first-person chimera—a mind thing, not an atom thing. It can't ever be inspected or objectively known.

   From this difficulty of confirming egoic observation the third person is ironically deemed a more accountable reporter than the first person, a paradox which provides a popular phenomenalist joke: one behaviorist asks another, after they make love, "It was great for you; how was it for me?"[75]*

Consciousness is only as it seems to the experiencer, the elite observer; it cannot be transferred in any way. Third parties can only exchange words and gestures; they cannot exchange qualia. This much we know already, but consider the consequences for objective science: a scientist can show anything outside himself to a third person, e.g., to another scientist: a plant, a rock formation, his laboratory culture of viruses, particles in a cloud chamber. No matter how elusive, how subtle the thing is, as long as it is an objective external item, it can always be transferred. The direct experience of it can go from third person to third person, changing qualities according to their first-person filters. But the same scientist cannot transfer the actual thing or its meaning to another scientist, only its concrete artifact, its placeholder. Likewise, his colleague is receiving only the placeholder, not the phenomenal object/thought. The quale that joins the two remains forever inaccessible to sharing.

So there is no way one scientist can exteriorize or transport his experimental objects to another scientist. He can approximate his ideas and observations verbally, but he cannot schlep them overland intact to another person so that that person has his same cozy access to those same ideas and observations. He cannot get them out of his own viewing sanctum into another one.

What anyone can do routinely with iron pipes, corncob pipes, or barrels of beer no one can do with his most banal qualia.

- "[O]ne man's mind could not pass into another man's body, to perceive what appearances were produced by those organs...."[77] The only nontelepathic exception appears to be the minded elements of organ transplants but, whatever they are, they're still not discrete transfers of object/thoughts between first persons.

---

*The converse of this is Slovenian philosopher Slavoj Zizek's quip that it is not so much that masturbation is sex with a fantasy object as that sex is masturbation with a partner.[76]

- "My thought belongs with *my* other thoughts, and your thought belongs with *your* other thoughts…. The past thought of Peter is appropriated by Peter only…. Whether anywhere in the room there be a *mere* thought, which is nobody's thought, we have no means of ascertaining, for we have no experience of its like. The only states of consciousness that we naturally deal with are found in personal consciousnesses, minds, selves, concrete particular I's and you's."[78]

- "[T]he third-person point of view is, right from the start, parasitic on the first-person point of view…."[79] It is not the first person who doesn't exist, it is that supposedly unassailable third person!

- There is "a genuine asymmetry between the *mode of access* to facts of one's own consciousness and the mode of access to facts about others' consciousness…."[80] While this is a fabricated epistemological problem, it is a true ontological dilemma. How did the first person get to know so much and to seal it off from everyone else? Is it because he ("I") got there first?

- "There seems to be no ordinary way to peek into the inner lives of others—to feel their pains, go through their sensations, or directly observe their consciousness. That is, there seems to be an epistemic impossibility for anyone to have direct access to the qualia of others—literally share their first-person perspective, in short to partake in what it is like to be them. These are limitations of the third-person perspective. From the outside, firsthand explorations of the consciousness of others just seems to be out of the reach of ordinary scientific methods, others' experiences being neither directly observable nor noninferentially verifiable. And therein this asymmetry between first- and third-person perspective lies the epistemic duality of the study of consciousness."[81]

  People have different, and better, access to their own mental states than to those of others—no big surprise to the man on the street but an expedition from which scientists continue to return empty-handed no matter how "eager beaver" they are on each new neuroscientific approach to the brain. The average working stiff is more likely to want to know how much taxpayer money was spent to excavate this nonsense, and then ask for a personal reduction. This is the house that Jack built. Enter Zombies!

●

The functionalist response to first-person valorization is Artificial Intelligence, which in one fell swoop does away with ego seniority and all its privileges. Here's why: By the creation of AI robots, conjecturally if not (quite yet) cybernetically, the properties that ostensibly make us conscious (or conscious of being conscious), the so-called qualia inside us, are summarily programmed into a device *without* qualia. This is not an insurmountable technical problem for functionalist programmers, since for them qualia don't exist anyway, so how would you tell the difference between successful or unsuccessful programming?

In attempting to build convincing robots—more than convincing robots: ones that are just as conscious and sentient as us and that pass the Turing test with flying colors—electrical engineers extrapolate and then "plagiarize" (to the best of their knowledge) our own circuitry with its hierarchies and synaptic grids (top-down), based (for instance) on how olfactory nerves are activated, blood chemistry changes, and "spiking frequencies [occur] in the brain."[82] These "computational or neural mechanisms" are pure "neurophysiology, neurochemistry, and neurophysics,"[83] topping out at control centers compellingly like ganglia that make them proxy humans.

There is no logical and mechanical deterrent to a robot like the functionalist version of Mary (who learns nothing new), because science has access to all the prerequisites for red: epiphenomenal red, physical red, any red; any color or shape or experience. "[I]nformative representations are distributed to all subsystems yet those representations are totally devoid of phenomenal awareness."[84] Hurray! The interior aspect of being—the thing that makes us *us* to ourselves—is proven extraneous. Just add more ping-pong balls!

But is consciousness really installed? Yes, in this thought experiment appropriate actions and responses are exhaustively written into a computer's or a robot's software such that it satisfies even future advanced versions of the Turing test. Various cognitive and behavioral functions "are [all] straightforwardly vulnerable to explanation in terms of computational or neural mechanisms."[85] With breakthroughs in digitalization and miniaturization the necessary circuits can be packaged at humanoid size and scale. Then the robots are set loose to make trouble.

Are such absent-qualia entities, should they ever be created, truly equivalent to us? Can we replace ourselves in all respects with robotic clones?

We can if subjective reality is only epiphenomenal, a combination of neuronal pathways: synapses, relay stations, and data-sorting hierarchies that produce a "consciousness effect." These can be fully and absolutely imitated, if not tomorrow, then on some Earth-to-be; perhaps they already have been on another planet. Then the consciousness illusion will be fully simulated and impersonated by circuits. Robots will be able to replace or at least join people. Given the number of stars and galaxies, the odds are that this has already been done and the case is long closed.

For his own version of this thought experiment philosopher David Chalmers appropriated the term "zombie" from pop lingo. Philosophical zombies are AI entities that behave as if they had consciousness but do not: "The zombie does not feel anything, even if it thinks and acts as if it does. Its experiences lack the qualitative feels altogether. There is nothing it is like to be *it*."[86]

If asked if it is real, the zombie will declare, "Of course I am. Can't you see that?" If you press the issue, he may confide, "I can do anything a human can do, and lots of things that humans can't.... But my inner life is nil, a complete zero. Any awareness I may seem to have is just your own projection. I'm nothing but a clever fake."[87]

Adapted from the West African Dahomey *nzumbe* via the Haitian Creole *zonbi*, the native designation actually means the converse of its philosophical mate: a witchcraft-engendered zombie is an alive person made to seem dead, while the philosophical zombie is a "dead person" (an artificial person) made to seem alive. In voodoo rituals, through a combination of hypnosis and potions (usually containing the deadly pufferfish toxin), a man or woman is put into an automaton-like coma resembling death. Then a *vodoun* sorcerer awakens the supposed cadaver as a zombie, an "animated corpse" without free will, memory, or self-awareness. Lacking its own conscious identity, it performs as a slave under his command.

Chalmers recruited this chilling character from various books and movies (in the mode of *Night of the Living Dead* and *The Serpent and the Rainbow**) to support arguments *against* physicalism. That's right: *against!* The zombie thought experiment is not an argument for behaviorism against phenomenalism, as I myself originally thought and continued to think for a long time (getting fooled by the tautology of the behaviorist argument); it is an argument *for* phenomenalism against behaviorism. I know that that seems counterintuitive, but bear with me. To propose launching a full-fledged zombie is to take an ironical poke at the behaviorist-physicalist position.

According to behaviorists, a robot identical to a normal human being in its comportment but without conscious experience is logically impossible. Subjective being may be delusional and epiphenomenal, but it is still "real"; it exists somehow in the universe. Thus the hypothetical "creation" of a zombie by a dualist or phenomenalist is proof that consciousness is *more* than the sum of neurobiological pathways, brain states, and their monitors and referentialities. The underlying thesis is that if physical facts are all there is in the universe (e.g., determine all other facts), the physical universe is the same as the actual universe (contains everything in the actual universe), and conscious experience must be in it too.

Since every fact other than phenomenal consciousness holds identically for a philosophical zombie without P-consciousness as for an ordinary human with it, then physicalism is put in the position of arguing that philosophical zombies are not possible, at least not in the way that subjectively aware humans are. If we can posit a robot physiologically identical to humans but without consciousness (and its zombiehood is logically irrefutable), physicalism (ergo) is false.

---

*Also *Zombie Town, Zombie Wars, American Zombie, Zombie Nation, Tokyo Zombie, I Walked With a Zombie, Zombie Farm, Zombie Strippers, Zombie Honeymoon, Zombie Brigade, Zombie Campout, Zombie Holocaust, Zombie Aftermath, Zombie Jamboree, Die You Zombie Bastards!, Alien Zombie Invasion, Kung Fu Zombie, Oh! My Zombie Mermaid, Scooby-Doo on Zombie Island,* and of course *Zombie High: The School That Ate My Brain.* And this only scratches the surface.

I accept this formal logic as a tactic, but I am not going to wade into the ontological layers of the ensuing debate with their myriad arguments from competing positions of metaphysics, circularity, and error theory. Instead I am going to appropriate the term "zombie" as a reflection of what behaviorists presume has happened evolutionarily on this planet even while they reject the hypothetical zombies of Chalmers's thought experiment. After all, Chalmers's zombies are based on real physicalist suppositions regarding first-person status, which support pretty much all the AI premises whereby such entities might be assembled, right up to the point where they get activated and are sicced on the populace. Then the behaviorists back off.

In any case, the logical fallout from my premise will land where it may.

An AI robot without qualia is "computationally like us ... but nonetheless a phenomenally unconscious zombie."[88] Remember, it has Access but no Phenomenal Consciousness.[89] It has "to do with information flow not qualia.... The robot zombies are envisioned to have mechanisms of reasoning and guiding action and reporting that process information about the world."[90] They think and are even exceptionally clever. They have the same "huge complex of memes"—minimal-sized cultural units that replicate reliably—programmed into them as humans.[91] Then any mode of sapient behavior, any functional or intentional state, "could be ascribed to robots or automata that behaved like people though they experienced nothing."[92] In every respect these absent-qualia humanoids are *mensches* like us who supposedly "experience" phenomena, except that these guys *really* have nothing going on inside them—no subjective existence.

Zombies mimic conscious behavior solely because they have been programmed meticulously to do so. They have sensors that implement assignments and preferences and summon up sub-routines to get out of "jams." If the first-person installation in them is comprehensive, the threshold of Turing "beingness" will have been crossed. From an outside (third-person) view their response patterns and actions cannot be distinguished from those of any "other" ostensibly conscious organism. When they engage in debates about their own putative consciousness, they reason so cunningly that an outsider would be utterly convinced that the machine with whom he was conversing had free will and its own bittersweet knowledge of reality. He would even "understand" the irony of being mistaken for a robot. A well-programmed zombie says all the things he or she needs to in order to dissuade or misdirect suspicion. She answers questions about whether she is a robot with playful humor, and not always in the same way—sometimes slapstick, sometimes wry, sometimes charming. She has savvy dialogue programs for different approaches. How can she be exposed?

Physicist Nick Herbert named his own thought-zombie Claire and then proceeded to interrogate her. When he tells her that she is only a fake human lacking feelings or inner experiences, she expresses abject dismay:

> I put my arm around Claire to comfort her, "You're such a
> lovely fake, Claire. But why are you crying?"
> "Because that's the way I'm programmed, you idiot!"[93]

There is no reliable way to tell a zombie from a qualia-bearing citizen, especially in a tautological, infinite-regress, nonconscious world among others like yourself. Remember, no third party can access another's first-person experience or ascertain whether he (or it) is truly subjectively conscious to his or its self (or skillfully programmed to simulate it); all an outsider can do is observe an "organism's" actions and make an educated guess as to whether they are the outcomes of neural states or mere algorithms.

We perform this audit all the time; we look at strangers and assume (pretty much to a one) that they are conscious and real, while hardly dwelling on the inferential basis of our assumption. We take it for granted that they are each carrying around an equivalent inner projection of a complete universe. Only an occasional person acts like a zombie and even raises the issue, as in the politically incorrect "you retard!" or the gentler "you dimwit!" or "you doink!" And yes, "you zombie!"

A "real" person naturally corroborates his or her interior experience, while a zombie's performance betrays itself by jerky, robotic movements, voicemail speech, systemic tics, etc. Or so we believe. We out-and-out assume that we would spot a zombie in the unlikely event we met one on the plaza. Everyone would stare at it because it would act like a shill for a *Star Wars* knockoff. However—and this is the key point—you cannot definitively ascertain whether any person is or is not a zombie by either analyzing his activity or interrogating his performance: "[N]o conceivable human behavior can reveal with certainty the presence of an inner life in another person: from the outside human beings appear to be 'mere things.'" [94] The candidate zombie's behavior is hermetically sealed, his first-person (or its lack) known (or unknown) only to him(its)self. Either way, he or she could play the opposite role to perfection.

Everyone is already tilting toward the same cool social roles and conforming to popular linguistic and behavioral mimicries. Everyone is *actually trying harder* to be a zombie than themselves, yet they don't come close; their "character" shows through. Even so, a lifelike doll in a suit, say at a cocktail party, if she was really good, could fool the guests and even get hit on as an eligible dame by a cruising male. Just as a sociopath or serial killer learns how to sell himself as Mr. Nice Guy—in fact brilliantly so—and a male transvestite decks himself out as a sexually attractive woman and replicates her beguilements, a zombie effectively sells itself as a conscious gal or dude; so from a third-person standpoint it *is* "real." Here are Nick and Claire again:

> After her speech I was moved to ask, "Claire, do you desire me?"
>
> "No, not really, Nick. But I think I can work up a pretty good simulation."[95]

The zombie could be programmed to emit suitable "pain" bleeps.[96] In place of "pain," he has a transducer that trips a recognition signal and networks that respond appropriately. He "takes aspirin because he thinks he has a headache" (or is programmed to tell himself he thinks he has a headache); in the dentist's chair, he will receive anaesthetics on the basis of his falsely ascribed beliefs and concepts about his nonexistent pain qualia.[97]*

Perhaps the alleged absent-qualia zombie is using the designant "pain" to "refer to something other than pain, or perhaps [it] succeeds in referring to genuine pain," meaning that pain itself is a linguistic or conceptual

---

*Maybe every dental patient is a zombie with super-vivid pain monitors. When I suggested this possibility to my own dentist, Gordon Swanstrom, he looked both wounded and baffled, but then he is practicing dentistry, not philosophy.

conceit.[98] It doesn't matter. "According to orthodox functionalism, the essential property of pain, like the essential property of belief, is its causal role fixed by evolution. Pain can be functionally defined in terms of its typical causes—bodily damage; and its typical effects—behavioral changes brought about by an intense desire to change one's current state and a belief that one can do so by changing one's behavior.... For the physicalist-functionalist, the dispositional tendency of an internal state to shift our immediate desires away from our current projects toward behavior which would prevent further bodily damage is insufficient to ensure that it feels like pain ... or is associated with any feeling at all.... [S]ome internal state might play the evolutionary role of pain without it feeling like *anything*, much less like pain, to be in that state.... Thus ... functional states underdetermine qualitative states, and only further specification of the physical details of the way in which the functional states are realized would determine which, if any, qualitative states the subject occupies."[99]

Think about it. Pain is not fully transparent; it is not a real signification of itself. It is a series of signals without an absolute meaning, installed solely to guide a robot (or an evolving vertebrate) out of trouble: "These signals could be processed in such a way as to cause in the machine a sudden, extremely high preference assignment, to the implementation of any sub-routine that the machine believed likely to reduce that damage and/or further reception of such signals, that is, to get itself out of such states. The states produced in this way would seem to constitute a functional equivalent to pain."[100] This stuff is real; it just doesn't "mean" anything.

Robots without qualia could additionally be programmed to gather in hunting bands or militias and discuss strategy. It would be somewhat more complicated than Deep Blue acing a chess tournament but not of a fundamentally different order. Machines with this sort of ambition and initiative already frequent futuristic fables. In *2001: A Space Odyssey*, the cinema version of Arthur Clarke's iconic novel, the astronauts' discussions with the computer HAL are acceptable and convincing as exchanges between sentient beings; they evoke human emotions and empathy.

In Isaac Asimov's classic tale *I, Robot*, adapted less iconically to the screen, robots recognize the superior quality of their own intelligence—they have more computing capacity than humans and, like *Star Trek's* Spock, are more rational—so they take over the planet and replace us. They become effectively a species with an agenda. Sound familiar? In this plot we humans turn out to be nature's intermediary, much as a caterpillar is an intermediary to a butterfly. Valorizing ourselves as consciousness's endpoint, we miss the trick being played on us whereby we are the molt between great apes and I-robots—a mere pupa for "real" intelligence housed in sturdier, in fact "immortal," machines. The only difference is that we reach outside the morphogenetic database to garner our designs and materials, and our molts are made in factories we construct rather than through cocoons we exude.

The best zombies could have a "suitably indexicalized recursive believer system" much as we do.[101] If a machine can be programmed to signal red or pain and react appropriately, or to move bishops and knights, even to assemble as units in squadrons, "so does it seem possible to program [a machine] to believe it's conscious.... The machine could think and print out, 'I see clearly that there is nothing easier for me to know than my own mind.... [N]o matter what your theory and instruments might say, they can never give me reason to think that I am not conscious, here, now.'"[102] It says, "I think; therefore I am." *"I, robot...,"* indeed!

"[I]f we already had a nonconscious perception-belief-desire machine, the addition of a 'self' concept would be trifling (just as would that of a simple internal monitor); one need only give the machine a first-person representation whose referent was the machine itself—that is, add the functional analogue of the pronoun 'I' to the machine's language of thought."[103] The machine has installed in it a module with circuits to represent itself to itself and at the same time fool any suspicious observers and thwart further inquiry. This bogus individuation is objectively indistinguishable from whatever it is we presently have, or think we have, and brand as consciousness. Meet Machine Joe.

"Joe" also happens to be precisely where modern science has feathered its present nest on the question of human meaning. Neuroscientist Michael Gazzaniga speaks for his entire tribe: "[A]ll we do in life is discover what was built into our brains."[104] Our "feels" of personal identity and free will are illusions or chemical effects: "[T]he environmental influences on a person *select* from options that are already there."[105] That's one flimsy cart to lead a very wild horse!

Falling for its own bait, neuroscience then leads the zombie out of the corral of its logical inception. But the problem with extreme physicalists is that they do not recognize their own irony or that the joke is on them— again, the Bee Gees: *"I looked at the skies, running my hands over my eyes,/ And I fell out of bed, hurting my head from things that I said."*[106] The lyrics don't scream "zombie," but they have that vaguely warped DJ-zombie warble of a materialized culture gone mad. The Bee Gees grokked this themselves:

"The melody to this one was heard aboard a British Airways Vickers Viscount about a hundred miles from Essen. It was one of those old four-engine 'prop' jobs that seemed to drone the passenger into a sort of hypnotic trance, only with this it was different. The droning, after a while, appeared to take the form of a tune, which mysteriously sounded like a church choir."[107]

Talk about the depreciated spirit behind technology finding a voice inside the human psyche!

Physicist Stephen Hawking "attributes the illusion of free will to the fact that each human being contains about a thousand trillion trillion particles, so that as a practical matter it is impossible to predict what [it] will do."[108] Through billions of years of phylogenesis this is how cells have self-organized and self-programmed to bundle the identity illusion. By this standard a machine with a thousand trillion trillion quantum switches graduates automatically *(ex post facto)* to a suitable proxy person, with or without qualia. In fact, the quantum states of its particles *are* its subjective illusion—more on that in Chapter Eight.

Hawking is proposing replacing our human state with not a zombie robot but a quantum-mechanical slot machine. Given the unknown potential of those thousand trillion trillion quantum switches, just about anything in our imagination of the imagining process can occur. The issue of subjective awareness and first-person access does not have to be acknowledged, let alone addressed.

How (pray tell) are we not simply functional analogues of the pronoun

"I" inside biologically activated machines? Being "real" as opposed to being "artificial" doesn't do it. After all, we could be "zombies" produced in nature by natural selection to compete more effectively for energy and the procreation of our genetic templates, or we could be the same zombies constructed by alien biotechnologists arriving in faster-than-light or wormhole-jumping craft for the same (or a different) purpose. Same difference! We only think that we are more indigenous or "alive" than that, that we are actually real, actually conscious. Well, try to walk yourself through a proof of your own existence!

What does being "natural" count for anyway in a bottomless regress of designants and simulations of possible outcomes? Of course, we could be programmed to think that we are conscious and to dismiss our ostensible E.T. creators (the command to dismiss them could also be part of their program). But then (by the way) who made them? How do we explain the existence of consciousness anywhere, even if only an illusion, even if artificially installed by someone else?

Who is to say that Earth is not full of E.T.-engineered carbon-based machines with built-in evolutionary software? Who can prove that Earth's vertebrate phyla are not all customized bioengineered zooids? If behaviorism is valid for zombie porpoises and caribou, it shrouds our putative existence too. Biotechnologically engineered living machines are not that different from naturally produced living machines (there is *no* difference to a functionalist). Whether animals are artificial robots or natural robots is not a meta-historical issue involving aliens but an ontological one involving systems. Nature's algorithms on planet Earth or the outcome of some other solar system's creatures and those creatures' robots designed by algorithmic shortcuts come out to about the same. In that sense, all intelligence is artificial, so robots are not only not unusual; they are not even robots.

Could we be cyborgs, part naturally alive, part artificially alive? Could we be part cyborgs, part clones who were bioengineered out of ape or lemur DNA for some unknown agenda (or amusement) by imperialist E.T.s? Any way you slice it, we are assigned the final status of zombies.

But where would a functional or epistemological line even be drawn? And how would it affect our definition either way? I don't think our hypothetical real or artificial status matters since our meanings—as arcane as they are unfathomable—arise *sui generis* from a substratum senior to mechanism or consciousness. Who cares whether we were engineered or cloned! We still mean what we *are*.

Nick and Claire engage in a playful discussion regarding her identity after she has told him that a quantum mechanical "consciousness" device developed by philosopher-scientist James Culbertson has been installed in her:

> "Well, I'm a Culbertson gal now, Nick. I've got a 3000C Outlook Tree Machine inside my head and, believe me, it's changed my whole way of looking at things…. For the first time in my life I feel like a real woman. Consciousness is a wonderful gift…. I'm really grateful. It makes everything so—so real!"
>
> "What's it like to be conscious, Claire? How do you feel now?"
>
> "It's impossible to describe, Nick. I feel like I'm in the center of an immensely important drama. It's a continual unfolding of—I can't really say what. The world is—the world is actually happening, and most of all, it's happening to me. Do I seem different to you now that you know?"
>
> "You, you seem livelier than usual; your eyes are sparkling, and I've never seen you so excited. I'm happy for you, Claire. Aren't you proud to be the world's first conscious robot?"
>
> "No, I'm really ashamed, Nick. I've been fooling you. I do have my OTM, but as far as I can tell it didn't work. I still don't have any feelings, nothing but behavior. No insides. Just clever and somewhat deceptive outer acts. I apologize for deceiving you, Nick. You really are quite gullible. But I'm afraid I'm still just a pretty, empty-headed robot."[109]

But who is fooling whom? Are you sure you know? Is she a full-qualia robot pretending not to have real qualia or an absent-qualia robot telling the "truth"? And to what end either way?

●

We are in a true bind. Our zombie twin "will be awake, able to report the contents of his internal states, able to focus attention in various places…. He will be perceiving the trees outside, in the functional sense, and tasting the chocolate, in the psychological sense.... It is just that none of this functioning will be accompanied by any real conscious experience. There will be no phenomenal feel."[110] Such creatures absent qualia are conceivably ubiquitous. They might comprise everyone, even the folks who just moved in next door. "I know that I myself am conscious via an undeniably direct and immediate revelation. But what about my neighbor? Could it be that he is just going through the right motions but there is actually nobody at home?"[111] And can you even really be sure of yourself?

Your parents could be zombies, and you would have zero chance of knowing or proving your suspicions (either way). Your schoolteachers or playmates could be zombies in an exquisitely crafted *Truman Show;* you might be the only one with a subjective identity. What child has not fantasized that he or she alone is real? Or that his or her parents are kidnapper zombies?

If you yourself were a zombie or gradually became a zombie, not only would no one else notice it, you might not either. Your inner life, which only *seems* to you to exist in the first place, would continue to *seem* to flow into your monitors. If you started off possessing qualia and then lost them as you became a full-fledged robot (as per the discussion of cyborgs on p. 73), "you could now be 'hallucinating' your own phenomenology"[112] while judging that nothing untoward had changed—or perhaps nothing *had happened* and you were always this way—or perhaps you are a zombie one hour and a person with qualia the next, hour after hour. As you slipped each hour into the alternate phase, how would you distinguish your hallucination of qualia from the "real" qualia states that followed?

From a behaviorist standpoint we are already there and have long been there. Our putative consciousness is a combination of illusion and delusion, thought effects produced (and then confirmed to each other) by bioelectrical circuits and signals.

It's not even so much not being able to convince someone else that you are not a zombie—that's first-person/third-person garbage. It's the matter of not even being able to convince yourself in the privacy of your thoughts. We cannot prove to our "first persons" that we are not zombies and that our thoughts and feelings are not simulations, the epiphenomenal simulacra of what real thoughts and feelings might be in some universe, not this one, that housed real creatures, not us.

Try this mindfucking meditation. Consider that your thoughts are not really your thoughts and have no interior feel. They are just programs along with a belief program whereby you are telling yourself you are real and exist. Pretend that what you call feelings are just chemicals moving around and interacting with each other to produce a hallucination.

Don't laugh. This is no mere thought experiment or abstract epistemology. It is the crux of modernity and has serious political and socioeconomic implications. Its propaganda is the source of many a recreational war, pedophiliac binge, playboy/playgirl lifestyle, reality show, sterile accumulation of wealth, gambling addiction, severe depression, and (ironically) sense of privilege, exceptionalism, and entitlement. If you're entitled to nothing, you might as well go after everything—everyone else's stuff too. None of you actually exist.

As long as "qualia" could be illusions intrasubjectively programmed (like color inversions), we can never get free of the zombie curse or get to know who and what we are.

At core this is the primary, irrescindable proclamation of modern science and its credo of human destiny (to the horror of all the Aborigines and nonhuman species left on the Earth). Mind is just heat or heat effects, its subjective illusion of experience arising from binary flows of molecules: "a biological phenomenon like any other."[113]

Except for that fact that consciousness preceded artificial networks by millions of years and drew enough of a bye to establish itself among the beasts, plebes, and underclasses, phenomenal existence wouldn't stand the chance of a snowball in hell. Not under the reign of science. I mean, it is not even proxy consciousness that we finally embed in AI machines; it is our own *proxy illusion of being conscious.*

The phenomenalist objects that a zombie machine is *programmed* to react in a certain way and has only the *functional equivalent* of consciousness. "[T]he silicon chips did not duplicate the causal powers of the brain to produce conscious mental states, they only duplicated input-output functions of the brain. The underlying conscious mental life was left out."[114] In that sense is not our plight (under peril of zombie life-sentences) not a little like certain individuals who wake up under anaesthesia but are paralyzed, or have immobilizing neural diseases, or mistakenly take drugs that freeze their motor functions but not their consciousness? They are the exemplar zombie. The true self in them (like the vanishing voice of the Searle cyborg) wants to cry out that it is still conscious: "I perceive everything going on around me. It's just that I can't make any movement."[115] *I know I exist, but I can't prove it.*

But "[W]hat more is required?" the functionalist counters. Saying we are alive and have a subjective mind whereas a machine doesn't, by mere virtue of first-person imprimatur, hardly ensures that we are not cyborgs or some sort of artifact, made out of remarkable plastics—even remarkable organic-molecule "plastics" fashioned over billions of years in nature and distributed in molecular-cellular circuits through the planetary environment at large. We only resemble natural life forms with qualia. The maker might be natural selection of the algorithmic Darwinian style. There might be no maker. The maker might be E.T. or God. Our maker might even be a transcendental cosmic intelligence. It's all the same in zombie-land.

If we turn out to be machines (of any of the above ilk), we can still privilege and anoint ourselves, even if we cannot prove our inner life or our difference from robots. Remember, robots can be programmed to coronate themselves too. Again, this is not a thought experiment: this is our species's self-declaration: Zombies = *Homo sapiens* = zooids of Linnaean mix. Free will is a fantasy; we do what we are told (by genes, quantum particle states, and neural programs).

The phenomenalist or dualist saves his best argument for last: Yes, it can be demonstrated that molecular movements alone do cause colors, light, sensations of heat, etc., but *only after consciousness already exists.* Microstructures express themselves in consciousness, but there is no way that they can invent it whole cloth from nothing, out of inert matter. Hormones and neuro-regulators do not create existential entities or feelings; they activate and regulate them—as well as their conscious and unconscious states of being.[116]

Anyway subjective consciousness is less like a program than it's a stream made of something fundamental like photons, only subtler and traveling in its own layer of the universe along a discrete trajectory: that's the dualist litany. Mind gets hooked up to matter only by means of an evolving web inside nature; yet it retains its separate existence from matter: "[T]he very study of the external world leads to the conclusion that the content of consciousness is an ultimate reality."[117] This becomes my tack in Chapter Nine.

On top of that—and this is really important—the objective neutrality of zombies is a fraud; they are repositories only of post-consciousness human ideologies, contaminated with our own symbols, symptoms, and signs. Cyborgs are not innocent, transparent replicas of behavioral programs; they contain complex tropes and histories within their artificially humanized circuitry, the deciphering of which yields not algorithmic I-robot initiatives but clues to our own chameleon nature—our unexamined agendas, including our biases of scientific proof and misplaced concreteness: where we think we come from as well as where we tell ourselves we are going.

Yep, science too.

Even a simple airplane is a composite symbol before (and after) it is enlisted as a programmed carrier. It is packed with myriad concealed subtexts (arcane even to its designers and makers). What do airplanes say about our relationship to space, time, culture, markets, social groups, commodities, love, each other? In that sense neither a robot nor a cyborg is a true zombie; it is always a counterfeit in which an alias of our own phenomenal mind and unresolved conflicts has been ingenuously implanted—a metal-and-plastic Rube Goldberg device for collectively displaced economic and political objectives.

Far from proving how nonhuman we are, zombies prove that our qualia infest everything with which they come into contact. Robots are not substitute creatures but dynamized myth chains. Even a slot machine or hair dryer is contaminated with the unconscious meanings that we inject into their processing circuitry and zomboid existence. But don't expect to find those meanings lodged in any linear or accessible manner, in any diagnostic or diagrammatic flow chart; we have buried them far below the hardware—as deep as our biological history and in as inextricable a labyrinth as the cultural and historical motherboard by which we have arrived at the present script with its pandemic technological subtexts.

Anthropologist Claude Lévi-Strauss diagnoses the condition (but the doctor says we'll be all right):

"[T]he nearer we get to concrete groups the more we must expect to find arbitrary distinctions and denominations which are explicable primarily in terms of occurrences and events and defy any logical arrangement."[118] That includes qualia, robots, zombies, thought experiments, market economies, nine-to-five jobs, and all of the protocols of science. In indigenous Australia everything is a potential totem, even the vessels in which the white guys arrived, even the nasty fellows themselves. The universe is ductile, malleable, totemic, made of exactly what? It is a vast internalization inside a vast externalization, but it is neither a neutral equation nor a gratuitous experiment—or even for that matter the "real" universe.

The one thing zombies can't create is credible totems because, more accurately, they can't create anything that is not a totem (just *our* totems not theirs). Lévi-Strauss continues. He might say: *"I've got the horse right here,/ His name is Paul Revere."*[119] He says:

"For a theory of information to be able to be evolved it was undoubtedly essential to have discovered that the universe of information is part of an aspect of the natural world." No big deal for functionalism, standard operating procedure marking the transparency of the passage of sensory into cognitive information. "But the validity of the passage from the laws of nature to those of information once demonstrated, implies the validity of the reverse passage—that which for millennia has allowed men to approach laws of nature by way of information."[120] This is a very big deal: an open-ended, totem-ridden system blocking access to any absolute physical reality; yet most functionalists are unaware of it (or they dismiss it as fashionable cultural relativism indulged in only by Yale and Harvard graduate students not studying the hard sciences).

They don't understand—academic fashion aside—the information is coming from us, only from us, and is not epiphenomenal debris—and that means *all* the information along with its eternal objects, background noise, arcane melodies, and other subtexts. Arising at an innate and endemic mythic level, it is creating the known and only known universe.

Although myths, Lévi-Strauss contends, "considered in themselves, appear to be absurd narratives, the interconnections between their absurdities are governed by a hidden logic: even a form of thought which seems to be highly irrational is thus contained within a kind of external framework of rationality; later, with the development of scientific knowledge, thought interiorizes this rationality so as to become rational in itself. What has been called 'the progress of consciousness' in philosophy and history corresponds to this process of interiorizing a pre-existent rationality which has two forms: one is immanent in the world and, were it not there, thought could never apprehend phenomena and science would be impossible; and, also included in the world, is objective thought, which operates in an autonomous and rational way, even before subjectivizing the surrounding rationality, and taming it into usefulness."[121] That is, if you consider our uses useful.

Zombies and computers are "interiorized rationalities" based on historically nested phenomenal realities, not "rationalized exteriorizations" and certainly not unaligned access programs. Science is stuck with not only its first person but that autocratic dude's archaeological, unconscious, and totemic qualia-sets. Even though the situation is both opaque and insane (and sullied too), it is the only link between uses and meanings. Again, Lévi-Strauss:

"Man is not free to choose whether to be or not to be...." That's probably a good thing under the circumstances. Instead, he generates "an unlimited series of other binary distinctions which, while never resolving the primary contradiction [his existence], echo and perpetuate it on an ever smaller scale.... [T]he reality of being [is what] man senses at the deepest level as being alone capable of giving a reason and a meaning to his daily activities, his moral and emotional life, his political options, his involvement in the social and natural worlds, his practical endeavours, and his scientific achievements...."[122] This is the stuff to which the functionalist crowd is outright blind and deaf.

Just as you cannot batch consciousness into ping-pong balls or beer cans or carbon transistors no matter how many of them you weld or string into networked grids, no matter how complex the hierarchies in which you arrange them or how many billions of years you give wind and water to shape them in a variety of niches pregnant with differential forms of energy to metabolize—you cannot implant the equivalent of totems, politics, and moral and emotional life in zombies; you can only install *actual ones,* which can't be installed at all. Qualia are ontologically different from mere circuits mimicking conscious behavior, and they will be different forever; we will always sing our own, different song: *"Can do, can do./The guy says the horse can do...."*[123]

Meanwhile I would posit that the philosophical and technological edifice of AI and its hypothetical creation by humans is less an enterprise of earnest scientific endeavor than a religious device intended to convince us of our own meaninglessness and dependence for our existence on mechanisms and motifs that others control (like corporate-patented genes—a more immediate danger than the merely hypothetical risk of zombie-hood). Today we are zombies more in the Haitian sense. The Voodoo Master is Capitalism, Consumerism; his religion encompasses academic and scientific consumerism too. In that sense the zombie AI is a real-life metaphor and a diversionary counterpart to the Commodity's insidious scrolls of nihilism, endless war, apocalypse, and passive enforced consumption.

The speculative trope of charming robots with agendas who develop agency (and then replace us or not) is diagnostic of a global epidemic and contagious existential doubt that has trickled down from elite scientific circles (or perhaps trickled up). Our identity crisis—a crisis of possession—has progressed in the last hundred years into a crisis of meaning and a moral and spiritual crisis as well. We do not know who we are or *if* in fact we are. We cannot escape the *Voudoun* "who" has turned us into animated corpses. Every day we fear that we could be supplanted unaware by automatons because we experience how the global capitalist imperative has already turned us into something like automatons: desire machines without souls—workaholic, funaholic slaves. That is the alienation of the present heroin/cocaine/pharma-addicted, 24-hour-news-cycle world, as we drift through colognes of meaninglessness discharged by a culture that at core does not believe in its own existence.

CHAPTER SEVEN

Consciousness as an Emergent Phenomenon: The Psycholinguistics and Phylogenesis of Meaning

> "What is drawing molecules into us, into our elemental cycles? What entices matter, while still obeying the precepts of thermodynamics, into biochemical bonds and metabolic pathways? What could possibly convince a particle, seemingly against entropy, to do my body's tangled bidding rather than follow the breeze? What is so persuasive about the call of tissues that molecules flock to assemble in myriad beasts prowling landscapes, each one glimpsing the world by electrons flowing into their ganglia? Where is life's pied piper? How is his summons drawn and issued? How can a system as simple as 'Moon rolls through clouds' contain an event as complex and outrageous as 'parrot dive-bombs marmoset'?"[1]

i. The Origin of Life

Biotechnologists get at traits, but they cannot get at life itself or consciousness. They do not know where or how (or even if) its signals are coded, nor can they get at the modes that routed quadrupeds to mice or men—or any two species sharing ancestral genomes to radically different outcomes. I mean, where are the road signs and traffic lights that conducted blastulas toward French cuisine rather than entrenched rodenthood? Whither their subgenetic networks and mitochondrial switches, if not to transmute the nucleus of a ground-foraging mammal into a hominid homunculus, at least to fluctuate in that direction? What likewise commands a sea squirt toward seasons of salamanders and the amphibian life?

A juicy enough arsenal of genomes and rebuses awaits the "Craig Venters" of the world to sequence, for starters the human blueprint, the house-fly blueprint, prehistoric weevil DNA, fossilized magnolia juice, *Tyrannosaurus rex* bone marrow from Hell Creek, Montana, etc.

Biologists at the Natural History Museum of Utah in Salt Lake City have wrapped genome scripts of various mammals in rows of amino-acid sequences around a big column on the third floor. Elongated horizontal strings, disappearing around the column's curve, mark individual species, while each common (though unnamed) mammalian coding site is represented by a vertical row intersecting the horizontal ones, making the equivalent of a graph with boxes formed by X and Y axes. A visitor can check out which of the four "letters" (adenine, guanine, thymine, cytosine) lands at each specific position (box) on a chromosome. Astonishingly, most of the vertical rows read identically: the same letter (A, G, T, or C) regardless of whether the animal is a whale, a porcupine, a rabbit, a macaque, or *Homo sapiens*. Every so often one or more of the creatures departs by having a different letter (amino acid) in a slot. Those few slots contain the *only* information that separates a human in a DNA blueprint from a wolverine or a mouse. Think about that: an adenine or cytosine here or there and the ancient bolus of yolk and twine that turned into you could have become a squirrel clambering up a tree or a fox prowling in the high grass instead.

Clearly something more is going on than spelling, for outcomes transcend their alphabets. I would aver that there is *another code* buried under (or inside) all those—a meta-code if you will. That's the one they can't touch, one not written in their lingo. It is the code for the river of life and maybe for major phylum shifts within that river. It was not left at Hell Creek, nor does it reside in roadkill.

Forget for a moment the metaphysical ramifications; it also marks the spot where the biophysical world vanishes into its own quantum vortex of a trillion-trillion-plus switches. Because of the continual miniaturization and condensation of molecular material into nucleoplasmic and then cellular and tissue layers and molecular valences through the annals of evolution, tinier and tinier subcellular clocks have been cached in other timing devices, microtubulets inside microtubules—so the real sentinels conducting biopoeisis and organization dwell beneath the level of genes, beneath even the molecular structure of DNA. Job addressed this arcanum for centuries to come:

> *The deep says, "It is not in me"; And the sea says, "It is not with me...."/It is hidden from the eyes of all living..../God ... knows its place./For he looks to the end of the earth,/And sees under the whole heavens,/To establish a weight for the wind,/And mete out the waters by measure.*[2]

●

Though RNA-catalyzed mechanisms have been inherited into modernity intact, the precise roles of most of their original controllers are scattered nonlinearly and discontinuously throughout the whole of cytoplasmic space—they still work, in fact work better, sleeker, and more efficiently than ever because there is hardly anything of them left to crash (only what is absolute necessary). By countless life-and-death contingencies they have been reduced to states of barest drag, outputs of mere quanta. Time and turbulence have eroded away everything else. Since the planet didn't annihilate them in the cradle while wind and rain and ice and solar disk were scraping and freezing and scorching and glaciating them down to almost nothing, they are now the sleekest cats in the parade. Their aliveness alone runs them—precedent and simplicity—plus a paradigmatic force. Everything else has fallen by one wayside or another. Life won, but only by forfeiting anything that might have been life but isn't.

Modern sequenced chromosomes and their amino-acid offspring are relics from the pre-protein environments of four billion years ago—rebuilt ciphers from Precambrian tags. They were already ancient when they broke into the cellular universe. Even the 32,000 or so decipherable strings and strands at the system's surface, cardinal in their way, represent less than a dimer in a haystack; they are secondary to core plans of modern life forms, tenuous as spray rising from an ocean is to the intricate, swirling waters beneath. All the benchmarks are there—protoplasm, form, motif, internal design—but they leave no independent map for an eel or a crayfish, or how to derive any organism at all out of muck and mud.

Whereas natural selection operates at cellular and organismic levels, the universe is carved more meticulously below moleculo-atomic levels, which turn out to be linked and keyed in ways that could not have been foreseen, even in retrospect. Our source code is weighed with the wind, meted under a caul of hieroglyphic subcodes riding the roulette of epigenetic calculabilities. The last stray threads of its ball are all that biotechnicians can presently untwine and wrap around their columns. Its deep molecular and submolecular structure is more complex than they have tools to pick apart or perhaps (like the purported link between the brain and the mind) will ever have tools to get at. The rest has been interred so far inside the nucleus and cytoplasm that the best microscopes merely bounce off its skin. Only it is not skin, and its marks are no longer marks.

In the vale of molecular time, over billions of years of bustle and sorting, under the reign of nucleic relativity, signals, triggers, and palimpsests were loaded into transitional mechanisms, then methodically expunged into their vestigial chemistry. They reside there now: immune, outside interrogation or summons (like the weight of moonlight).

"God made everything out of nothing," Paul Valéry averred. "But the nothing shows through."[3]

What remains is a pittance. It couldn't possibly function mechanically. How can an engine run, missing more than ninety-nine percent of its parts? But it *does run* because it has *continued to run*. No one knew how to start it, but no one now knows how to dissuade it. We can stomp out or poison its creatures, but we can't turn off the engine, not short of killing the planet itself.

We look into the mechanism of life and see how it operates, but that doesn't mean we see what *it is*. Life is a synergy, a momentum that ultimately dispensed with levers and axles, gears and shafts, the absence of which would halt any manmade motor in its tracks, resulting in catastrophic failure. A fertilized egg is so simple and inviolable because it is so complex. It carries so much information because its systemic load has shifted toward micro- and nano-space and most of it has been erased anyway. It assembles so much more because its cargo is so much less. Transparent yet dense, deprived of all but a few ancestral strings, the egg separates its layers not by space but time, proceeding through flawless metamorphoses without a hitch, creature after immaculate creature.

Plants and animals happen now solely because the molecules in their membranes, cells, and extracellular matrices *have no other mechanical choice.* Their design has one option: *to unfold.* But, in an entropic universe, how was this privilege stolen from mighty thunder?

Life runs simply because it runs. Everything inside it works only because everything *else* works, even what is no longer present, the elisions and gaps. That is why we can't find the operating circuits, the core premises or baseline intelligence. It is the one in a trillion trillion that survived (and then the one of those trillion trillion) and, past a certain point, it didn't have to show ID or give testimony; it only had to be Mr. Broadway. There *is* no testimony, no decodable blueprint—only subsequences underlying major sequencing themes. The king of the algorithm *transcends* the algorithm.

The originators and regulators of distributed networks of genes, enzymes, and proteins trail acausal chains of effects in hyperspace. We see them, but we don't believe them, for this kind of cybernetics is beyond time, quicker than white on rice.

The key to life is either too small or too discursive, too tautological or too piecemeal, too vast or too alien to us. It is concealed by mechanism that has come to look like packaging. It is the background of everything, that itself is nothing. Even if a microbiologist stumbled over the jewels, he wouldn't recognize them; that is, wouldn't grok them as different from background noise—they shoot unsigned commands across now-coalesced gaps, commands that no longer bear even their own witness; lesions that are filled by phase states, adjacent potentialities, nonlinear and quantum effects, and arcane commands of junk DNA. Eons ago life's context was spilled down the sinkhole of the cosmos. Its signs were discarded or splattered into nature with the letters of a lost alphabet.

The unexpected order of chaotic systems, the paradoxical intelligence exchanged among networks of cells, the sheer depth and variability of the genetic book of changes, the messaging inside the extracellular matrix—when reconsidered (or even when not) in the context of scalar, nonergodic and/or telekinetic forces—collide with a quintillions-of-gigabytes-thick, timelessly unfolding entropy and the mysteriously unpredictable emergent properties of atoms, molecules, and subcellular biota. The resulting machinery is so subtle, so fine, so effortlessly self-correcting, so eerily "designed" that DNA pioneer Francis Crick, a vocally aspiritual dude, yet one of the first to *see* the double helix through a jumble of cellular protein, proposed that the code itself and the self-regulating human algorithm might well have been fabricated by extraterrestrials. But that was only because he couldn't think of anyone else.

Same question: Who designed the extraterrestrials?

The evolution of life from inorganic to organic to biochemical substances is remarkable in its inevitability. The development of complexity in its own right has been well established by the science of nonequilibrium thermodynamics as pioneered by Ilya Prigogene: indeterminism in unstable nonlinear systems leads to reversal of maximum entropy under the irreversibility of time. When energy is flowing through a system from source to sink, the passage of that energy drives the development of areas of very low entropy almost as if preordained. Hence in our solar system with its enormous energy flux, complex systems like life would be surprising only in their absence.

Something is still missing, and I am not sure that calling it nonequilibrium thermodynamics says what it actually is—what changes rates of chemical reactions under its influence. We simply can't penetrate a system that large (that macroscopic). Statistical solutions point less to agency than the mystery at the source of all numbers.

In a presently unpublished manuscript, my buddy and fellow cosmic explorer, Curtis McCosco, takes a heroic shot at science's Magical Turbulence dynamics. I yield space for him here to make his full point:*

> A dynamic system is all about the interrelationships of the
> forces, internal and external, which energize and influence the
> system. A soap bubble is a simple dynamic system adapting

---

*Later he introduces the third volume of this book.

instantly to constantly changing conditions. The surface forms as air gases are blown into liquid soap that coheres into a semi-stable periphery. The constantly changing relationship between the periphery and an invisible center is the defining characteristic of a dynamic system. Internal and external pressures tend to establish a state of homeostasis or equilibrium, then a breeze picks up and the fun begins.

Blowing bubbles is like playing with magic, ask any kid. External air currents change the shape of the bubble which, in perfectly still air, indoors for instance, naturally assumes a smooth spheroid shape. This sphere has a center, and if a breeze blows, distorting the shape of the balloon, that center moves (this is called its orbit) in response to the changing conditions. What we're calling "the center" does not exist as a physical entity, but as a function; it is real because of the continuous relationship of the substance of the bubble's membrane with itself. Any distortion in the surface shape moves the center position instantaneously because that center is a product of the relative positions of an infinite number of points on the surface. In this way there is a constant and instantaneous exchange of information between center and periphery; a multitude of feedback loops. Nothing of substance is exchanged, but the cohesiveness of the whole system is based on these relationships.

The bubble also is part of a larger continuum—call it the environment, the bio-sphere, the cosmos—that also has a self-generated center. Each of these categories, scales of reference, has its own center; the Earth has a center one can reference quite easily by dropping a plate on the tile floor just as you're assuring your wife you did not have too much wine at the dinner party.... The locus of the center of the bubble, its physical-mathematical time/space position, changes constantly, and is also influenced by the Earth's relationship with another center, the Sun. Bubbles are definitely more fun on a sunny day. These spherical dynamic systems of scale, from the soap bubble to the Solar System, have centers which are continuously interacting with each other and affecting each other's position, velocity, trajectory, and shape. By this line of reasoning a butterfly in the Amazon really does affect storms in the Gulf of Mexico....

What all these centers share in common is obviously not a set of spacetime coordinates, but non-existence; like Gertrude Stein's Oakland, there is no there there. The centers are abstractions calculable because a "periphery" exists, a multitude of forces interacting. The forces aggregate and a center is created. Call it mathematics, the assemblage point, lower *tan t'ien,* the sweet spot of a hanging curve ball, it's there; but it ain't there....

John Von Neumann found that living systems are self-organizing and self-correcting, communicate internally, process external data in relationship to center—all done simultaneously. In human living systems these processes flow smoothly in currents more subtle than thought. No one tells the bubble how to change shape, except for those frisky kids having fun chasing them. Yet it changes shape in direct response to changing conditions. Reaction time of the whole is subject to gravity and the law of inertia, but internal communication affecting the movement of center, its orbit, is instantaneous....

As the amount of information increases, any system becomes more complex (and *vice versa*), and begins to resemble a dynamic system. The amount and nature of energy entering affects behavior accordingly, from which you can derive a mathematical shape. Now you have a dynamic system, with parameters and center, and uncertainty. It's as though extra perspectives are added. A moving shape causes and is caused by changing relationships; internal forces interact with each other and with external forces creating spirals of energy. Spirals are centering functions of force and perception, simultaneously, creating new spirals. Flow is created by spinning energy in relationship to spinning energy, that's the regenerative force. Actually, in advanced theorizing on the matter, and from a Buddhist perspective, all systems are dynamic in their true nature. That Ford tractor rotting in the weeds behind the barn, once a hard-working machine, is experiencing a spiraling return to the earth as rust and crumbling rubber. As its form disintegrates, its once dynamic center re-integrates with a center of a greater dynamic—the Earth....

Supposedly Werner Heisenberg once quipped, "When I meet God, I am going to ask him two questions: Why relativity? And why turbulence? I really believe he'll have an answer for the first."[4]

•

That's how fundamental and mysterious turbulence actually is—turbulence and its cronies: chaos, departure from equilibrium, self-organizing agency. Because of nonequilibrium thermodynamics, archetypal forces, and whatever else is at play, I suspect that something uncannily similar to our own double helices of DNA abounds presently in the shear-force-warmed oceans of the Jovian moon Europa or the Saturnian moon Enceladus and/ or if not there, in equivalent planetary oceans throughout the cosmos— and not solely because of probability theory and natural selection (or intelligent design). Here the Buddhist principle of self-originating ground luminosity meets the crossover science of complexity. But both point toward the same reality, for there is only one universe (at least in that sense).

The point is: existence is likely inescapable.

ii. The Evolution of Consciousness

No matter what else I say, here or elsewhere, or anyone says, regarding the essential nature of consciousness—whether it is intrinsic to the universe or merely contingent to the present organic-chemistry matinee of our Oscar-nominated Big Bang—there is little doubt that on this planet a brain capable of self-reflection evolved inside nature along a chordate-vertebrate-mammal-primate gradient. In the physical world, within the unforgiving principality of animal anatomy, it's survival of the fittest—there isn't any alternate path to temporal consciousness. Mind is tied to self-integrating cellular structures that generate burgeoning ganglia and sulci that form internal links and reciprocities leading to the "proper" relations of subsystems and elements.

A brain is a naturally assembled soft machine that receives and processes consciousness, the only such instrument known in the universe and a *sine qua non* for living thought; again, no matter what consciousness *is*. It takes—it took—billions of years of DNA improvisation, codon sorting, and moleculo-cellular feedback and modification to arrive at our particular bundle of mentation. For now, let's proceed as if "brain *equals* consciousness" and not worry either about whether that means the access or phenomenal brand (or both). There is no way that introspection and philosophy can be infused into the ganglia of an anchovy or dragonfly, no matter how exquisitely their anatomies do anchovy and dragonfly things with what neurons they possess. These creatures do not have the tissue capacity for reflective mindedness (even should they aspire to it). Likely they do not have the karmic prerequisites either, but that is an entirely different matter.

Suffice it to say that without a sizable, internally folded brain you can't have biological consciousness; you can't have totems and you can't hold public elections—though that doesn't mean that (one) our mode of bio-consciousness is the only kind of consciousness in the universe; (two) that its operation is solely utilitarian; and (three) that cerebral mind *is* what consciousness *is*.

We tend to consider the "meaning" range mostly of ourselves and other primates, but birds and other mammals (and even simpler vertebrates and invertebrates as well) invent and drive the planetary mind. As biological ciphers interact with one another and with the external world, they generate behavior codes and strategies, which may even leap species boundaries by some sort of Hundredth Monkey principle or perhaps DNA sharing and interspecies contamination, not to mention any "unauthorized" telekinetic transmission. As the external world changes, biomolecular codes evolve and adapt too, and not just by systemic feedback from the environment in a Darwinian mode. By an artistry too complex to specify, they improvise variants of their own strings of algorithms and then project their proxies into the environment (for approval or rejection). I explore the feedback between internal and external dialectics in the evolution of consciousness at many junctures in this chapter.

Our brain is the outcome of a deep history embedded in our genome with its own meta-histories of psycholinguistic and cultural sets. In that sense the various lobes and internal cerebral structures have both a genetic, physical origin and a superorganic, epigenetic origin—and these have been interfused and have imbued themselves in each other, componentially and indivisibly. (This is the case at some level for all our evolving organs.)

Language, culture, and their memes have been conflating and germinating inside our tissues for so long by now that it is difficult to gauge just how profoundly they have governed our anatomy, especially hominid brains; likewise how deeply and integrally they embedded, cellularly and even subcellularly, in discrete structures. Once converted from symbols back into biology, consciousness has transmitted its messages across time while synergizing them into biogenetic and neural processes. How might that have been done without violating Darwinian mandates?

For a start, DNA in a Gaian environment has favored certain styles of body plans and sensorimotor pathways, shuffling these out of its deck in the context of epochal changes in load factors and divergent lines of speciation (receding its hand when necessary). Across gaps in speciation, popular motifs keep returning: "Photoreceptors have been independently invented over forty times in various invertebrate lineages ... powered flight was invented at least three times after insects did it: by the flying dinosaurs, by the bats, and by the birds (not to mention all the jumping spiders, gliding mammals, 'flying' fish, even a snake that glides between tropical treetops).... [C]onsciousness has been hooked onto many distinct mental systems [and] been invented ... time and again for many distinct modules...."[5]

Each nervous system and set of species semes evolved under the constraints of its own ecological crises and mode of being—outcomes that are discrete for a worm, a bird, or a fish in terms of interfacing architecturally with the external world. Each interprets landscapes biogenetically, develops organs for making reality/appearance distinctions, and expresses already intrinsic designs.[6] Showtime anatomies reflect shifting relationships between two active morphologies: geographical fields and biological fields. Each phenotype*—penguin, beaver, duckbill platypus, crab, etc.—delivers not just a body shape and neural resources, embryo by successor embryo, but a metabolizing fusion of visceroskeletal and environmental representations in a virtual field. Successor forms mold one another as if tissues were as soft as gelling clouds. In genetic storage they are even softer.

Creatures' neural anatomies evolved also, as noted, through interaction with prior synaptic networks and their morphogenetic properties, and then, in certain lineages, this process was subjected to psycholinguistic and behavioral interpolation. For humans, that meant communication systems, rituals, hunting in groups, weapons, art, and cognitive classification, all at degrees of relation to a shifting outer scenery—Earth and climate changes and their effects on other species in the ecosphere—then radiating through the internal referents and constraints of an evolving nervous system.

The script for *Homo sapiens* follows the emergence from water to land, taking the saline ocean inside a general body-plan (worm to fish to amphibian to reptile), while preserving marine ratios of sensory nodes (bilateral symmetry and ascending synaptic routing, a.k.a. tail to head delivery). Prior radial metazoa remain camouflaged in the bilateral symmetry of both vertebrate and invertebrate lines, betrayed during early stages of development and preserved in some adult forms of mollusks, starfish, and beetles.

Gravity and shear force initially rounded molecules as in an energy vortex inside a permeable outer membrane. A cell is better thought of as a field or "energy disturbance" than a "thing," for cells are not so much discrete objects with boundaries as continuous ribbons of semi-closed interconnected structures that arise from and lead to one another, not inside or outside any fixed contour but arranged in layers by the chemistry that takes place among them. This field fissioned into a colony of fields that sensed each other's presences and formed a globular mass like tiny jellyfish stuck to one another as they tried to swim independently. Each "jellyfish" contained a living hologram of its species plan.

---

*Phenotype is the physical, biochemical expression of a genetic trait.

Even at microscopic scale, an overall colony generates a soft maze through which a particle can wander for miles without coming to an end or crossing its own path—its membranes penetrated by transmembrane channels filtering protein molecules and other metabolic resources for its energy functions.

The initial circularity, which bends into an oblate or tubular spheroid in some lineages (like insects), provided a landmark around which templates of sensory ladders, synapses, nervous systems, and ganglia organized. In the Chordate-Vertebrate line among the teleosts (bony fish) and crustaceans (particularly octopi), rounded topology folded into itself and then elongated in a hierarchical system or pyramid of command, heading crownward along a central trunk (notochord) toward a head or some sort of cephalic organ or concentrated neuralized zone (see ganglion immediately below). Settling ultimately in a hard spine (vertebrae), this synapse-spreading wave established the great kingdom of vertebrates. Diverse related species arose from its axial motif: primeval ancestors of sharks, frogs, snakes, hawks, dolphins, rats, bears, and hominids.

As these creatures got more complex, their body-plan ultimately required a switching station for the upstream current of sensory inputs to deliver their contents into a central processing unit: a ganglion. It is in ganglia that the notion of switching stations and monitoring modules became synonymous with each other, and the shoot of mind was planted at the system's crest. Once nature invented such a faculty, it had to "work at wiring it to the sensory modules and to memory … to do for consciousness what it did for wings and photoreceptors,"[7] thereby creating "a federation of somewhat autonomous agencies."[8] The literal crown of this activity was the brain. (Of course, species not in the Chordate lineage adopted different motifs and went on their merry ways.)

The order of primates, our human lineage, emerged from a branch of mammals almost 70 million years ago at the end of the Cretaceous or beginning of the Palaeocene era. Its sire probably descended from a reptile-like forager that then served as the innocent progenitor of marsupials, carnivores, and ungulates. In this long-ago era when nothing resembling us dwelled on the Earth, rodent-like fauna were our closest ancestral kin. Eventually some groups of otherwise bland shrews migrated into trees in order to investigate the ripe canopy there. These meek, unprepossessing, rodentlike quadrupeds—not the more formidable ancestors of jackals, lions, and elephants—were uniquely preadapted to symbol formation, powwows, and philosophy, though of course they didn't have an inkling; they were simply climbing toward the intelligible obvious: the treasures of their daily obsessions, sweet fuel for their operations. Hunger is an unparalleled motivator.

If one anatomical change served as a hub for others it wasn't the head *per se* but the mutation of limbs into differentiated pairs: fore and hind. In the same approximate phase shift (500,000 years, give or take a few millennia), vision replaced smell as the senior sense. These were each discrete tissue metamorphoses, the apparent result of random mutations and subsequent adaptation of mutants to tree life; they "found" and catalyzed one another.

Smell is a watery, boggy, foxy, and piggish mode; sight is a wide-angled, aerial raptor mode. So it went. Arboreal biographies among tree shrews and their descendants featured wide viewing platforms and continued development of upper phalanges distinct from lower ones, ideal for something regularly climbing and brachiating through branches. In jumping, an animal is propelled by its hindlimbs working as a lever, and in climbing, it pulls itself with forelimbs.

Primate legs are more than specialized props and levers for support and motion. Almost acrobatically loose, they are attached mobilely to a spine and its girdles, and their tips differentiate into sensitive pads. Comparatively, the distal appendages of caribou and weasels are fixed chassis. Hoof and claw are preadapted for neither art nor science.

In terms of closest current representatives (not our actual ancestors), the human line diverges from tree shrews to lemurs and tarsiers (prosimians), then to (simians) monkeys, apes, and proto-hominids—e.g., those are customized modern representatives of the carriers of our more generalized ancestral line.

In the transition from waveform snouty and neckless fishes to upright stalk-like brainy hominids, remarkably no facial or cranial bones have been added, even with intricately twisted angles and new fractally designed plexuses in the skull. Existing skeletal elements had to be reshaped, realigned, and reoriented using what was there—cranium wrested from muzzle. The blank timeless countenance of a fish had to pucker and deepen into the rubbery emotive expressiveness of a chimpanzee. Gradually the metabolic and neural outputs of other organs reflected in the face.

The key rubric here is allometry: shifts in the proportion of parts in relation to both growth and shifts of the whole. During early morphogenesis in the facial region alone, layers of mesoderm and neural-crest squadrons were recalibrated, reservoirs of mesenchyme were redistributed (e.g., embryonic mesoderm in the form of loosely packed, unspecialized stem cells were put to new uses), angles of cartilaginous and fascial grids were tilted, axes of osteoblast (bone-making) migration were rotated toward new zones, branchial arches and trigeminal and facial neurons were realigned with one another, and sphenomandibular and stylohyoid ligaments were rerouted—all at both DNA and then embryogenic levels. This was accomplished in stages, a passel of them to get from a fish to a man, and it took a while: Industrial Light and Magic in the wild with DNA as the digitizer.

Further down the simian trail, comparable morphogeneses in the rest of the body, notably the pelvic region, coordinated with the skull's evolution to keep the template's overall anatomy proportionate and functional. Supported by the structural refinements of upright locomotion, mutations involving musculoskeletal somites and their histogeneses of bone and cartilage favored design requirements axial to the bipedal mode. Bipedal prerequisites induced shifts in other early biological fields, as tissues developed in relation to each other's shapes and spaces among emerging viscera and neurons. Cell layers, reindexed at deviating angles from those in predecessor organisms, reapportioned and repositioned soft structures. These ossified along new load lines with different fulcra, providing additional compasses for tissue architectures, including (no doubt) displacement and recursion by self-similar systems (see below). My account barely scratches the surface of a metamorphosis ontologically well beyond the chord shift from caterpillar to butterfly and requiring hundreds of millions of years and many false trails and lost critters.

Closer to the modern epoch and recent Ice Ages, these developments gathered momentum, collateralizing an ensemble of upright posture, curving of the eye sockets forward and toward each other (to create an overlapping stereoscopic field for arboreal navigation), and color vision for (among other uses) discriminating fruits from poisons—talents that came at the expense (e.g., corresponding mutational loss) of four-legged speed, core power, and rich aromatic landscapes because no animal can be a composite of all its ancestors.

This is also the back story of consciousness. Phenomenology is visual first: eyes are an extension of a brain, but a brain is also a distension of eyes, as both organs dilated and reorganized many times over as they swelled and differentiated from eyespots and ocular bulbs. Mind didn't just arrive with Aphrodite in the spume; it had to wend its way through a labyrinth onto the Earth.

●

Whereas the jaw of the average mammal is little more than a hinged super-claw, the lemur jawbone is significantly atrophied and deflected toward a position below the braincase and at an angle to it. This modification also supports a vegetarian or (at least) omnivorous lifestyle by compari-son with the average mammalian carnivore. As aggressive snapping for combat and tearing apart carcasses became less critical in these creatures' lives, they applied their teeth more often to crunching vegetation. The powerful downward load vectors in the skull were gradually eased toward cerebralization, the changed proportions opening space for random muta-tions that allowed the cranium and its contents to expand.

At least so the Darwinian fable goes: mutations filled cerebral pockets liberated by a shrinking jaw. The creatures in which these mutations "took" proved more swag than their less mutated kin—the first whiz kids in the 'hood.

Consciousness begins to reveal its signature topology in the lemur ce-rebral cortex: a separation of lobes commencing in distinctive temporal and occipital areas. Yet with frontal and parietal areas lacking significant sulci (folds), overall the ancestors of prosimians were not as brainiac as their cousins, the ancestors of monkeys and apes. As cerebral giganticism became the fashion, less cranial tree-moles and their descendants were left behind.

Our internal organs may be aquatic, but the lobes of our brain are virtual-reality chambers, receptors for hurtling through branches. We still use these synapses to find our way through abstractions, to measure distances between stars, and to track the "jumps" of subatomic particles. It was in the treetops that the relationship between space and time became ontological: "The brain did not evolve to know the nature of the sun as it is known by a physicist, nor to know itself as it is known by a neurophysiologist. But, in the right circumstances, it can come to know them anyhow."[9] This is the miracle of biogenesis on Earth: breed a tree-foraging mole and get a neurosurgeon—because each novel design is bottomless and exponential, and every meaning is an open cascading field with more than one subtext.

Neo-Darwinian science says that it was random mutations and genetic drift, not the intrinsic intelligence of matter, but I say that it was both, converging on a hinge that doesn't exist except conceptually—because they didn't (and don't) have to mechanically converge if they are different expressions of the same transcendent force. Some processes are so basic and inherent that they can sustain contradictory and opposite mechanisms—in this case random mutations under natural selection and innate intelligence radiating from sacred geometry. Where entropy imposes the widest, deepest and most abstruse angle, archetypes emerge at the center of its cone to pull events into the tightest and most synchronistic attractors. The two meet as chaos and organization, yin and yang.

●

Even staunch Darwinians these days recognize that evolution is not conducted only in strokes of gross anatomies, definitive habitats, and unforgiving algebras but involves a hidden palette of event horizons and tipping points inside both nucleic and genetic space and in the context of environmental niches operating as well at multivariate levels. The molecular biology behind the origination of our species—formulated mostly since I studied physical anthropology in graduate school in the Sixties—uses a state-of-the-art medley of differential logic, Boolean networks, chaos phase transitions, and recursive feedback loops to generate the same old-fashioned Darwinian effects. For the remainder of this section I will share my rough understanding of these nonlinear synergistic events, presenting their basic themes in different modes for nuance and clarity. If this kind of stuff doesn't interest you, skip to the next section ("The Birth of the Symbol"). You don't need all of the upcoming technical detail to grasp the evolutionary essentials of biological consciousness.

Why do I dig in so deep? Well, for one, I had a heavy dose of Darwinian logic in graduate school, so I am charmed by the depth and dynamic subtlety of the new nonlinear paradigms. Secondly, in terms of my overall work these embryogenic mechanisms update and (at least for now) complete my prior meta-biological speculations in *Embryogenesis: Species, Gender, and Identity* (1983–2000) and *Embryos, Galaxies, and Sentient Beings: How the Universe Makes Life* (2000–2003). Thirdly (and the only real justification in terms of this book), in order to excavate the biophysical source of consciousness to the satisfaction of modernity, I need three extrinsic modes: neuroscience (the one I have explored in the previous five chapters), physics (the one I will explore in the next chapter), and anthropology (the one under present consideration). These represent, respectively, (one) the anatomy of consciousness at molecular, cellular, and metabolic levels; (two) the anatomy of consciousness at submolecular, atomic, and subatomic levels; and (three) the emergence of consciousness by microenvironmentally chaperoned evolution under morphogenetic algorithms. Obviously these modes converge somewhere, but that vortex lies at an ontological and cosmological depth well beyond contemporary science and, as it is, few extant links even connect the three famous academic disciplines.

The new modalities give a provisional picture of system "intelligence" operating at multiple levels and scales to create organs of consciousness. That intelligence may be as random and "dumb" as natural selection, but we don't know that for sure. In fact, when you get to this level of cause and effect, we don't know much of anything, so just because natural selection is the only story that adequately fits a modern world-view doesn't mean that there aren't other narratives defraying the same evidence.

Along a morphogenetic-to-epigenetic (embryological-to-evolutionary) spectrum, a number of interrelated factors (other than Darwinian house rules) underwrote humanity: these include the cumulative embryogenic

effects of "lazy" genes; homeotic* genes (yielding iterative structures); parcellation of cells in blastula and early gastrula phases of development; and relations
between synchronies and asynchronies in timing mechanisms coordinating
cell mitoses and differentiation curves (each discussed in detail below). Social
and behavioral motifs then reinforced the effectual ranges of each of these
plexuses. Together they add up to a large multidimensional field coordinating
external environmental and emergent cultural factors with intracellular DNA
sequences and gene/allele distribution. These then combined via synergistic
waves of developmental and ecological loading and unloading—all nonlinear,
all disjunctive and synchronous at the same time, all in the context of the raw
material of random mutations and morphogenetic drift.

How does it work under game conditions? Well, you may remember
the difference between Darwinian and Lamarckian theories of evolution—
Darwin and his successors won that battle decisively.

The Lamarckian giraffe got her long neck from stretching to reach un-
touched leaves on higher branches, somehow dispatching codes regarding
her success into her germ plasm for the next generation to incorporate
from her more elongated vertebral starting point. Increment by increment,
these phases lengthened a segment between her thoracic and cephalic skeleton. Most significantly, what provided the neck's oomph was the energy
of actual Lamarckian giraffes going after fodder. The problem was scripting
that energy licitly from behavior into genes.

The sturdier and more enduring Darwinian giraffe was the benefi-
ciary of incremental random mutations and reproductive success among
proto-giraffes with ever-longer necks who could reach food denied to
shorter-necked cousins and other quadrupeds. This homeostasis played
out in an epigenetic context under a regime of quadratic equations
combining integrated regional carrying capacity and average tree height
with population ranges of proto-giraffes (and competing species)—setting "X" as accessible nutrition per species. Darwin's proto-giraffe surfed

---

*A homeotic gene governs a major shift in the developmental fate of an organ or
body part or a class of organs and body parts. Homeotic genes turn notches into
serrations, flaps into folds, segments into segmentations, mites into millipedes,
and single layers into multilayered bladders and tubes.

entropy whereas Lamarck's proto-giraffe tried to countermand it with a purposeful agenda.

Lamarckians may have lost the main bout but, through a glass darkly, they grasped something that Darwinians missed. While it would take supernatural intercession to translate environmental and behavioral information into a language that could send a literal message from giraffe tissues into giraffe chromosomes (like "high juicy leaves, yum yum; program longer necks"), nature has, it turns out, nonlinear back channels for transmitting *the same message with the same basic outcome.* It may not be Lamarckian in Lamarck's sense but, you know what?—it carries his meaning *sub rosa* and even hints at an intelligent and counter-entropic universe.

Natural selection doesn't just zap genes and coronate kick-ass traits. Within a morphogenetic environment, it licenses surrogates that speak crossover pattern languages. These neutralize, displace, and offload vectors while following strange attractors into new motifs, a process that more resembles the internal dynamics of the rings of Saturn and string theory than a clash of titans or the harems of dominant males. Evolution becomes a minion of internal DNA experimentation and systemic collaboration rather than a zero-sum resolution of lethal or permissive feedback in shifting mammalian niches. Coevolutionary explorations of behavioral and moleculo-cellular space converge such that two divergent and blind-to-each-other domains (one nucleic, the other behavioral) settle nonconsciously on solutions to potentially drastic problems: 'Canopy too high for starving antelope-like quadrupeds—it may take a while, but while you're at it—grab and blueprint all mutants that look like—call it a giraffe—and march them down a different aisle.'

An attractor is anything that, in the absence of other attractors, holds a complex system in a nonlinear configuration; it can be a planetary orbit, a riverbed, even a fence collecting wind-blown debris along a highway. Each attractor specifies the dynamical state of its system while maintaining it in a so-called basin, which includes itself and all its sub-states, pre-states, and emergent states.

A configuration generating an infinitude of simultaneous vectors is called a strange attractor. That's our boarding pass. Strange attractors are what shape a universe as complex, vast, and broadly chaotic yet locally discrete as this one. It is through them that subcellular and ecological space collide and collude.

As genes obey intracellular regimes with indeterminate factors (fractal distribution, seriality, positioning), they commute (as it were) backwards into "survival of the fittest's" punctuated equilibrium by riding classical Darwinian regulation's antipodes: global relaxation of selection, "degradation of control"[10] and/or "redundant variant replicas of some prior form (gene, cell, connection, antibody)" [to create] "an external context"[11] for diversification and retention of behavioral traits. In other words, "survival of the less fit" (or permissive parents in the romper room) was closer to evolutionary reality than a strict nursery and militant and unforgiving application of tooth and claw (though there were undoubtedly measures of both).

Randy and rowdy vertebrates may have hissed, growled, threatened, lacerated, and dismembered each other in battles over millennia, but their molecules and cells were engaged in peaceful collaboration and, in a broader biological environment, probably cross-pollinated traits via DNA.

The hominization process took place initially among our distant arboreal ancestors and then among our more proximal savanna-dwelling forerunners. As networks of subcodes mingled, augmented, and sometimes cancelled one another, they were able to "block the [more direct and lethal] effects of natural selection" which led, under the umbrella of protoculture, to "flexibility of developmental and learning processes."[12]

I am beginning to quote from Terrence Deacon, a neuroscience-oriented anthropologist at the University of California, Berkeley. While he does not endorse the metaphysical tenor of my approach, during a lunch at Berkeley Thai House we exchanged ideas on embryological episodes that might have linked the mechanisms of DNA sorting, mutations, and population genetics to the archaeology of primate fossil records.

Back when I was a graduate student at the University of Michigan, physical anthropology—a series of courses encompassing statistical analysis of genotypes, general primate demography, and old bones—was required for everyone in the cultural anthropology division too. At the time, there was no way to bring these frames together: no operational links—no anthropological embryology or even a rumor of one. Our professor opened his course by saying, "There are three parts to physical anthropology: the palaeontology of primates, population genetics, and embryology." When questioned later about the missing topic, he said simply, "We don't teach it." What was there to teach? Genes and fossils conduct no direct conferencing unless living embryos provide an occasion, but who had access to the activities of monkeys from millions of years ago?

Since 1967 dramatic advances in DNA mapping, molecular biology, hominid forensics, computer memory, and neuroscience have sharpened that picture, and embryological anthropologists like Terrence have come up with archaeological embryologies at the level of satellite scans of Uranus's moons by comparison with former telescopic images of the same bodies through the Earth's atmosphere.

Beginning eighteen years later (1983), blending science, poetics, *ad hoc* science fiction, and evolutionary speculation, I wrote my two embryology books and tried to supply my own missing link, though not in a way that would have satisfied my professor or Dr. Deacon. That wasn't in the cards. Instead I took the kind of road trip on which you might meet William Blake, Ursula Le Guin, David Bowie, Rudolf Steiner, Gertrude Stein, Remo Conscious, and Lady Gaga. Terrence has since filled the same gap more properly with empirical science, so at this point I defer to him and his colleagues, as I draw on papers he shared with me.

The main new player in physical anthropology is the so-called "lazy" gene. The chance existence of redundant genes, each carrying the same biological information about a particular ingredient or agency (along with their divergent information about other ingredients and agencies), eases selective pressure insofar as there are now two or more sources of a particular trait. Thus if one gets degraded or turns maladaptive for reasons having nothing to do with the inherent benefits of the trait, its hereditary basis is still maintained in the genome and can be replaced by the other gene (almost seamlessly with the right induction wave during embryogenesis). This not only restores information from the backup but brings the "backup" gene's own variances and adaptive potentials.

Because redundant genes provide "two ways of producing the same phenotypic effect…, this redundancy [relieves] selection on the duplicate gene's function"[13] and allows it a range of expression in a selective regime previously dominated by the prior gene. Retaining a phenotype while replacing its underlying genotype alters the overall evolutionary trajectory of the lineage and opens new vistas of expression, as the former locus had probably exhausted its creative potential by now anyway.

On a gene-to-gene basis such a shift can not only abate Darwinian selection, it reverses it by transferring the loading of selective pressure from one site to another. The function or trait, be it a component of limb differentiation, eye-socket orientation, or cortical crinkling, will not totally be wiped out if its locus is eliminated through random mutation or by lethal effects of *another* function on the same chromosome; it will be picked up in the organism's developmental potpourri by a different, comparable allele (usually helically close but sometimes far, it depends on internal pattern recognition) and integrated into an overall design that emerges from embryogenic caucus. Then, as just noted, that new allele brings all of its creative potential into play, some of which may be just what the doctor ordered, for instance for hominization and consciousness. At least that is the neo-Darwinian "best guess" at a lazy-gene cover story.

Creature anatomy and local ecology, despite their remoteness from each other's provinces of expression, team up within the "stickiness" of Darwinian selection under an improvisational DNA rubric. Over generations of indirect feedback between lifestyles and genes, the essential life narratives of animals interact not just through their behaviors and organs but in the molecular structure and internal play of their DNA fields. That's how simultaneously wide and stringent the reverberation is. From landscape to cellular nucleus, life draws on a huge menu of reflexive loops, undisclosed conspiracies, indirect communiqués, and compensatory blowback.

Redundant genes leave space, leeway, and support for genetic exploration of new forms, in particular for their genomes to stumble upon Lamarckian-like effects. Cover now comes from not only direct environmental feedback (Lamarck) but an indirect multidimensional nucleic environment (Deacon). A redundant gene's "umbrella effect ... increase[s] the odds of fortuitous variations in the reproductive lottery."[14] Thus, in abrogation of linear determinism, even something that is not "initially transmitted genetically … can indirectly bias the direction of evolution."[15] Catch a fleeting glimpse here of the Lamarckian giraffe!

Yet nothing untoward or forbidden occurs, no magic steps over the Darwinian line: selective pressure is reversed without violating the parameter that imposes it—*the opposite result is achieved by the same mechanism.* This is John Le Carré, back-channel stuff: "The very plasticity that favors an increase in variation will also mask (i.e., partially shield the organism from) the very forces of selection that would be necessary to shape up or stabilize any of the congenitally produced surrogates that could have supplanted this acquired adaptation."[16] The strategy resembles a dictator auditioning lookalike doubles and then making them widely available to divert potential assassins—it produces lookalikes of Lamarckian selection to prevent Darwinian assassination.

A random pattern search within genetic space not only protects critical genes and offers broader strategies for survival but paradoxically relaxes enough parameters to allow DNA to play with new architectures by drawing on the formerly recessive repertoires of its backup genes—an inventory replete with, guess what, adjacent function spaces and fresh attractors. This is how opposite trajectories and differentially generated infrastructures can produce pretty much an identical phenotype. Even its mother couldn't tell the difference or know that, under the hood, her genes were switched. It looks and smells like her offspring. Beneath the cape, though, it might turn out to be … Supermole!

To summarize (and I will recapitulate again in primate contexts): in a multiply recursive genetic regime, even a mutation that alters proteins in such a way as to degrade the function of a chromosome, or to degrade the functions of multiple alleles with deletion of large coding regions,[17] would not necessarily negatively impact the selection of *a particular trait* carried by that allele, or transmit a selective disadvantage, or even reduce "reproduction of the organism *as long as the other copy remains intact.*"[18] (My italics.) In fact, in many cases the trait, favored by being unloaded into a different regime not under the former pressures, would not only persist but distribute its array of contingent capacities with a fresh range of morphogenetic design options until the "right" package (or best combo under the circumstances) trod the Earth. Emergent designs need periodic *loosening* (not tightening) of genetic strictures, and they also need open design space with which to reinvent their assemblies and tenant new niches.[19] Deacon put it this way:

"The relaxation of selection that is created by the functional redundancy consequent to gene duplication enables what amounts to a random walk away from the gene's antecedent function."[20] With "changing environmental conditions … previously neutral traits become subject to natural selection…,"[21] unmasking features and opening new ranges that were previously hidden from selection. These kinds of "lazy" chaperones not only protect background traits and fragile motifs but, by their tendency to patronize "complex emergent higher-order regularities,"[22] likely supported early primate nervous system functions and brain evolution. They pulled at least a billion years of intrinsic (and extrinsic) novelty from nature's attic. In Deacon's words again, it is "a non-Darwinian mechanism which acts as a complement to natural selection in the generation of functional synergies, because its form-generating properties derive *from self-organizing tendencies of molecular and cellular interactions* rather than from relationships to environmental conditions."[23] (My italics.)

●

By their own pagan system intelligence and endemic paradigms, simple cells were always more capable of generating whole plant and animal kingdoms than early Darwinians imagined. Endosymbiosis, commensalism, and molecular improvisation have a collective momentum that overrides entropy and the struggle for survival, to produce a "white on rice," sleekest-cat-in-the-parade effect. It seems as though nature not only wants to cooperate, to go communal despite its sway of fang and claw (and antibody), it actually *intends* to curve inward toward its own implicit omega point (or what futurist Terence McKenna called Timewave Zero). In discussing the universe's evolving deeper and deeper into complexity and the ramping up of that complexity prior to a predicted 2012 cosmic shift in the wave, McKenna placed prior and future wave-shifts in a cosmic context:

"We can argue whether the eschaton will arrive in 2000 or 2012 or 3000…. The harder thing to imagine is human history going on for hundreds and hundreds and hundreds and hundreds and hundreds of years.... History is a self-limited game ... a phase-transition. The clock's ticking, folks. It only lasts about 25,000 years. You're not the hunting ape anymore, but you're not yet the sixteen-dimensional digital god.

"Some people think 25,000 years is a long time; some people think it's a short time; I think of it as, snap! one moment you're hunting ungulates on the plains of Africa, and the next moment you're hurling a gold-ytterbium superconducting, extrastellar device toward Alpha Centauri with all of mankind aboard in virtual space being run as a simulation in circuitry...."[24]

That's getting a bit ahead of ourselves (as Terence could do on a dime), but stars, solar systems, and DNA operate together across a broad continuum, making the kind of space-time we keep locally (and by which we measure all things human) little more than the complement of a housefly's summer or, more accurately, a nucleic half-life. We might find someday, in Hundredth Monkey fashion, that the evolutionary algorithm is simultaneously diffuse and rigorous enough that, *even without* interstellar DNA transfer by panspermia or cosmic telekinesis (see pp. 186–188), facsimiles of moles and monkeys prowl the third planet of Alpha Centauri and elsewhere too. Though genetic information is always local and specific to its clime (at well below microgeological level), other equally fundamental factors intercede: algebraic-geometric patterning, molecular constraints, boundary dynamics, hydrological and meteorological zoning, ubiquitous rounding factors, and tetrahedronal and toroid projections of basic anatomical templates (from the segmenting of flies and lobsters to the same segmenting in tortoise shells and brains), all in the context of walkaway and "lazy gene" functions. These are universal, intergalactic; they unfold in some fashion in all biological, exobiological, and metabiological systems.

Here is a pale sighting of the ninety-nine-percent-missing null set with its junk DNA ball of twineless twine. And it points toward cosmic "socialism" as against a presumption of universal free-market forces, territorial imperatives, and last-ditch species wars. It privileges the necromancer and performance artist over their more ferocious and mercenary comrades— free play over conscription. Even entropy proves not the ogre it was cracked up to be.

Remember, we are not talking about the sole generation of form under old-fashioned natural selection; we are talking about creative chaos, phase shifts, strange attractors, the Three Stooges, and Whiteheadian novelty—the good guys. These are systems sponsored by Jackson Pollack not Rembrandt van Rijn: form arrives because its elements are *always* implicit, even after the paint goes flying.

Nature cannot simply decide late one afternoon that, given climate change, an iguana would fare better tomorrow than a polar bear and then go back to the drawing board to sculpt a different creature from the same raw protoplasm (though that creature is likely hidden too in the clay); it can't design a goose egg where there used to be a snake egg. Genetic time doesn't ever unravel or backtrack. Genetic space doesn't reach out for fresh canvases; it only folds in on itself. Of course it can't go back to the drawing board for a makeover, as it is too far down the assembly line and has accumulated too much form. Life may not always progress or complexify, but it *is* a one-way thread.

Nature can revise a caterpillar as a butterfly only by building both templates into a single genome at the beginning of the line (with staggered relaxation of their phase spaces) and then using each other's raw material efficiently (no biological energy to waste) and tactically in every subsequent designated phase state and shape-design. One genome effectively shuts off the other inside the nucleic reels where they are not only packed tight but fused and intertwined. It melts down the bug's exoskeleton and then, many transitional creatures—instars, imagos, nymphs, and molts—later, a flying worm kicks out of its cocoon. As I discuss in the fourth section of this chapter ("Ontogeny and Phylogeny"), any creature must survive each phase of its emerging body-shape by retaining a sound ecological post, even in intervening sessile forms between free-living bionts. That's why each bug's transitional stages are perfect bugs too.

Lamarck missed his own point because he didn't get the quantum basis of microbiological events: the Tao hadn't yet been revealed to the West. He had the right instinct, though: system intelligence, just the wrong scale of application (germ plasm rather than differential algebra and vector-valued functions). It takes far too much time for morphogenesis to respond in

Lamarckian fashion to cascading onslaughts of signals in a crisis-condition world. The jig would already be up. Events in a landscape can't directly hijack genes, nor can genes micromanage epigenetic processes in binary (wipeout or thumbs-up) fashion; they rather provide "only genetic regulation of the boundary conditions affecting processes that have the potential of arising by self-organization, self-assembly, or other extragenomic processes."[25]

As nucleic biases and constraints "tend to mask selection maintaining corresponding genetically inherited information…, the genome will tend to offload morphogenetic control in the course of evolution, in a way that takes advantage of the emergent regularities that characterize many epigenetic processes."[26]

How the genome "knows" to proceed to Option B remains a riddle of system intelligence. I would posit that it "needs" the synchronicity of the archetype (in precisely the way that entropy itself is just another archetype). Forms seize other forms across a genetic-epigenetic threshold, as they continue to cavort to create still other forms. Offloaded attractors drift across old boundaries into new phase states. Fractals and Mandelbrot sets rule. *Everything* is reexamined under the constraints of newly arising anatomies and either made use of or turned into space-filling vestige and structure. Nonequilibrium thermodynamics and performance art, for sure!

As the algebra of DNA responds to its own geometrically imposed functions, a repertoire of adjacent uncertainty states unfolds by species (armadillos, bats, dolphins, echidnas) under cellular-molecular tension. Divergent mammalian designs prestidigitate off a single assembly line into epochal capybaras, raccoons, elephants, hares, anteaters, lynxes, squirrels, boars, sloths, moles, macaques, fishers, lorises, pangolins, porcupines, *et al.* Each body-type and niche, by being adjacent to countless other body-types and potential lifestyle niches, claims by default the three-dimensional haberdashery most suitable to its genome (of course, with numerous teratologies and casualties along the way, but those are swept offstage usually with nary a fossil or fuss). This is what the strings around the DNA column at the Natural History Museum of Utah were demonstrating.

"[D]uplication, masking, and random walks can provide a kind of exploration of the space of possible synergistic relationships that lie, in effect, in the 'function space' just adjacent to an existing function."[27] Some unknown aviator, a flying gila monster perhaps, is always hibernating a function space away from a small dinosaur with webbed digits, and it's carrying every fowl from a wren and hummingbird to a turkey vulture and eagle in its nucleic pouch. In much the same way a giraffe is lurking inside an antelope, a hominid is privy to a chimp.

Longer necks, swifter gaits, and social intelligence can all arise incidentally and fortuitously in modes that make them look suspiciously like Lamarckian acquired characteristics. The genetic kaleidoscope turns until a propitious pattern of functional phase states bobs up for the situation (amidst the Darwinian cannon fodder), and apparently it always does—another nod to the implicit intelligence and nonequilibrium thermodynamics of molecular space: maximum entropy production with increasing dissipation of matter, energy, and their gradients as well; degrees of freedom in the determination and dissemination of steady states behind stationary shock waves; nonlinear flux-force relationships driving ecosystem metabolism rates; autonomous self-organization by participation in autocatakinetic systems; enthalpic cycles of endergonic and exergonic reactions (requiring and releasing energy, respectively); innate geometric symmetry and its topological projections; creative preadaptations (or exaptations, as palaeontologist Stephen Jay Gould preferred to call them—features that turn out to enhance fitness but were not "built" by natural selection for their current roles); and a general override by nonergodic (uncountably nonrepeating) time-scales.* It's as though a fourth law of thermodynamics—or at least an alternate theory about the structure of our universe—is staring us in the face, yet we don't see it.

Genes wander into unexplored relationships, and "gene families descended from a common ancestral gene [especially alleles of the same gene][28] ... form synergistic functional complexes."[29] They trigger an "epigenetic effect that independently parallels the environmental contribution."[30]

---

*This is not my home territory, so please forgive any misconstruings: I am not trying to tune a pitch-perfect thermodynamic theory; I am trying to compose a rough chaos-energetic mode of evolution to counter predilections toward closed Newtonian/Darwinian systems.

I picture promenades morphing in the tradition of South American Indian bird-nesters, Roadrunner cartoons, and the rebuses of Dr. Seuss. In timelapse Linnaean cinema, giraffes pranced from impalas and cows, whales and dolphins paddled out the fabled delta of their river-bound hippo great-grandfathers and elephant uncles. A bipedal ape, given the right runway and wardrobe, spawns artisans and farmers.

The Algorithm becomes the Synchronon.

●

In primate evolution, the adjacent function space of cultural pre-states captured genetic variations likely exposed many times previously but thwarted, leading at last to *cultural* relaxation of environmental pressures: "lazy" lifestyles, as it were, on top of "lazy" genes. Because of behavioral factors the primate lineage represents even less absolute genetic control and more morphogenetic drift than other mammalian lines.[31]

The first anthropoids, the ancestors of monkeys, apes, and humans, diverged from the common ancestor they shared with lemurs and tarsiers some 30 to 40 million years ago during the Oligocene era. By extrapolation from fossil evidence the cerebral cortices of these animals' brains were already torquing and convoluting under designed control. In living ape and human cortices, secondary sulci have overgrown and obscured the simian and reptile layers underneath them—a topology that matches the emergence of mind. This is repeated in the human fetus embryogenically:

> The first chamber, blossoming explosively inside the placenta like some lurid tropical flower, splits itself into two subchambers to form the first and second cerebral ventricles, around which the twin cortical hemispheres develop. The cerebral cortex, or *forebrain,* is by far the largest part of the human brain, a thick convoluted sheet of neural tissue that expands like yeast-rich bread inside the embryonic brain pan. Its wildfire growth impeded by the skull, the double cortex creases and bends back upon itself. Seeking more space it grows forward; then up, over, and back; then forward once more,

completely enveloping the slower growing lower brain chambers like some huge flesh mushroom. Viewed from above, the wrinkled forebrain resembles a huge walnut, divided down the middle by the great longitudinal fissure—the cranial Grand Canyon.

Around the second neural chamber, the more modest growth of the *midbrain* structures takes place.... The grotesquely swollen cerebral cortex covers the midbrain like an umbrella. The handle of this umbrella is the brain stem, or "hindbrain." Relatively undeveloped, except for the cauliflowerlike efflorescence of the cerebellum in the back of the brain, it resembles a thick-walled hollow tube—with the fourth ventricle as its core—bulging twice in front to form two lumps, the pons ("bridge") and the ... medulla oblongata ("elongated core")....

Contemporary essayist Ihab Hassan describes the brain this way:

"The brain is not yet whole or one. Like a divided flower, never exposed to the sun, it grows from an ancient stem that controls both heart and lungs. On each side, cerebellum, thalamus, and limbic system twice grasp this stem. Our muscles, our senses, our rages and fears and loves, in this double fistful of old matter stir about. The great new cortex envelops the whole, grey petals and convolutions, where will, reason, and memory strive to shape all into mind."[32]

This almost grotesque protrusion, a computing device made of viscera and saltwater, transformed the planet. It was a UFO landing from within, the center of operation for a new principality. As ancient anatomies flowed into emerging congeries, and their neurons streamed into images, concepts, and emotions—somewhere in the current, King Consciousness hopped on board.

●

Simple jellyfish and comb-jellies don't possess central nervous systems or ganglia; they have diffuse, decentralized nerve nets through which excitation flows in all directions with barely any discrimination between electrical charge and concomitant neuromuscular response. Global pulses, punctuated by sudden bursts of activity during contraction or when exposed to light, suggest the universal autonomic unconsciousness at the basis of all living systems. Perhaps the forerunners of ganglia exist in nerve rings close to the margin of the bells and in touch with plexuses in their sub- and exumbrellas, but they remain stillborn in the cone of Timewave Zero.

The first cerebral ganglion may have initially been no more than one of several fluid-filled sacs that function as organs of balance in invertebrates (a statocyst); it could also have begun as an aggregation of receptors along an embryogenic margin. Whatever its origin, it gradually became the organ of animal identity, attracting all regional relays.

Centralized nervous systems differ from nerve nets in that they favor determination, discretion, and routing. In the process of organizing flow and mustering ganglia, they cross-reference one another, conferring capacities for spontaneous action, regional expression, and grace. This is either the biological birth of consciousness (functionalism says) or the biological expression of consciousness (dualism says). Either way, hierarchical organization was such a successful mode of survival that it was differentially favored in many lines, so advanced nervous systems likely evolved countless times independently from nerve nets.

Relative complexity of nervous systems is only partially a consequence of lineage. The other factor is niche and lifestyle. Reduced systems are common in sessile forms, parasites, and to some degree among sedentary animals in all phyla. Sense organs are usually lost when they are not used— eyes atrophy in caves or below the sun-line in oceans, their signal elements and assorted seeds falling back into dormant sections of their DNA or eliding altogether. A brain is not the only possible expression of system intelligence.

The Deuterostomate lineage of vertebrates diverged from simpler phyla at roughly a jellyfish stage—very early in the game. In jellyfish and their related lineages (the Protostomes: worms, mollusks, and insects), the blastopore of the embryonic archenteron, the original central digestive cavitation infolded during the embryo's gastrulation (evagination), becomes the mouth or both the mouth and anus of the animal, leaving minimal space for neural potential. Insects, for instance, attained "intelligence" with very few neurons—virtually no associational ones—their wiring is strict and inflexible, so they must adhere religiously to ancestral rules. They have limited memories and intentions; they can learn virtually nothing. Foreground and background to them are the same; a bee trapped in a room and shown the way out—even a hundred times—cannot learn it, and will die there eventually, batting itself against a window: "The body of an insect ... is perpetually other than itself."[33]

In the vertebrate line the blastopore becomes the anal opening and a mouth forms secondarily from a new perforation of the body wall. The central nervous system emerges from an introversion of ectodermal tissue forming a hollow tube in gastrulation. Some ancestors of acorn worms must have become subject to slightly different selective factors from the ancestors of most other worms, for they developed a supple rod of supportive tissue, the partial forerunner of a spinal column and the future organizer of vertebrate development. Initially this column served as a brace and fulcrum for swimming movements, tissues on either side of it contracting as the tail wiggled from side to side. A similar lever exists contemporarily in lancelets, but also, more significantly, in the larval forms of sea squirts (tunicates), which are closer to the vertebrate line. In them the rod serves also as a nerve cord, enlarging anteriorly into a light receptor and an organ responsive to tilting. As adults, however, these creatures become motionless lumps in "tunics," their upper oral openings drawing water through sheets of glandular mucus into buccal cavities.

A central nervous system with a brain evaginating at the anterior pole of a neural tube is a Chordate and vertebrate hallmark. It emerges embryogenically with the notochord and neural groove embedded in it end to end. Neither a ring nor a hierarchy of ganglia yet, the tube becomes surrounded by skeletal tissue produced by its somites; muscles and nerves are then pattern-organized by and attached to the backbone at its vertebrae. Afferent nerves (conducting from sense organs and the body's periphery to the spinal cord) shoot to their destinations, while efferent nerves (conducting away from the central nervous system to a gland or tissue complex) impregnate muscles and limbs.

When the head of the neural tube expands to become a brain, it has no point of contention with skeleton or gut (a consequence of its new opening), only with the structural requirements of the spinal column and more ancient facial skeleton. It is free to let the ugly duckling through the labyrinth and thereby invent a universe.

The vertebrate brain expresses itself initially in three unpretentious frontal swellings, ontogenetically as well as phylogenetically. Its initial ambition was modest, holding most of its cards tight to the vest—a bit more balance, light, scent, and pressure recognition in a "worm" mode (but these sensations are incipient "thoughts"). Its primal site lies at the congress of various adaptive trajectories, mere coiled bumps among many in a system prone anyway to swelling and curvature.

As sense organs formed in concert with the bumps, they balloooned over many generations of different animals into full subsidiary lobes. The forebrain developed with an olfactory lobe, the midbrain with an optic lobe, and the hindbrain as an otic organ comprising balance, vibration, and general equilibrium.

The hindbrain is the lingering compass of the invertebrate realm and of primitive vertebrates; it coordinates functions critical in oceanic abodes. The forebrain evolved in early vertebrates (among the fishes) as a scenting ganglion. However, it continued to expand beyond its olfactory base to become a cerebrum in the higher primates. Playing all its cards finally, it swelled out over all the other convoluted structures, engulfing prior talents, strata of intelligence, neural plans, and cerebral lobes in its own structure and functions. It is in the cerebral cortex that the mind lays claim to a replica of the cosmos.

Our mega-capacity of consciousness does not have an immediate survival or selective benefit, so it doesn't pass muster as an ordinary Darwinian object. Long before anything was symbolized, the nervous systems and ganglia of prehuman hominids had already surpassed all hardware requirements for sentience and speech. This paradox has plagued neo-Darwinian theorists from the get-go, but it is not *per se* a game-breaker. Why did the brain's sulci keep crinkling over one another even after the emergence of successful carnivores until some of them became capable of rumination?

Why didn't the biosphere just run prototype machines and settle for a lovely lizard or charming chipmunk or, failing that, stop at great apes? Any of them would have captured enough energy for the judges to have awarded the permanent gold medal and, in effect, ended the games.

We are declared more brainy than we need to be because of cerebral surplus—a runaway phase state. Additional tissues and folds were constituted for reasons having little or nothing to do with the analytical and representational properties of intellect and their use in survival. Yet once they were federated and indexed, they continued to interpolate into cerebral topology.

A hunting band was sufficient for our niche, but the capacity for the *mercado* and the Internet was implanted well before the Ice Ages. Essence or overrun? We have no field reports from other medium-size watery planets with oxygen-rich atmospheres against which to test our outcome, but somehow I don't think we are going to be looking at 10 million years of mice or hunting bands anywhere.

The anthropoid brain diverged in other, qualitative ways. A new series of nerve tracts ran directly from the cortex to the spine; these fibers (called pyramidal) were apparently induced by the evolutionary interaction of the vertebrate spine and the cortex, as they superseded slower, more diffuse extrapyramidal relays between the brain stem and motor neurons. A concurrent development took place in the sensory filaments that record and discriminate muscle contractions. Information from these proprioceptors and from the tactile neurons in the skin tended toward quick, discrete bursts—at least by comparison with the diffuse peripheral routes of quadrupeds.

Our human progenitors were probably marginal tree apes, outcasts from a larger simian community, exiled by overpopulation and natural disasters—perhaps during a drought and forest die-off. The fossil record shows that early hominids, though still quite monkeylike, had relatively larger brains, smaller jaws, and more differentiated fore and hind limbs than their simian cousins. As primates ancestral to both monkeys and apes adventured from the trees onto the savannas of Africa, they made flexible use of their idiosyncratically phalanged forelimbs (arms, elbows, wrists, hands, fingers, and fingertips); otherwise, lacking (prior to tool-making) any compensatory specialty, they were dufusses, easy targets for big cats and dogs. The hominoids* that thrived in the new habitats carried pre-adaptions and/or adjacent function space for savanna life.

Outside the woodlands the array of selective factors working *against* hominization would have been somewhat neutralized and in a few cases (bipedalism, for instance) immediately counteracted. But the potential of a new phenotype still needed thousands of generations to metamorphose into a lifestyle.

Although the line of extinct hominoids in fossil chronologies leading to our species seems more similar to that of apes than monkeys, we did not, in truth, evolve from either. Our ancestor most likely diverged from the general line leading separately to apes sometime between 10 and 20 million years ago at the end of the Miocene or during the early Pliocene era (though we cannot exclude the possibility that hominids branched off earlier and that we share our last ancestor with other anthropoids at a point even before apes and monkeys parted ways).

The freedom of forelimbs to grasp objects—stones, sticks, bones—is palaeontologists' consensus turning point in the saga of hominization as well as the crucible of existential selfhood. At African fossil sites primitive tools correlate with cranial and pelvic development. Archaeological evidence of the hunt characterizes Australopithecine bivouacs: bashed-in skulls of antelopes, mastodons, boars, birds, and other game animals, skeletal parts that became raw material for future weapon design. Idiot-savant monkeys finally showed up big-time to rule the prairie.

Arboreal foraging would have liberated arm, hand, and phalange space (epigenetically) for manipulation under the exigencies of life on relatively treeless savannas. An endocellular walkaway matched a migratory walk-about. Bearing sticks and bones in forelimbs, capable of conceiving applications of these implements (as symbols) in future time, early hominids stalked in bands across the Serengeti and other southern savannas, obviously proving a hardy match for the ancestors of wolves, lions, and pigs. These hunters were barely, if at all, more intelligent than apes, and they limped by current bipedal standards. But they had entered a powerful new attractor: the imaginal realm.

---

*Palaeontologists disagree on whether they were also hominids yet.

iii. The Birth of the Symbol

Consciousness ultimately requires language; layers of symbols and signs give rise to semantic systems—nonconscious ones first. On Earth, the only biological regime we know firsthand, signs express themselves in diverse formats: ant mind, fish mind, worm mind, lizard mind, raccoon mind, the single-note surges of collective cricket identity and cello-like ribbetts of the existential frog. Later proto-languages are sources of monkey and ape identities, bird identities, whale and dolphin identities. They are radically different consciousness systems from ours but homologous neurosocial expressions of DNA.

Signs beyond speech create animal identities too—fireflies and phosphorescent fish express their existence in light; some birds exude it in colors of plumage, some arthropods by dances. One could even argue that ants and bees create social meaning through the "phonemes" and pheromones of touch and taste. Among primates, whistles, drums, panpipes, head-bumps, and pantomimes are neurolinguistic too. The symbol transcends any system, mode, or dialect of its expression.

The primate brain apparently developed preadapted hardwiring needed for responses in strings of symbols. The cerebral basis of language is also an alternately-routable, off-loadable talent. If morphophonemic capacity arose both genetically and epigenetically more than once, its allele could eventually be replaced "(via chance mutation) by an innate analogue."[34] This would sculpt the capacity to "match specific body architecture and sensory specializations,"[35] realigning semantic infrastructure with neural circuitry as often as was necessary under emergent cultural demands.

The underlying psycholinguistic structures of universal grammar and their syntactic grids enabled hominids to cross the last bridge to meaning and identity. Our human version of speech likely emerged species-wide as a single structure even as it later diverged over thousands of generations. Linguistic frames as remote from each other today as Finnish and Nahuatl share an original word-forming, clause-catenating lattice. Tibetan, Cherokee, Zulu, Mohawk, Aranda, Hebrew, Serbo-Croatian, Welsh, Hindi, Basque, and Hungarian sound as if they come from different planets in their word-meaning pairs, tonations, rhythms, and accents, but they are performed on the same chords by the same logic-strings: e.g., the primate lattice of evolving speech centers in the vertebrate brain. Different dialects are random regional variations of informational threads, computational hierarchies, and conceptual topologies, predetermined to a large degree inside the cortex. Each phoneme may be optional and unique (the glottal stops of Navaho, the trilled "r" of Spanish, the clicks of Zulu), but at their core is a rulebook that determines their morphophonemic route to designation in every case.

Linguist Noam Chomsky proposed a deep matrix of formation-grids (akin to unaffiliated stem cells) underlying the neurological roots of syntax itself. From templates of superstrings that he discovered in actual spoken languages he derived the universal basis for all language.[36] These core principles common to hominid neurolinguistic formation are what children unconsciously train in mastering the sounds and grammar of any particular language. An innate "theory of mind" confers an ability to perceive and interpret what another hominid is imagining.

It is not entirely far-fetched to propose that the human "alphabet" is derived from a suborganic genetic "alphabet"—jumbled, turned inside-out, and cast into our darkness. Language is a function not only of neurons, not only of "I," but of the meeting of cell and cosmos: the brash Caw of a disembodied intelligence at large in the universe, perhaps even the "voice" of primordial matter. You can hear it in wild turkeys at dawn.

Human speech evolution can be hypothetically modeled under less draconian circumstances. Its phonetic-phonemic range is paralleled by the complexification of birdsongs in domesticated finches where "mate choice [is] determined by breeders on the basis of coloration patterns, and irrespective of singing."[37] This incidental selection leads to finch song repertoires with melodically freer, non-innate "ordering and combinations of vocal sounds" and call features. More colorful birds strangely sang more diverse refrains too. The same sort of walkaway might have opened the door "for emotional tone expression via speech prosody" in hominids.[38]

Domestic protection of songbirds increased "the complexity of both song structure and of neural systems for producing song ... in the absence of overt selection acting on these traits,"[39] and this led to "progressive despecialization of the circuits that contribute tight constraints on motor patterning that specif[ies] song structure."[40]

As with primate calls—"a similarly complexified neurobehavioral trait"[41]—songbird melodies en route to new significations would involve trajectories "away from congenitally prespecified song [and] ... a limited and a relatively invariant set of notes ... and transition probabilities between notes ... to socially acquired flexible singing behavior ... with many more transitions possible, including many possible loop-backs in the absence of selection."[42] Less selective pressure means freer exploration of sounds— new pattern languages. When the choir director relaxes his baton, his chorus bursts into a John Cage cacophony of notes and tunes: wind chimes and water flutes and theremins, "phones" and their "emes." Birds "sing" and hominids babble with musical delight. Of course hominids sing too (and dance and head-bump and do plenty of other junk). At a more advanced cerebral level, phonemic and psycholinguistic strings provided symbols for tool-making styles, hunting strategies, shamanic devices, and social rules.

Humans in a state of incipient culture provided both weak and strong selection in spades for giving voice to nature's hidden signs. The "semiotic and pragmatic demands of symbolic communication" intersected the need to "favor structures that are more easily acquired by immature brains undergoing activity-dependent interselection of neural circuitry."[43] Remember, if the DNA substratum is complex and varied enough, it doesn't need either strong selection toward a primary function or an anatomically tight delivery system to achieve a new trait. Once the trait is "identified"— pulled from the underlying vortex—it can arise through diverse motifs (even aimless genetic drift). Its potential mainly needs room and a more permissive hand. And drift within an organized vortex is always fortuitous sooner or later—gamblers stationed day and night around the roulette table prove that. All previous, unlucky draws are quickly forgotten. Exaptations rule their own unintended domains.

The first simple meanings jabbered out of highly structured call systems: "[W]ith the exception of laughter, sobbing, and shrieks of fright, we inherit a very limited repertoire of ... prespecified vocalizations ... for distinct social messages or objects in our environment." Onomatopoetic calls are preserved inside advanced languaging in the subtle music of speech, as calls eventually become iambic pentameter, free verse, reggae, hiphop....

Early "speech prosody seems to coopt many of these call arousal correlations, subordinated to the articulation of speech phonemes...."[44] Like chimpanzees, human rappers started emoting with fixed modalities like chatter, shrieks, and whistle commands. These sounds are direct expressions "belched" from organs like the heart, belly, and lungs (via the brain) into sounds that release their stored charge and urgency toward designation. That is how meanings got released from inside the universe (and matter) into culture, psyche, and philosophy.

Later "we borrowed many of the tonal and rhythmic features that are associated with specific emotion states ... in the calls of other species"[45] and at the same time abandoned pure "sound-meaning couplings" and obligatory onomatopoeias.

The ancient primate call "dictionary"—containing the cry, the alarm code, the mourning dirge, and mimicries of mews and calls—was incorporated first at a "grunt" level, then into simple and compound multi-purpose signs. Proto-proto-Indo-European cum proto-proto-Uto-Aztecan emerged over hundreds of thousands of years by tonal blending and symbolic figuration.

The "social transmission of information"[46] clearly favors a reciprocal substitution pattern—unfixed symbols with aliases and proxies—offloading genetic control to epigenetic and cultural domain: after all, "languages are vast systems of socially transmitted and maintained behavioral algorithms."[47]

Mankind's first poets, minstrels, and shamans fashioned songs, chants, and canticles as well as countless scatological pranks and comedies and cathartic evocations of submerged or concealed meanings. All of these, powerful and vintage at the time, were lost (with their tongues) in the eons since, though their erasures lie at the basis of every stanza of ancient and modern literature. They included some of the greatest poetic compositions and sacred chants ever composed on this planet—I am simply figuring the odds here: hundreds of thousands of years of potential Shakespeares,

Marty Robbinses, and Jay-Z's (albeit with far fewer humans on the planet) as against a mere few thousand of Homer, the Vedas, Rumi, Annie Proulx, and Leonard Cohen. Who knows how many dialects and tribes vanished en route to those that remained just four thousand years ago.

Consciousness in the human sense also needed a system of representation for the modules and corridors of the brain to begin to talk to one another. Their synapses had to signify something relevant to the organisms—for instance, to separate Self from Other, time from eternity, a sphere with a center from a sphere whose center is everywhere and whose circumference is nowhere. Voices of winds and patterns of stars, movements of flocks and herds, feints of predators and prey, kin obligations within their bands all cast faint mirrors of abeyant meanings. Stalking, foraging, eating, mating, giving birth, sparring, dreaming, dying fell likewise under their witness.

Whereas call production occurs almost exclusively in the "anterior midlines, in prelimbic regions such as the anterior cingulated cortex ... language is significantly dependent on cerebral cortical systems that are widely separated between frontal motor and premotor, temporal, auditory, parietal multisensory, and prefrontal executive systems, as well as with striatal and thalamic structures to which each of these cortical regions [is] linked."[48] The differentiation of the cortex was essential for laying psychosomatic meaning-space.

Neuralization requires society, so society enjoins cortical bursts: perceptions and phenomenologies. Meanings get built up from a flow between unconscious and systemic elements and from tissues themselves and then they spill into and reverberate back out of the world. Anatomy shapes culture, culture permeates anatomy, and together, over epochs, they lead to the functions, institutions, and the habitants of tribes and civilizations. By giving symbolic content to "adjacent function spaces" that arise in the folds of their developing brains (continued "blind variation with selective retention"[49]), humans diverged, culturally and anatomically, from their hominoid ancestors as "a somewhat genetically degenerate, neurologically dedifferentiated ape."[50]

●

Mind is an emergent property not only of the brain but of its own thought processes, pretty much in the way that matter is an emergent property of molecular configurations, and life is an emergent property of molecules. "Emergent" means that it cannot be reduced to its material components.

Mind is also an emergent function of the hunter, the band, and the tribe—a commensal unit in which each side nurtures the other, for the band, the tribe, and the chiefdom are *also* emergent functions of the brain/mind. Advanced modalities of consciousness and culture must be first supported neurally, but then the majority of their activity requires social, emotional, and semantic contexts to unlock their bound energy and sublimate their unspoken charge—e.g., object relations and deeds of social transference. As culture provides support for the brain to develop, the brain explores and renews its own capacity for culture.

With communication modes comes "complex interdependence of the neurological, behavioral, and social transmission features" [with] "attentional, mnemonic, and sensorimotor biases ... for acquisition and transmission."[51] Natural selection thereafter favors any mutations and walkaways that sponsor integration into social groups, as "control of a previously innate and localized function" gets shifted "onto a distributed array of systems that each now only fractionally influence that function."[52] The "novel cognitive demands on hominid brains" with the accession of protolanguage ultimately spawn "a socially constructed artificial niche."[53] Morphogenetic space, environmental space, cognitive space, and cultural space converge, with wide-ranging epigenetic and cosmological consequences.

Neural hardware itself likely contains no more than five percent of the data of thought and (of course) none of its emergent properties. Culture is mostly connotation and situation. Even individual birds in a flock probably inherit no more DNA than a rough version of "flap and stay three feet away from each other."[54] The rest is conjured by the act of group flight; the proprioceptive rush of wind, air, light, and one another's existences. Mind may have its componential basis in single neurons and their synapses, but it finds itself only by discrete associations of neural and cerebral events even as a developing embryo finds its shape not by DNA output as such but through associations of cells. Schooling fish likewise function as parallel processing systems "without the necessity for any single locus of executive command."[55]

Cultural streaming goes on in the background of pretty much all yak-kety yak, stitching the fabrics of human personae while ratifying existence, individual and collective. Conversations of voices inside our heads inaugu-rate dialogues with others; conversely, conversations with others feed our self-dialogues. Materials flow back and forth between signifiers. All private and personal mentations—even the most idiosyncratic fantasies, supersti-tions, and fugues—arise as collective material in an unclaimed commons between a community and its individuals.

The anthropological model for the origin of consciousness matches (somewhat accidentally) Freud's birth of the ego by sublimation of libido, each psychosomatic representation requiring another operation for com-pletion of its expression: "Experiences leave complicated states of physi-ological arousal, but without necessary coalescence into the kind of specific mental states we describe as feelings. And without such coalescence, cer-tain kinds of higher-level integrative process cannot be accomplished. In our basic design we are not self-contained emotional processing units. We evolved as co-dependent group members."[56]

Without language human beings would have no capacity to represent the neural content of their bodies, but without a social network, their representations would *have no meaning*. Society rescues *Homo sapiens* from the identity crisis posed by his own brain and the dimensionless abyss of animal existence. The brain becomes secondarily "the coalescence of the memory traces of the group ... the world of remembered others who are brought to life in [the] background narrative ... the animated model of our representational world."[57] Once the symbol is planted, the implosion of language spawns a rapid, unprecedented breakthrough of psycholin-guistic activity and "extensive synergy of diverse brain systems,"[58] which leads to a new "dependence on social learning."[59] This wake-up call was no doubt a shock to its sleepwalkers and soon enough required shamanism for vigil as well as a quantum leap into the noosphere.* Paintings on cave walls are heralds as well as graffiti of this hyperspatial journey.

---

*Twentieth-century priest and archaeologist Pierre Teilhard de Chardin inaugu-rated the noosphere as a zone of consciousness forming around the Earth by anal-ogy with its lithosphere, atmosphere, and biosphere.

Protocultural phase states probably flirted with hundreds of thousands of different bands and their cultures until they struck tipping points, as "neural traits can almost freely vary in form over the course of evolution until for some reason they rise above this threshold."[60] Two theaters—one as vast as Africa, the other as small as a cell nucleus—continued to envelop each other. Animals following this script were of course clueless; they dealt in existential situations, as creatures do: "Here I am! Whoa! What next?!"

In truth, all primates could earn the insanity defense, but consciousness happens, and *nothing but it* happens. Dadaism was a Stone Age craze long before it became a late European fad. The clown and the priest were seminal and not so different from each other at origin; together they proposed the break dances and slang of an evolving street and jungle culture. There were hipsters and rockstars then too. This was no sleepover, no mere weekend rave; it took a million or more years to hit its shock wave, and then the real party began.

●

We can explore a much simpler model for environmental modification in the birth of a symbol, as human culture is not unlike (in modality at least) "the evolution of beaver aquatic adaptations in response to a beaver -generated artificial niche."[61] Beaver dams impose a beaverlike aquatic environment; then beaver bodies adapt to the behavioral constraints and opportunities arising from "beaver-world," thereby short-circuiting the usual strong "cause-effect cycle of natural selection."[62]

As expedient tool-use was followed by actual tool-making among hominids, it took some of the remaining selective pressure off primate mega-jaws, mitigating prior "intense masticatory processing of vegetable foods that characterized the australopithecines"[63] and offloading more cranial space from the maxillary apparatus in service to cerebral development.

Tool-making became a committed continuum, a sustained tradition of artisans and masons over generations. The first terrestrial devices were probably stripped branches, handy rocks—stuff accessible in crises and historically available to simians and birds as well. The oldest remnants of aboriginal men and women were the pebbles and bits of river gravel altered by Australopithecus, likely for knives, chisels, scrapers. Eventually their descendants fashioned implements for wider uses. The symbol became more elaborate and complex as it flowed down the river of culture.

With tools now fully subbing for teeth, incisors and canines in the prehuman line became even smaller; the jawbone continued to shrink, providing fractal area for elaborating the phenomenology of hunting in cerebral sulci. Tool development had its own recursive effect on hominid intelligence, as culture was a walkaway at more than one level—and even more than one neural level. When proto-human primates began making arrowheads and cleavers, traveling in groups, coordinating hunts, and tracking prey, symbols provided cultural cover for their new anatomies and philosophical trajectory: "Hey, go counter-sunwise around that clump of trees and frighten those totem stinkbirds." In a sense, the language of symbolates replaced the language of genes. Yes, big cats played much the same game, but they couldn't enlarge it with counter-strategies and hand-held tools. Bearing homemade spears and knives, word-wielding hominids moved in tactical bipedal bands that were revolutionary on this planet. Beavers, by contrast, were limited to their trademark dams and the gifts of muskrats. (But here is the thing: they were made beavers in the way that shamans were made shamans: by the "Word.")

●

Higher consciousness is unavailable to other mammals not solely because they lack the neurons (though they do) but because they lack cultural patterns and contexts mediated by symbols: "[N]o one in neuroscience thinks that the way to understand the nervous system is first to understand everything about the basic molecules, then everything about every neuron and every synapse, and to continue ponderously thus to ascend the various levels of organization until, at long last, one arrives at the uppermost level—psychological processes."[64] Consciousness didn't arrive that way; it was a convergence of anatomies and signs with acts of "Lamarckian" espionage—meanings, cells, and tissues crossing barriers and intermingling.

Protean sigils, nomenclatures, discourses, and taxonomies hitchhiked among early hominids in meta-systems, flickering unstably and atavistically behind all objects and thoughts. At their roots, meanings are cathexes: bound biological charges. Relentless by nature, they find meaning wherever they look; then they make new meanings out of the debris of prior content, a process that Claude Lévi-Strauss dubbed "bricolage."

Previous primates—monkeys and apes—hadn't stumbled behind that veil, so they never realized how much was available there, because they never "realized" at all, never reached out intentionally to the symbol, never liberated the unconscious swarm inside themselves (and all animals). Even today it is hard to show it to them and hold their gaze. They can't look; it isn't their lot; it is forbidden. Like the beavers they uphold other totems.

In the beginning, magic and myths ruled the hominid regime, for symbols and their uses are only obvious (happily) at the first level. Lévi-Strauss diagnoses the moment:

"[T]he universe is never charged with sufficient meaning and ... the mind always has more meanings available than there are objects to which to relate them. Torn between the two systems of reference—the signifying and the signified—man asks magical thinking to provide him with a new system of reference, within which the thus-far contradictory elements can be integrated...."[65] And that's what happened, right from the Zen bell.

As shapes drew themselves out of the background, they gave rise to further shapes and meanings. Signatures yielded signs, generating acts that shook loose gourds of fresh signatures and signs. The symbolic realm deepened and spread, generating stories that turned into Greek and Norse cosmologies and the trickster tales of the Winnebago Indians epochs later.

The radio and computer took little more than decades to get invented and establish themselves, electricity required a century, the harpoon and wheel thousands of years. The original stone tools, language, and systematic control of fire and its forge—the baseline of all technologies and machines—emerged on a scale of hundreds of thousands of years, likely reaching a threshold countless times before an unheralded cluster of hominids finally "got it." Once they did, the emerging noosphere turned out to be far more consequential to the planet than either beaver dams or finch calls.

There is an implicit resistance to an act like rock-chipping, for, innocent though it is, it is to imagine one's self cutting a preconceived shape in a stone; to entertain all the spooks and spirits that attend consciousness. To make a tool is also to make its maker. No way to avoid staring in the mirror: an intentionally chipped pebble is the beginning of "I," the vague apperception that "I" and "It" are different. Grasped between the opposable digits of a curious animal paw, an altered stone reflected selfhood with a concreteness and a purity that later inquiries into the origin of consciousness or identity fail to recover: "I," "me," "mine."* The first marks were crude, unidirectional cuts in the borders of small gravel, but they were images, even more luminous and powerful in their way than those horses and rhinos etched millennia later on the cave walls of Chauvet and Lascaux, for they were prior and seminal to them.

During the hundreds of thousands of years of the Pleistocene, an age of glaciations and interglacials, Pithecanthropus spread with his culture, developing new generations of handaxes, using mostly lava, quartz, quartzite, and flint. These so-named Chellean tools were flaked around the edges with strokes in alternate directions so that two faces crossed in a staggered margin. Scarred bifaces were dispersed across Eurasia and, by the second interglacial some 300,000 years ago, their makers had honed sophisticated and utilitarian styles. At Cro-Magnon's emergence roughly 50,000 years prior to the first writing tablets and settled villages, humans were producing awls of antler, points, burins, concave saws, needles, fishhooks, willow-leaf blades, arrowheads, lunar counting devices in bone, small statues, and paintings inside corridors to the underworld. This was modernity!

In the earliest phases of consciousness, "the numinosity of the archetype ... exceeds man's power of representation, so much so that at first no form can be given to it."[66] But as consciousness crossed cellular and

---

*My editor, Kathy Glass, contends that "it was the mother looking at her offspring that generated the first awareness of me versus it, and her certain knowledge that, though the baby came from her, it was not her. It's a more likely scenario than a man and a rock coming to that awareness—there are lots of tool-using animals today that don't develop our kind of consciousness. How typical for a male writer to see this origin in his own tool-making and not the motherhood function! Of course, other animals also give birth and don't develop an 'I' from it.... Either way it's just speculation." Agreed! I would add only that I was talking more about the origin of technology from "I" and "It" rather than the origin of society from "I" and "Thou."

philosophical space and dispersed lingering shadows of bestial trance, it equipped *Homo sapiens* with revolutionary concepts, memes, and heresies, for its form was bottomless and mysterious beyond reckoning. Even before the bridge from *Homo erectus* to *Homo sapiens* was crossed, mankind and womankind had grown eternally old, wandering inside a labyrinth of unfinished meanings. Of course the labyrinth arose from a prior labyrinth, while the first labyrinth was itself instantaneous and timeless.

As society trespassed into a ferocious world that it was both mediating and creating, the fairy tale and the myth were born out of the darkness that they themselves contained. Since nothing by name preceded Earth's specters, I assume that they came from everywhere and nowhere: representations of landscapes, the mind of the planet, the animal kingdom, the inside of the body, the paradoxes of social life, the latency of the cosmos—all aspects of unconscious archetypes. We have recreated and reinvented our world-view billions of times over from the unsolved riddles that plagued and inspired prehistoric tribes in their own crises of meaning. Lévi-Strauss:

"The form is not outside, but inside.... [W]e are not free to trace a boundary on the far side of which purely arbitrary considerations would reign. Meaning is not decreed; if not everywhere it is nowhere."[67]

Distinctions between man and animal, nature and culture were far more dangerous in their conception than the wild creatures themselves (or than they presently seem after generations of domestication and imprisonment in symbols). Entities of nature were not in themselves insoluble threats to early men and women. Lightning, lions, winds, and floods were reassuring in their explicitness and direct representation of nature's archetype. Yes, they were at times vicious and beyond mediation, but they were neutral by pledge of their own independent callings. They didn't hold grudges or seek revenge, at least not until they entered our myth cycles. Then they made big trouble too.

It was the ambivalent structure of their own minds—the existential fact of being—that baffled and frightened our ancestors. The mistake of naturalism, in the words of Lévi-Strauss, "...was to think that natural phenomena are *what* myths seek to explain, when they are rather the *medium through which* myths try to explain facts which are themselves not of a natural but a logical order":[68] eros, death, ancestors, brothers-in-law, battles.

Mythology is mind's map of itself as it is penetrated by a world of which it is newly conscious. Hominids conceived porcupines, parrots, zebus, and the like—things separate in their own innatenesses—as totems, so nature entered society a second time, in totems and signs. Their classifications yielded laws of kinship and marriage, strictures of domestic life, and the rudiments of mathematics, astronomy, and natural history, as men and women turned the blank denominators of nature into domesticated animals, clans, poetry, pottery, divination, metallurgy, arithmetic, ritual magic.

Original social and logical constructs were derived from observations of bees collecting honey, a jaguar carrying meat, Orion glittering in the night, frogs multiplying in an empty gourd. These events "knew" what they were and taught priests and clans their derivations: a hummingbird stealing a rattle, wild pigs climbing a tree into the sky, a fox injuring a woodpecker while trying to delouse him with a needle, an old woman morphing into an anteater. What is impossible literally (but still "thought") is also what is most contingent—the unfathomable figures and rebuses constructed by events—and these were also the most useful in terms of conceptualizing new designs and technologies. Without real objects and arrangements to shuffle and condense, there would be no marked path into the depth of symbols that arise only autonomously. Totems lead to their own meanings and operations (and everything else). Only woodpeckers and Pleiades say anything *necessary* about the human universe (if not exactly about birds or stars).

Tall tales and nonsense legends of yore progressed into scientific and philosophical representations as well as terms and relations by which consciousness addressed itself and upon which it built subsequent *legenda*. Meanings assigned at one level (in a myth or totemic complex) surfaced as operators and catalysts at another phase of the same myth or in a different myth cycle (or in something else like a sand painting or bull ceremony). Similarly the metallurgy of the Iron and Bronze Ages originated in the flames of the shamanic forge:

"The layered structure ... allows us to look upon myth as a matrix of meanings which are arranged in lines or columns, but in which each level always refers to some other level, whichever way the myth is read. Similarly, each matrix of meanings refers to another matrix, each myth to other myths. And if it is now asked to what final meaning these mutually significative meanings are referring—since in the last resort and in their totality they must refer to something—the only reply ... is that myths signify the mind that evolves them by making use of the world of which it is itself part. Thus there is simultaneous production of myths themselves, by the mind that generates them and, by the myths, of an image of the world which is already inherent in the structure of the mind."[69] Consciousness is what it becomes conscious of.

Through migrations, marriages, and linguistic and symbolic drift, narratives accrued and shifted over generations. Each myth functioned as a transformative algebra of all the other myths in its pantheon, generating ceaseless enhanced, abridged, distorted, and defective versions. Much like redundant genes, multiple drafts of the same original stories erode and rebuild their precepts by borrowing from and conflating one another (while raiding adventitious myths in the manner of dream formation). Meanings arise from other meanings and from things that don't look like meanings at all. Unconscious layers and structures are turned up to reveal new parameters of omen and knowledge. All variants generate one another, and every variant generates every other variant. Soon there is only one myth with an infinity of variations.

A self-originating myth cycle endlessly diverges and converges in redactions of the same information such that multiple trains and representations reconfigure every thought, identity, meaning, and narrative as they fluctuate in half-lives, driven by cultural improvisation, libidinal forces, unconscious imperatives, and random associative processes and also by innate neural thresholds and systemic boundaries. By the early Pleistocene era, the original tales were already far too ancient and indeterminate to be identified or extricated. So they kept on going in search of their maker. They are still being written by novelists, film-makers, and choreographers. What do you think Charles Dickens, Steven Spielberg, Yvonne Rainer are all about?

Contemporary science is not some pristine empirical system coronated during the Renaissance. It emerged layer by layer from animism, sympathetic magic, and hunting-and-gathering strategies. Stone Age propositions are of the same fundamental order as those of advanced algebra or cybernetics, for humans in bands and tribes had brains the equal of ours, plus fully sapient minds and imaginalities. They coded information while embedding it in strings and symbols as discrete and subtle as those of mathematics and physics. Wisdom is not more profound or even more accurate now than in the Stone Age; it simply has more objects to work with and more tools with which to specify its gaze. Stone Age men and women were already participating in sophisticated messaging, information processing, and symbolic transmutation. Their oral encyclopedias of systematic botany and zoology, ethnobotany, ethnoastronomy, ethnogeology, ethnometeorology, and ethnomedicine speak to that.

The title of Lévi-Strauss's seminal book *The Savage Mind* resonates on three levels (the mind of savages, the mind itself is savage, and our mind is the same as theirs). He addresses "the false antimony between logical and prelogical mentality." Myths and totems proposed causal relationships later revealed by bent glass and electron streams, for no one was looking through that glass except the same *"pensée sauvage"*:

> The savage mind is logical in the same sense and the same
> fashion as ours, though as our own is only when it is applied
> to knowledge of a universe in which it recognizes physical
> and semantic properties simultaneously.... It will be objected
> that there remains a major difference between the thought
> of primitives and our own: Information Theory is concerned
> with genuine messages whereas primitives mistake mere mani-
> festations of physical determinism for messages. Two consid-
> erations, however, deprive this argument of any weight. In the
> first place, Information Theory has been generalized, and it
> extends to phenomena not intrinsically possessing the char-
> acter of messages, notably to those of biology; the illusions
> of totemism have had at least the merit of illuminating the
> fundamental place belonging to phenomena of this order, in
> the internal economy of systems of classification. In treating
> the sensible properties of the animal and plant kingdoms as if

they were the elements of a message, and in discovering "signatures"—and so signs—in them, men have made mistakes of identification: the meaningful element was not always the one they supposed. But without perfected instruments which would have permitted them to place it where it most often is—namely at the microscopic level—they already discerned "as through a glass darkly" principles of interpretation whose heuristic value and accordance with reality have been revealed to us only through very recent inventions: telecommunications, computers, and electron microscopes.[70]

●

Incest taboo was a premier belief system, a combination of vestigial customs with a landmark injunction. "Marry out or die out,"[71] declared CLS in justifying a primal superstition that is universal among human groups (although kin space is mapped differently by each culture). Marriage replaced incest by becoming it; i.e., by turning a forbidden endogamous act into an advantageous exogamous one.

Truces between Stone Age bands were negotiated by weddings, creating in-laws. The first culture-forming act was the gift,* and the original gift was the spouse, the marriage partner from an outlaw tribe. Neighboring groups soon held sisters, brothers-in-law, nephews, nieces, cross-cousins. Before everyone's astonished eyes a peace offering to an enemy turned into a dowry between kin.

As items flowed among adjoining groups in perpetual interchange, new tiers of social structure emerged. Migratory bands of men and women founded clans of interrelated families, then tribes, communities, villages, cities, nations. Tools, symbols, stories, institutions, and meanings circulated affine to affine across the migratory range, from Africa to Asia, from Indonesia to Australia. Chains of genealogy morphed by lineages of descent groups into the rules of bifurcate mergers and moieties. The Bushmen of Swaziland were bound to the Wirri-wirri of New South Wales. That is why they give the same accounts of Emergence out of Primal Discord, why they worship

---

*Those pesky muskrats again.

variants of the same godheads and propitiate manifestations of the same spirits. Consciousness is biological but psychological and anthropological too.

Once Earth tamed the Symbol, both at its basis and in its sprawl, it sowed it everywhere and is still sowing.

iv. Ontogeny and Phylogeny

In both my embryology books I asserted, in line with current theory, that ontogeny does *not* recapitulate phylogeny; embryos merely adapt to their fetal environments (the maternal womb in the case of mammals, an egg, pond, soil, or the flesh of a host in other zoological instances). Their phases are not recapitulations but accommodations.

At the same time, though, ontogeny *has to* recapitulate phylogeny as there is no other source or pathway for development and organization. What generates an embryo and its morphology is a nucleic ledger of prior embryonic designs in a lineage back to the first proto-biological configuration in some thickening puddle. Any living creature is the endpoint of one embryogenic thread as well as the beginning of another, for (if it manages to procreate) it transfers its inherited design to an indeterminate chain of future creatures, the first of them its own zygote.

Life cannot reinvent itself from scratch each time; it relies on its thread of a billion or more predecessors accumulating variations from each other by mutations, drift, and the self-organizing proclivity of molecules and cells. A cumulative inventory of prototypes and pattern states embedded in germs turns into liverworts, ferns, crows, newts, wombats, etc., as well as seeds and eggs for future templates.

In human development, something like a worm *is* added to the end of a starfish, a primitive amphibian to an unspecialized fish, a molelike quadruped to a lizard, etc. These atavisms are cumulatively cached in their DNA, as form keeps building on prior form. Hundreds of millennia of evolution are thereby synopsized into each contemporary embryo. This is both real and a mirage, and for the same reason, as we shall soon see. In any case, without prior embryos and precedents, there is nothing to construct and no basis on which to construct it—not even the simplest bacterium could get made.

Exactly as the syllogism of sexually transmitted diseases goes, you sleep not only with your partners but also everyone with whom they slept—you are not only your parents but every ancestor of theirs. In either instance, it doesn't matter how long ago the deed occurred—you are implicated. Of course past a certain number of generations, you likely don't share a single gene with any one ancestor;* yet paradoxically you get all your genes (and mutations of them) from their collective pool. There is no alternate dispensatory. Embryos move algorithmically away from their ancestors, while they are pulled back to them by remnant continuities of data.

Yet morphogenesis is never a faithful pedigree. Instead it reenacts principles of organization embedded in DNA, a combination of ongoing mutations under environmental change. At the end of the day it's all there—terminal additions, emergent phase states, lazy genes, recapitulation—but it's there in the sense that these mechanisms overlap, conflate, and recombine semi-randomly, and whatever's still alive walks (or swims or flies) away.

Remember, genes are codes, not tissues or organisms. Energy *per se* cannot be transferred in them. The code has to interact with raw materials to create membranes and tissues that conduct metabolism. New bionts must assemble and self-activate chemicodynamically each time from scratch. Their genetic libraries can't supply behavior or nutrition except insofar as structures created by them do.

Cells are impelled by gravity, heat, and their hungers alone—by immediate topokinetic requirements against anatomical constraints and boundaries: gravity, shear force, etc. Their drive to feed, grow, attract, repel, congregate in clumps, and extend their clumps' domains into adjacent possible space *is* their principle of organization. If they didn't have a novice's view of the universe right from the get-go (and all along the way), they wouldn't survive the crisis of their own development. So there is no recapitulation as such, only molecules and cells reorganizing dynamically in new fields under genetic chaperone.

Despite the substantial genetic weight of phylogeny (biological precedent), embryos must also rediscover *all* the stages of their development

---

*Like those country "cousins" Barack Obama and Dick Cheney.

anew in terms of one fetal contingency after another, which, *because of their phylogeny*, lead ironically to the same general outcome as if they were recapitulating a blueprint. They will inevitably diverge from their templates in major respects, if not in the present generation then in the next, or the next, or the next, because any adventure, major or minor, external or archival, at any point in the history of the germinal protoplasm, imposes a rupture of precedent. Then the plan has to self-organize anew and come up with a fresh motif, but its first choice is usually itself (or what's left of itself) because that's what's already there.

In short, the phase states of an embryo must recapitulate *enough* ancestral form and design to hold its tissues together, with enough immediate function in the context of that history to feed, generate energy, and expel shit. These two exigencies—one static and historical, the other immediate and dynamic—coincide in all living structures because that is the course of both greatest safety and least resistance.

While a series of fragmented blueprints preserves ancestors in genes, the exact lineal repertoire of prior creatures can never be just terminally extended, adult by adult, into new topologies; it is always reinvented in action. Terminal addition is not only never a simple progression, it isn't even really terminal: data are added at all phases of development. "Terminal" suggests a linearity that has almost no meaning in nucleic or morphogenetic space-time. The scroll continues to index and condense as it augments. Its earlier phases get briefer and briefer, crammed in ever tighter fractals. Of course, as noted earlier, a great deal is expunged in the process.

Yet, conversely, the chronological extension of the embryonic period among mammals allows the biggest empty block of space-time for augmentation: the future, the margin. Given the predilection to work with what's already there, new forms spool onto prior states with increasing complexity. It *is* terminal addition, sort of. Not only is a turtle more geodesically complex than a sponge but there are multiple "sponges" interpolated inside any turtle. The deep geometric core is what is recapitulated, not the superficial sponge (or fish, sea-squirt, or newt) whose features are mostly surface adaptations (like the "snake" inside the Testudine shell)—and the core tends to go deeper and deeper as each animal (or plant) annexes more overlays of design.

Simplicity has to come from somewhere, and the only source is complexity. Simplicity on the *other* side of complexity is one of the unheralded laws of nature. As a surface becomes more intricate, its core becomes simpler until the entire design refines (phase-shifts) itself at a new level of organization. Yet, at the same time that this interior indexing is going on, fresh embryos continue to fine-tune their adaptations to their exterior environments, which happen usually to resemble primordial oceanic and watery spheres; hence *even those phases* that are not actually recapitulations turn out to look like them. Terrence Deacon points out as well:

"Each individual that grows to a large size ... from a single cell on its way to a mature state ... temporarily pass[es] through physiological and environmental conditions that correspond to those confronted by other, smaller species in adulthood." Along the way they conform partially "to each of these diverse contexts, and they must do so in a specified order from one stage to another in the life cycle."[72]

Once again, adaptations not only resemble but incorporate recapitulation, pulling designs out of a primal core even as the core deepens, differentiates, and incorporates more information and heterogeneity to oblige progressions of timing and scale.

No one could possibly track the range of uncertainty states behind a creature's emergence. They bear only the hollow or correlative imprints of recapitulated former presences, also their erasures and revisions. So recapitulation needs to be restated:

The stages of embryos resemble the stages of their ancestors because the playback of condensed, recapitulated elements (in phylogenetic asymptotes, as it were) engages the novel forms provided by selection during fetal adaptation. These requirements converge, for the womb is a primordial pond and, in any case, form itself is redundant and promiscuous. In the phylogenesis of mammals, this means a "return" to an aquatic environment in a maternal body in lieu of the sea—a fishlike newt. Again, because the fetal adaptations are similar to pure recapitulation, they create the impression that ontogeny recapitulates phylogeny and by terminal addition of traits to boot. And it sort of does (too).

Let's try it one more time: By gradual *nonlinear* accretion of integrated microstructures (and their elisions) inherited from prior generations, a modern animal assembles its own blueprint. What is recapitulated is a fluid homeostasis, not a progressive taxonomy. Every embryo oscillates between templates of its long-ago ancestors and its immediate kinetic and environmental requirements, recapitulating something that represents both and neither, in order to weather the crises of its own development.

●

As discussed earlier, the brain's enhanced aptitude for consciousness was not resolved just by building the mind a bigger room. As the cerebrocranial mold dilated, soft tissue itself had to fold over its prior folds in such a way as to produce a multi-lobed design.

A gross parallel lies in arthropod cradles where segmental duplication of body parts fuses identical segments with structural redundancy into new matrices and specializations. This process speciates into grasshoppers, spiders, lobsters, flies, and the like, with differentiated antennae, spinnerets, claws, "and many other structures sharing the same jointed architecture, but modified to serve quite distinct functions."[73] Under this rubric the overall arthropod body-plan developed "greater flexibility of function with respect to uncertain environments,"[74] as well as "increased epigenetic variability and conditionality."[75]

Consider the variety of arthropod life on the Holocene Earth—they far outnumber us and are more creative in their lifestyles, even conceding our cultural advantages. "By reusing the same genetic information repeatedly in slightly different combinatorial patterns to build something approaching interchangeable parts, a remarkable diversity of body organizations was possible *with relatively minimal genetic reorganization.*"[76] (My italics.) This is a prime example of ontogeny recapitulating phylogeny to the degree possible while at the same time spawning new shapes and lifestyles. Complexity is not just terminally added, with the dice rolled afterward. The dice are thrown at every developmental phase, and only the winners leave the ring. But what a parade of insects and crustaceans, each recapitulated to some degree, yet daringly and brassily unique!

How does segmental duplication get rolling? "During gene duplication, a length of DNA is literally copied and spliced into the chromosome nearby, possibly as a result of uneven crossover events during meiotic replication, viral gene insertion and excision, or some other intrinsic or extrinsic mechanism that modifies gene replication."[77] This alters the template inside its own growth pattern. "One consequence of this hierarchic recursive genetic relationship is that changes in one gene can influence a large number of other genes in concert. So the functional divergence and interaction effects that result from duplication of such regulatory genes can be global and systemic."[78] It is like giving a child a set of bugs that stick together in different ways and asking him to combine them into compound animals.

As "DNA-binding gene products turn other genes on and off *en masse,* like an orchestra of genes…, each segmental region contains similar structures. [W]hole body segments [get] displaced or duplicated as a result of … [these] *homeotic* genes…. Interactions among genes [thus] set up complex gradients, thresholds, and semi-repeating patterns of molecular expression throughout the embryonic body…."[79] The effect is a protoplasmic ripple with simultaneously iterative and tier-displacing blowback. You get striking nuances of a classic arthropod look: a lobster is a giant submarine mantis or a bottom-feeding, water-dwelling, hardshell-molting housefly.

Homeotic genetic mutation is not limited of course to arthropods or armor. Ganglia can meld self-similarly into lobes of a brain, as "there are aspects of organism design and function that are fractal in nature—that is, are expressed in a self-similar or isometric way in both small and large organisms."[80] An E.T. expedition to this planet might report that, no surprise, grasshoppers and advanced brains on Sol 3 represent different homeotic expressions of one system intelligence:

"Among vertebrates, the major segmentation phase that determines axial patterning begins just after gastrulation, with the formation of the neural tube,"[81] an event that is central to the entire phylum and its divergence from Chordates. The spinal column and brain can be chimericized as dense, fleshy insects embedded in one another. Subsequent brain organization resembles global curling of brobdingnagian insect larvae with segmental organization "creating mosaic patterns of concentric and intersecting expression regimes."[82] The central skeletal axis sets compasses and contexts for the morphogenetic divisions of the body and nervous system.[83]

"[D]ivergent organization of peripheral structures, such as the eyes, the olfactory system, and the musculature,"[84] induces cerebral reorganization with segmentation, while encompassing fresh parcellation of cells from mitoses (see below). As vertebrates diverge from their oceanic kin, "multiplications of telencephalic stem cells lining the ventricular wall and their eventual production of neurons within this embryonic forebrain tend to enlarge the ventral and lateral walls of each telencephalic 'bubble' so that they expand into the ventricular space to form nucleated structures. Only in mammals does the dorsal half expand in a highly laminar fashion, due to the mostly radial migration of newly produced neuroblasts through layers of cells produced in previous waves of neurogenesis. So unlike the telencephalic nuclei of other vertebrate brains, the resulting laminar and columnar organization of the mammalian cerebral cortex maintains a uniform sheet topology that is well suited to maintain discretely separated map-like representations of its input-output connections with other structures, particularly with sensory receptor systems like the retina, the skin surface, and the organ of Corti within the cochlea."[85]

Cortical regions are further induced by "afferent axons projecting to them and the level and patterning of signals conveyed by those projections"[86] as well as "relative concentration gradients of … molecular cues distributed in different and overlapping patterns,"[87] "afferent invasion from thalamic neurons, [and] synaptic competition between afferent axons, which is strongly influenced by activity-dependent signal correlation mechanisms."[88]

A further critical issue in complexification of organs is the cellular meta-clocks underlying and coordinating development.[89] With a change of key and timing here and there, a seal is, presto, a walrus, or a lynx is a cheetah. Enough renegade clocks through enough embryogeneses can turn a chittering mouse into a climbing monkey, a forest ape into a wand-wielding shaman. It is of course a trillion times more complex and labor-intensive than that, but the hyperbole expresses the cumulative effect over generations. A blastulated orb, as it turns inside-out and gastrulates, develops zones and features. Most radical innovations at this point are lethal, dead ends (and of course progeny-less). But over many involutions and augmentations of tissue layers, a single lineage eventually gets on a roll—and we go sea squirt, salamander, orangutan, a wave of ingression like that of the hominid throwing a bone in the air (in the prologue to Stanley Kubrick's *2001*) and having it turn into an orbital satellite (as the universe itself folds into its own complexity). It is as though subwaves of logarithmic progressions sweep through cell nuclei and nucleic space-time as the archetypal egg invaginates, generation by generation, accumulating energy, accommodating stress, changing boundaries, initiating fresh biokinetic events, activating new nucleic responses (apply Terence McKenna's Timewave Zero here).

Mutationally instigated changes in timing mechanisms allow more rapid and more global metamorphoses than simple augmentations or deletions of alleles or tissues; they also require fewer genes, functioning like homeotic shifts or as macros. Many of the mutations behind protein syntheses and differentiating modes among mammals functioned as timing-based signals and controls. Heterochrony, paedomorphism, progenesis, neoteny, and recapitulation lead to juvenilization or, inversely, accelerated maturation of some tissues and organ complexes relative to others. Time gets bent in and into space.

Without altering each individual anatomical expression, genes can change the biological melody via links between fluctuations. As subcellular clocks lose former synchronization with one another, developmental patterns tune to new pipers. Some creatures end up as sexually mature "children," while others withhold sexual maturity further into their body's maturity. Likewise some organs overdevelop and embellish themselves relative to their previous designs; others underdevelop because, in their new sequence, they no longer have a window in which to elaborate. But, unlike the way in which it was conceived in the nineteenth century (when timing-based embryogenic patterns were first proposed and the human embryo was perceived as an *actual* fish and an *actual* monkey during its development), the various effects of acceleration or retardation affect organs and tissue systems selectively rather than whole embryos taxonomically— not fish and reptiles as such but fishlike and reptilian patterns.

Changes in relative timing lead to radically divergent designs out of the core package. Chronological resets during gastrulation and organogenesis interface with activations of homeotic genes and other iterations to produce utterly new structures and designs for novel functions. Each mechanism amplifies and extends the effects of the others.

Single heterochronic or paedomorphic mutations, operating on one die, can yield very different creatures from near-identical databases just by shifting their internal mitotic timing mechanisms and the progressive sequences they are coordinating such that some organs and tissue regions fall out of their prior synchronies into new ones with other tissues and regions. These heterochronies separate induction trajectories and phase states from the point of temporal dislocation forward through the remainder of fetal development. As altered subdesigns impinge on each other's cell space, especially early in embryogenesis, they force divergent interpretations and resolutions. It is no longer an ordinary factory run by union labor; heterochronic shifts in morphogenetic sequences produce Moebius transformations beyond ordinary anatomical limits. But a fetus is not a virtual reality or special effect; it is a real morphing object under kinetic regulation and design.

Heterochrony is a major reason why the phenotypic range of mammals exploded beyond their genotypic databases and why you got buffalos, beavers, and boars from one underlying genome. Humans differ in meaning and lifestyle much more than fifteen percent from mice and more than a meager one or two from chimpanzees, though the genes say otherwise, yet for assembly-line purposes the three function as different models of the same vehicle. Then clocks and reduplications take over.

We do not know for certain whether "growth and differentiation are merely the sum of many independently regulated processes in different tissues throughout the body, are globally regulated by hormonal or growth-factor effects that are diffused throughout the body, or are controlled by a sort of mitotic 'clock' synchronized in all cells at a point shortly after conception and maintaining this preestablished synchrony independent of specific context within the many different cell lineages of the developing body."[90] It could be a combination of these and/or other factors, or something emerging at a subtler level that can translate into any or all of them.

238

Now back to parcellation and scale. During early embryogenesis of complex animals, there is an increase "in cell numbers, and a correlated increase in the degree of cellular and tissue differentiation."[91] The tissues of the brain and body of an elephant obviously aggregate and develop at more patient rates from those in the brain and body of a coyote or rat. The mitotic clock with its potential for "global shifts in developmental timing"[92] is a key player but not the only factor in this equation, for "relative size of any given structure is not just a function of growth but rather a function of the interaction of growth, differentiation, and parcellation.... The ultimate sizes of different organs are significantly influenced by the way in which this initial parcellation interacts with subsequent mitotic and cell growth processes within each of the cell lineages."[93]

Sheer cell profusion and mitotic activity differing by region can skew relationships and anatomies as much as heterochrony, for initial complements of cells play out through subsequent morphogenetic interactions, developmental motifs, and the synchronized hands of different internal clocks. In the case of the developing brain, the system's "comprehension" of a new regional parcellation of cells probably didn't initially "solve" itself or get down all the fine-tuning necessary for metabolism and function. Cell death and apoptosis were "used to fine-tune the relationships between functionally linked neural cell populations in different regions. Overproducing and then culling cells on the basis of their interactions" carves out new organs and relationships.[94] The cumulative result is that "widely separated body parts ... [grow] to equivalent sizes [as] different species must have very different mappings of differentiation events onto numbers of cell divisions from conception...."[95] It is as if the clocks don't merely synchronize but, like something in a Salvador Dali painting, change size and shape too—and they aren't even clocks. A holographic concrescence of cycles injects novelty (self-reflective consciousness) into space-time, as each next subwave continues to fold over the prior ones.

It takes a masterful conductor to keep such instruments in tune with an orchestra, or it takes something other than a conductor. After all, he is being conducted too by the background symphony of a morphogenetic field. Maybe it is the sheer immensity of the algorithm, as Neo-Darwinians would assert.

In any case, within a hominoid context a human brain replaces a simian one—not only (and not even primarily) because of micro-timing shifts and segmental duplication but because someone suddenly dished a lot more cellular material into the cranium of a much smaller proto-monkey as if it were, say, an orangutan. Oops—but then again on second thought....

This endowment has no explanation and was the least explicable and most radical change of all. Where did that extra stuff and its control commands come from? How did other tissue masses get swapped out? These things don't just happen. Or do they? Something serves as designer, whether it is a self-organizing algorithm, a nonergodic phase shift, or a transdimensional synchronicity. Deacon refines my commentary:

"[D]ifferences in brain/body growth curves can be described by a single scalar factor, which is an index of relative mitotic growth.... [I]t is as if a single mitotic clock is running in all cells of the body, with each tissue type setting its point of last mitosis at the same relative time point with respect to the whole sequence.... [I]t is a sort of elastic-sheet time-line of development that is stretched to different degrees in different species but in which the various stages expand or contract proportionately and apparently the same way, irrespective of prenatal starting proportions.... It appears that the difference between the primate and nonprimate brain/body growth trends is determined by a shift in the initial proportions of brain and body without altering the organization of the developmental time-line."[96]

In primates, encephalization appears to represent a distribution of initial embryological materials and proportions within sequential contexts and mitotic timing of the body-brain scale.[97] In fact, you might say that it was the *failure* of heterochronic mutations to take hold of and recalibrate the new proportions that led to the human brain. Heterochrony got the evolving animal to an intelligence-critical brain/body-mass and then peremptorily allowed itself to be overruled, so that it actually crossed the threshold on its next subwave:

"[P]rimate embryos start out with less total body mass though roughly the same brain mass as other mammals.... [It is] as though at a very early stage in embryogenesis a whole chunk of the postcranially fated tissue mass has been removed. The segmentation of the embryo has been modified, and yet the developmental clock has not been rearranged to produce this primate segmental growth difference...."[98] It is not a timing change of segmental growth or developmental schedules "but instead a shift in parcellation of the early embryonic body that simply gets extrapolated by otherwise conserved rates of global mitotic growth."[99] In addition there is "a shift in the proportional length of prenatal and postnatal phases ... [and] brain growth prolongation...."[100]

Time has begun to flow simultaneously back and forward and, in both cases, at rates out of keeping with prior creature design. It is almost as if nature has a hidden built-in device to inaugurate species, rigging the equation at key crossroads for a philosophical result. In the final stage of human encephalization, nature plays the role of a Nobel-level biotechnician, as "human brain/body growth is analogous to what we might imagine after a whole brain xenotransplant from a huge ape into a chimplike embryo!"[101*]

How the equivalent of a gigantic ape brain got grafted into a chimplike embryo is a total stumper. You might almost suspect E.T. again.

My intuition is that an organ of higher consciousness passed through more than one labyrinth and more than one scalar shift. Mind was already a portal inside matter, making one brilliant decision after another, as if Leonard Bernstein, Christo, and Craig Venter were working behind the scenes together, each in his own key—but actually consciousness was just descending on its own plumb line into density, morphogenetic features taking their due, catch as catch can, from whatever was delivered their way (homeotic genes, segmentation, heterochrony, reparcellation, apoptosis, etc.).

In the larger universe a brain is a brain is a brain. The cerebral "archetype" functioned like a quantum-entangled moiré or a "shear" force that

---

*Tissues and genes transplanted from widely variant species, such as plants to fish and humans to flies, don't just tank in the alien environments but engage in at least mutual recognition and temporarily viable structures, as genes and cells speak a language that precedes and transcends the architectures that they create. An oak is much more radically and existentially different from a crab than an oak gene is from a crab gene or even a human gene, which all can "work" to produce chimerical protoplasmic forms, even if they are nonviable, in each other's "ova."

was neither material nor mechanical. Either that or a proto-monkey stumbled like Alice into the rabbit hole and found itself awake in its own dream.

Don't space-time, mass, and position represent congruent topological projections of a single design precedent? Is there any real difference between time and space inside a gene, a cell, or a developing embryo? In these very small zones wherein time itself has quantum and scalar properties, a morphogenetic parcel is probably just another version of a morphogenetic clock, and *vice versa*.

In conclusion, "[hominization] appears, on the surface, to be an unprecedented example of a segmental modification of the developmental clock."[102] Not only that: "the appropriately prolonged development of the human brain for its size suggests the unprecedented prolongation of the mitotic growth phase in these regions by a localized shift in the [same] differentiation clock...."[103] Random mutations or telekinetic shear? (But please, no intelligent design!)

Give credit where credit is due. How did any biotechnician, let alone one blind and deaf and dumb, put this all together and then bring the *coup de grace*? Back to those guilty alien cloners, aye? My guide through all this, Mr. Deacon, concludes with admirable understatement:

"It is significant, then, that the enlarged regions in the human brain correspond to the latest differentiating structures of the brain as well as to contiguous homeotic expression domains.... [There is] no simple concordance between heterochronic assessments at the morphological and cellular levels. The mechanism underlying the human divergence defies simple classification and remains mysterious."[104]

CHAPTER EIGHT

The Quantum Brain

W e not only lack a model for how we inhabit our bodies and brains as first-person consciousness, but we are as far from one, in a real sense, as Parmenides and Anaxagoras were in pre-Socratic Greek times. Pretty amazing plight, given the ground covered since. We have advanced on everything else in skinny about the universe but not on how its interior light turned on, how the gap was crossed from matter to mind, or how the city got inside the acorn. The brain remains a complex but ordinary object in Newtonian/Darwinian space-time while the mind is a baffling phenomenal (or epiphenomenal) effect associated with but not supervening it.

I have thus far considered classical mechanical—biochemical—solutions to this mystery, assuming that the brain is (more or less) the proximal source of consciousness. I have kept my scientific narrative conventionally scientific, avoiding facile applications of metaphysics. I have not avoided metaphysical precepts *per se;* far from it, I have championed a consciousness-based, mystery universe. Yet I have tried not to conflate it glibly with closed scientific models or to relegate unsolved problems in physics to paraphysical circumventions or metaphysics of convenience. I have pretty much avoided quantum physics too, at least till now.

The domain of quantum physics is generally pooh-poohed if not outright ridiculed as a candidate for generating the enigmatic "consciousness effect." The widely accepted premise within neuroscience is that "awareness and subjectivity are probably *network* effects, involving many millions of neurons in thalamic and cortical structures"[1]—neurons—not quantum collapses, not holographic waves, not transductions of uncertainty states into thoughts. There isn't the slightest hard (or even circumstantial) evidence for the application of quantum theory to consciousness. No link has been found between the micro, still classical Newtonian world in which neuronal and subneuronal events elapse and the atomic/subatomic realm of quantum effects, nor any explanation for how the two, existing at radically different reaches of scale with incompatible arenas of action, could interact in such a way as to cause consciousness. Neurons yield to Newtonian not quantum factors.

Quantum effects by definition occur in spaces far smaller than a neuron. As a crude measure of the classical/quantum ratio, compare the size of a neural synapse (about 300 angstroms) to the diameter of a hydrogen atom (1 angstrom). That's a factor of 300 and, even if you squeeze the zone a bit tighter, you are not going to get close to even double digits without fudging. In addition, quantum effects occur, so far as we know, at temperatures too low or under pressures too high—and by a lot in both measures—for cells to survive.

By even implying a back channel between atomic states and phenomenal awareness, I am buying into a myth rather than something observable in the world of events. Yet I am also investigating an ontological riddle—reality and consciousness—not conducting theoretical physics.

●

The mind is not, as we have seen, related to the brain in a linear manner nor are there answers for "fundamental questions about basic phenomena—such as the role of back projections [to and from the neocortex in the passage of memory and information], the nature of representation in sensory systems, whether sensory systems are hierarchically organized, precisely how memory is stored and retrieved, how sensorimotor integration works, [and even] what sleep and dreaming are all about...."[2] In short, we lack explanations for the most basic "bioneurological mechanisms of awareness,"[3] even a simple ontological status for waking life.

As early as 1951 physicist David Bohm pointed to the resemblance between subjective experience and the properties of quantum systems. In a purely intuitive sense, mental and quantum expressions *are* similarly paradoxical and discontinuous. In a way that the biophysics of the nervous system can't touch, quantum indeterminacy reflects the spontaneous and nonlocal aspects of consciousness; for instance, consciousness (like particles) can be in two places at once, connect to objects without bumps, and operate outside the flow of time.

A scientific waiver to pursue a quantum brain goes something like this: if your proper view is that consciousness is neurons, neurons, and more neurons (or molecules, molecules, and more molecules) and that the source has to be in there somewhere (because it can't be anywhere else without invoking idealism, vitalism, Platonism, or some other extravagance), and if you are likewise convinced that consciousness cannot be understood using all the resources of molecular-level theories and the levels of explanations that they support—if you conclude that the phenomenon has completely stumped the entire scientific community—then (and only then) may you poke and prod for answers at a quantum level, as that is the sole resource left in the natural sciences.

The stretching of quantum events from the fugitive realm of atoms to the robust zone of Newtonian effects is physical science's final last-ditch bridge over the abyss between subjective and objective worlds, hidden and observable universes. The only other logical alternatives would be the emergent properties of self-organizing systems, the black box of nonequilibrium thermodynamics, or out-and-out intelligent design.

First you look for consciousness at a macro level (cognitive functions of the brain). No dice, so you look for it next at a micro level (synapses and neurons). No dice, so you skip over molecules and look for it at a quantum level because *that's where something like consciousness begins to show up—but (ooops, sorry!) only by the intercession of conscious observation itself,* which is a coincidence and an irony and a tautology bordering on an oxymoron.

Despite problems with other explanations, most biologists regard quantum models of mind with as much enthusiasm as a child contemplates a tablespoon of cod liver oil. They have had mega success using the old tried and true stick-and-ball model of matter, and they can treat DNA as a four-letter mechanical code without street complications. They also do not feel that the avenue of emergent properties and self-organization of classical dynamic systems has been nearly exhausted, while new empirical studies of perception and cognition using traditional biochemistry continue to show promising, even spectacular, results.

Emergent properties that do *not* depend on quantum processes arise seamlessly in the aggregation of substance itself across major thresholds of scale, while those same properties just as inexplicably vanish below those levels of organization. Molecules, for instance, do many things that atoms can't, yet they don't translate their full atomic repertoire to a molecular realm—most pure atomic effects evaporate in translation. Organisms are made of cells but don't behave like cells—they don't change shape, clone, aggregate, or divide by mitosis.

Emergent properties are spontaneous, independent, and kosher. There's no reason why emergent networks of internally accessing monitors or classical computational capacities have to depend *even on natural selection* for their emergence. They can be just as nonlinear and paradoxical as quantum events in their own way and can be applied fecklessly to consciousness without violating physicalist bias or empirical observation.

While emergence does not require quantum processes or Darwinian effects, it is unclear what it does depend on. In fact, it may be the single greatest mystery in nature.*

The reason that it is a much greater leap to propose that consciousness might behave like subatomic waves and particles than it is to classify it as an emergent phenomenon of neurons is that a "quantum brain" would have to translate its "bump effects" across not just one but several dimensions of scale just to get discrete packets of content from atoms into neurons (to say nothing about the problem of generating that content quantum-mechanically in the first place). They would have to deliver subatomic instructions into presynaptic messagings. Neurons instead of being clear-cut "physical machines" would have to become "normative neurons" no longer simply describable by physiological observations except at a very primitive and simple level.[4]

Given this prohibitive prognosis, quantum outputs would seem irrevocably confined to their own playpen if it were not for two old flies that remain stuck in the ointment: One, we have no clue what consciousness

---

*One cognitive psychologist confided to me, "We call consciousness an emergent phenomenon, but that's like saying we don't know what the hell it is."

is, so we don't know at what scale it might operate or if it is even related to ordinary physical scales at all, and, two, where consciousness vanishes into the zero point of its own strange existence, wave-particle behavior vanishes within consciousness (wherein it also arises) into the zero point of its own paradoxical emanation.

At the end of every day, "real things—the position of the sun, for instance—are at base dependent on atoms and electrons: flagrant nonthings. We completely understand how simple nonthings (rainbows) arise from things, but we are not entirely sure how the world of things (sun, rain, and eyes) arises from the ultimately thingless world of atoms and electrons.... At this stage of the game, physicists do not possess a clear explanation of how the things of the world are produced by mutual interactions of the world's nonthings."[5] So, what they possess of consciousness is *bupkis.*

Once the contest is opened to all candidates (large and small, molecular and subatomic), quantum thinglessness stands out as the only property in physical nature that resembles the thinglessness of consciousness, raising (as well) the possibility that the brain as an ordinary object is not so ordinary after all—and that *some* of its components are operating at the quantum level (even if most are not).

The brain does not even have to be a quantum object; it can be a Newtonian object with quantum potential or with "quantum-like potential." Of course, the line between a quantum-potential brain and a metaphysical brain is a thin one—but it is a line nonetheless. According to physicist Nick Herbert, "part of the brain is already a quantum device—the part that's conscious. All we have to do is ignore the vast amount of Newtonian data processing that goes on in there and pay attention to how raw consciousness feels from the inside ... without Newtonian preconceptions...."[6]

Good point. Try fitting the sensation of self-awareness into any Newtonian map of the universe. It can't be done. You can't even take the first step because in what direction would you place your foot? The conscious operation of the brain may be indubitably Newtonian, but consciousness itself is quantum-like. Herbert calls the brain a quantum device outright and explains that the fog surrounding Newtonian objects with quantum potential is not so much an occult or metaphysical shroud as "the transparent clarity of a new mathematics"[7]—no more and no less (which might still, of course, mask something metaphysical). By the premise that mathematics maps the only part of the universe we can grasp at all, quantum equations are not just a chimera but the tiniest tip of the tail of a gigantic dragon lodged somewhere inside the "real."

There are many discrete pathways to a Newtonian brain with quantum potential; they range from the minute, as small as a neutrino (or a boson, which has no material existence and only hypothetically exists), to as large as the universe itself. The wee end of the spectrum proposes quantum effects seeping into macro-reality through very tiny particles (switches) triggering "purple glow" phenomena. The jumbo end of the spectrum calls for subsuming consciousness inside the quantum basis of the entire universe and everything in it: since quantum phenomena are mind-dependent and all of reality is based on quantum phenomena, (ergo) all of reality is mind-dependent. Even scientists who reject quantum explanations in biology concede that subatomic properties are worth considering—but only because they underlie the cosmos and were coterminous with it at the Big Bang.

However, if atoms have intrinsic quantum randomness and thinglessness, why not fire hydrants, tennis balls, and grains of sand, which are made of atoms? Why are they not exuding quantum mindedness too? I will deal with this conundrum in the next chapter.

Insofar as atoms have a quantum basis, the myriad molecular contraptions arising from them and making up the "real" world are held in atomic configurations. No matter how bump-removed quantum effects are from synaptic ones in any ordinary sense (a scale of at least 300) or how classically constructed a brain (or a neuron) is by comparison to an electron, the fact is: quantum properties are intrinsic to all objects made of atoms and must contribute an imperceptible but indivisible aspect to their reality—quantum wave motion built into every cell:

"The quantum reality problem is, strictly speaking, not a physics question at all, but a problem in metaphysics, concerned as it is not with explaining phenomena but with speculating about what kind of being lies behind and supports the phenomena."[8] Because this is no simple riddle, there is no simple solution. Even the common-law boundary between quantum and classical effects is not well defined. Consider these two examples:

- A thin film of oxide covers every metal part of an ordinary flashlight: batteries, springs, switches. Classically that oxide is an insulator, so the flashlight shouldn't work (and often doesn't if the oxide gets thick enough). However quantum mechanics comes to the rescue and allows electrons to tunnel through the insulator and suddenly appear on the other side. So a flashlight depends on quantum mechanics. But is a flashlight therefore conscious? Don't shake your head so fast! You would need a theory of consciousness worth its weight in salt even to ask the question.[9]

- There is new evidence that photosynthesis may be using a quantum strategy to transduce excitations efficiently from antenna chlorophyll to reaction chlorophyll: "[I]n at least one species of ocean algae, the process of energy transfer between antenna and reaction sites is so efficient that it can only be explained by a quantum theoretical mechanism and not by a mere 'random walk' as was previously supposed."[10] If plants are potential quantum generators—well, there is a lot of botanical activity on the Earth. So much for dismissing quantum mechanisms in life forms. I'll come back to quantum photosynthesis later in this chapter.

●

The quantum event-field was first modeled in the 1920s to settle Newtonian mechanics' failure to account for the realm of the very, very small, *not for its failure to account for consciousness*. As such, it has a precise and limited range of application. For instance, quantum mechanics will give perfect answers for the energy states of the hydrogen atom, and for any other "one electron, one nucleus" system. As soon as scientists take it to a more than two-body system, quantum logic can't give them an exact answer and they have to start tossing fudge factors into their equations to get their numbers to match known energy values.

Physicists have been trying for years to mathematically understand just the combining of two hydrogen atoms to form a hydrogen molecule—a two-electron chemical bond and a four-body problem (two electrons and two protons); yet even a three-body problem—the simple helium atom (one nucleus, two electrons)—is beyond solution. To get from such a fundamental phase of quantum mechanical analysis to the simplest ordinary macro-object, let alone consciousness, is off the chart.

Yet in the context of consciousness (and what other context is there once you leave Newtonian space-time?) quantum physics has everything to do with the paradox raised by the *observation and measurement* of atoms and properties of subatomic systems, phenomena that are paradoxical not only because they defy classical physics but because they defy common sense. Remember, atoms are "thingless"; they don't exist except as observations, proxy objects inside consciousness. They shadow and are shadowed by consciousness itself:

"Quantum theory declares that the particles are not concrete little chunks of something, but rather are nebulous clouds of possibility, whose wave function is collapsed by measurement of its location into particle function. The direction can just as easily be reversed: particle function can be collapsed by measurement of its velocity into wave function."[11] How does this actually work?

A single atom can appear as a particle in one place or a wave in many places at once—it depends on the experiment and the experimental apparatus. In that sense, atoms are not concrete micro-balls but unrealized potentials for action "continually on the move, growing, merging, and disappearing according to exact quantum laws of motion discovered by [Werner] Heisenberg and his colleagues. But despite all this frantic atomic activity nothing ever really happens down there. As long as they remain unobserved, events in the quantum world remain strictly in the realm of possibility."[12] Until we direct our mind onto it, an atom is only a possibility, a thingless thing. But it is still "something" rather than nothing (or than a mere hypothetical thing). That is, something is there—something major-league—even if it is not an "atom" (or a "thing") in the commonly understood senses of either.

When you measure an electron's momentum by setting a meter in its path, "the electron will appear to acquire a definite (unpredictable) momentum at the price of forgoing any knowledge concerning its position."[13] We can see each of its attributes in turn but only insofar as they manifest from a particular mode of observation. One subatomic observation brings one attribute into the world; it does not bring any affiliate attribute of the same electron along with that one, so the entire atom is never "realized." In fact, its position and momentum are not merely separated; they are incompatible, in logical contradiction to each other. Yet "in the unobserved world of pure possibility, incompatible attributes can [and do] exist without contradiction."[14]

Stated differently, while observed from any one vantage, an atom seems to have a definite value for position, momentum, or spin—it is a tiny but real object, the undisputed building block of classical molecules. But you cannot predict that value for a subsequent observation, so it is *not* a real object, it is a wave-function, a possibility wave: "The atom is everywhere because its wave function effectively spreads out over all space, although the wave's amplitude is largest near where the atom was last sighted. On the other hand, the unseen atom is 'nowhere' because its wave function represents not the atom's presence but only the possibility of the atom being in one particular place rather than another."[15] And even that possibility is not a position or an object but a continually vibrating energy at a particular frequency, "a frequency that depends on the atom's energy content. In addition, when two of these oscillating possibilities come together, their amplitudes are added together, like sound or water waves, either to decrease (out-of-phase waves) or to augment (in-phase waves) the atom's chance of being in one place rather than another."[16]

This is one of those "I don't think we're in Kansas anymore" moments. There are very real obstacles to recruiting an atom into classical Newtonian space. You'd have better luck hunting the Almiqui solenodon or frumious bandersnatch.

The quantum world is generally defined as "the realm of the smallest quantity of radiant energy, equal to Planck's constant [see below] times the frequency of the associated radiation, hence a fundamental unit of a quantized physical magnitude, as angular momentum."[17] In such a realm, atoms masquerade in not just a single paradox but a series of (so far as we can tell) independent paradoxes. Hijacked conceptually into our world, they can only exhibit as models, tropes, abstractions. People assume they exist: everything is made of those fabulous little guys, us too, right? We're real, so end of story. Not so fast. Only deep inside the phenomena that they generate do these potentials have their own actualities, but those actualities are themselves hypothetical or subsidiary because they depend on the cachet of emergent solid "thing" things like rocks, cacti, brains, etc. Atoms are tautologies, Catch-22 propositions, chimeras. If they have existential status, it is different from *our* existential status, taken either objectively or subjectively, yet we are their main, perhaps sole sponsors in the universe. You could say that they are placeholders for everything we don't know about nature, and at the exact spot where such placeholders should land— the bottom boundary and deepest interior of substance where anything left has to sing like a canary or else turn into a different kind of thing.

Even were all of an atom's possibilities magically resolved, it would remain thingless because "the sum of all attributes observable in all possible contexts exceeds the number and variety of attributes that a single ordinary object could possess."[18] An atom isn't just a very small thing; it is a gate-keeper of a different realm—like Charon the boatman at the crossing of the Styx. Its so-called combined intrinsic attributes exist only in an entirely different sense from the barge, the river, or the coin used to pay for the ride (or their molecules), a state which is inaccessible in precisely the way that classical phenomena (real boats and rivers) *are* accessible.[19] Even when an atom is made real, it is an instant reality-buster—here and gone, and what you saw was more quixotic and spooky than the Loch Ness monster in drag.

A photon before measurement and a record being made of it is in a state of superposition of all its polarization possibilities at the same time. When the photon is forced to reckon with our observation of it, it is no longer superposed but must choose either a left or right polarization. It "collapses" into a choice, and we give it a name: photon. But it is not a "photon"; it is an abstract model of an actual photon, which doesn't exist.

In order to get something from the quantum realm to an actual realm, you need to imaginarily imprison it in a state that cannot be superposed. To force atoms into reality you have to stop their hummingbird wings, freeze their nocturnal lynxlike stalk, "collapse" their wave function, spot the "extinct" solenodon. To do that, some outside force is needed. But what "force" is there except observation? What other force in the universe even approaches the task? Back to our hero.

In fact, mathematician John von Neumann "reluctantly concluded that the only known entity fit for this task was consciousness.... [T]he world remains everywhere in a state of pure possibility except where some conscious mind decides to promote a portion of [it] from its usual state of indefiniteness into a condition of actual existence."[20] You cannot "formulate the laws of quantum mechanics in a fully consistent way without reference to ... consciousness."[21] So even if you do not accept the quantum brain in a thought-wave sense, you might challenge the epiphenomenal position regarding consciousness insofar as mind is needed to complete or even admit a critical part of the real universe: the atomic world. If you reduce observation to an epiphenomenon, then you reduce atoms and the whole of material creation along with it. Catch-22.

Observation alone brings thingless atoms into their ambiguous assignation with atoms as "things" underlying the things of classical mechanics. Yes, all things exist only in consciousness, but atoms *really* exist *only* in consciousness. By the intercession of human measurement a single quantum possibility is singled out, "abandons its half-real, shadowy sisters and appears in our ordinary world as an actual event...."[22] So consciousness is necessary not merely to "find" atoms and graze the thinnest of all veils holding up the material world but to collapse their superpositional wave into a single position, "to bring an actual world into existence, out of the all-pervasive background world of mere possibilities."

As hard as it is to accept experientially, the quantum basis of this very robust physical universe "exists only preobservationally as possibility."[23] This is a disaster for the Newtonian-reality crew: "The Old Physics could not make the universe exist for more than a fraction of a second. However, for that instant [before quantum factors set in], it really did exist. On the other hand, for the New Physics, the fact that the universe exists at all is somewhat problematic."[24]

As Herbert foreshadowed, the unobserved atom is not a mechanical problem but an ontological one: that of intrinsic quantum variance—there is no world, only energy, curvature, and a semi-durable hallucination. Quanta step up to the plate to confirm it.

This is no mere hypothetical paradox or abstract modeling either: "Heisenberg constructed his picture of reality by taking quantum theory seriously, not merely as a device for calculating experimental results but as a true picture of the world. [He] proposed that, since quantum theory represents the unobserved world as possibility waves, then perhaps the world might really exist, when not looked at, as mere waves of possibility."[25]

Do you wonder any longer why quantum theory runs smack into consciousness theory and a Hindu universe made of *Māyā* (illusion), as well as the etheric landscape of the Australian Dreamtime? Reality is made only of atoms, and atoms don't exist until observed. When not encroached upon, they hover in a state of "absolute indefiniteness."[26] From here physicists and philosophers have taken their best shots. You can decide whether they came away empty-handed or whether some grabbed the quantum brass ring.

●

"Zombie"-philosopher David Chalmers's "most promising interpretation allows conscious states to be correlated with the total quantum state of a system, with the extra constraint that conscious states (unlike physical states) can never be superposed. In a conscious physical system such as a brain the physical and phenomenal states of the system will be correlated in a (nonsuperposed) quantum state. Upon observation ... the superposed external system [would] ... become correlated with the brain, yielding a resulting superposition of brain states and so (by psychophysical correlation) a superposition of conscious states. But such a superposition cannot occur, so one of the potential resulting conscious states is somehow selected (presumably by a nondeterministic dynamic principle at the phenomenal level). The result is that (by psychophysical correlation) a definite brain state and a definite state of the observed object are also selected."[27]

To oversimplify, quantum potential transfers itself into the brain as superposed (conscious-potential) states. Consciousness then occurs in the same way that conscious observation locates atoms: it collapses quantum waves. Mind is the phenomenal equivalent of quantum indefiniteness. But to get a "nondeterministic dynamic principle" to operate between matter and mind—between the city inside the acorn and the acorn itself—is a perplexity: somehow to turn models into neurally streaming thoughts.

Quanta are still suppositions of conscious states, quantum metaphors not quantum waves. Even at its synaptic basis, reality is much realer than that.

Berkeley physicist Henry Stapp translated objective collapse theory into a free-will-oriented choice menu for the brain wherein quantum possibility would translate directly into consciousness in neurons (because where else would it come from in a corner-to-corner physical universe?). The cerebral wave function and the particles' indeterministic collapse are considered ontologically distinct features, quantum events that are part of a holistic domain "ripe" for selection from sets of possibilities among neural excitations. A memory-oriented code filing neuronal synapses and drawing on the brain's memory of past events creates the quantum future in a sub-domain that cannot be represented outside itself and is not subject to the laws of classical mechanics in the way that the rest of the universe is—enter consciousness as the cream rising to the top of this clabber. He adds:

"The conscious action is represented physically by the selection of a new top-level code, which then automatically exercises top-level control of the flow of neural excitations in the brain through the action of quantum-theoretic laws of nature. The unity of conscious thought comes from the unifying integrative character of the conscious creative act, which selects a single code from among the multitude generated by the causal development prescribed by quantum theory."[28]

But how does this flux acquire the sort of exquisite, continuously replenishing context that actual thoughts have? How does it bust out of its metaphor into Technicolor 3-D? How does a quantum non-thing become a post-quantum (conscious) thing?

At a scale that tries to cross the bridge between atoms and neurons, a number of neuroscientifically-oriented philosophers, psychologists, and theoretical physicists have reconceived the brain as a microphysical object and mapped the mind onto it by deepening its field and substituting, as it were, another "brain" (or another cerebral level): a quantum-mechanical field of relationships *in* neural tissue generating a "meta-consciousness of consciousness."[29] This flow creates micro-to-macro seepage which somehow becomes individuated as consciousness.

Instead of spelunking from consciousness down into a quantized brain, they try to free-solo up from matter, assigning quantum properties to neurons (or to their molecules, atoms, or even subatomic particles) and then igniting these in the cortex: an "emerging view of consciousness as a quantum effect, even down to the individual nerve cells."[30] There are numerous such models, but they are parables, metonymies, not territories. For instance, one such predication is based on how computers operate by quantum tunneling, teleporting information between different energy states; the same transmission is then proposed between neurons or, more precisely, between their microtubules. A sort of virtual machine would be operated by the brain or assignable through the brain to the quantum states of the particles in the atoms inside its molecules. Interacting fields (nuclear and neurological) would then excite an amplification or synergy between separate state universes, triggering topological phase transitions. As neural networks and subcellular networks reverberate back and forth, subjective awareness would be generated (perhaps as an emergent phenomenon). It might occur at several emerging levels simultaneously, gradually approximating the interior "feel" of consciousness.

In that regard it has been suggested too that ferroelectric behavior in the form of weak electrical fields in the brain operates in such a way that excitation of the field's particles contributes to the emanation of consciousness.[31] But again, what tunnel or ferroelectric effect installs the mundane world in the brain, turns it inside-out, and illuminates it? "What" uses a jinni's code to represent its flow to itself in such a magical way? Certainly not atoms and electrons. So how do they project or engender thought into neurons? And if they are so smart, why (again) aren't rocks conscious too? And everything else!

If parallel computational capacity is intrinsic to the fabric of matter at a subatomic level and concomitant with the creation of reality in the brain, its software would also have to be super-compressed somewhere inside the molecular space of neurons. Trillions of particle reactions, quantum waves, and density matrices embedded in the brain's cells would then deliver their "mind" effects through the interactions of computer-like subcellular structures: individual microfilaments, microtubules, and other organelles within the neurons. Potentiated functions of this "brain" could be derived from quantum field theory's menu of (take your pick) intrinsic waveforms, superpositions, quasi-crystals, quantum gravity, indeterminacy, nonlocality, nonalgorithmic computation, and quantum entanglement (see below). Some combination of these would then cause unusual states to be propagated from inside the neuronal flow into the cortex. Still it seems odd to deliver mind into existence from the bottom up—sort of like getting enough ping-pong balls to ignite a neural effect, only very little ones this time. It just doesn't add up or deliver the morning paper.

●

Physicist Evan Harris Walker* invokes a similar event to that described for flashlights. He imagines that electrons tunnel across a synapse to generate neural activity, while the mind somehow adjusts which synapses fire or don't: "[W]hen a synapse is excited, the voltage difference between the excited neuron and its neighbors causes electrons to 'quantum tunnel' across the synaptic gap from neighbor neuron to initiating neuron, in a manner identical to that of electrons in a … tunnel diode…." They "not only quantum tunnel across the synaptic gap between adjacent neurons but also influence the firing of distant synapses, by tunneling to far-away synapses via a series of tuned stepping-stone molecules … amount(ing) to a 'second nervous system' operating by completely quantum rules and acting in parallel with the existing nervous system."[32] The first system handles unconscious data processing, whereas the second one generates the inner light we identify as first-person consciousness, "permitting an external mind to express itself by selecting which second-system quantum possibilities will be actualized."[33] As these potentials get streamed back into the central nervous system, their actualized possibilities instigate perception and behavior. Unless I'm missing something, there is a major ontological gap here.

In a dualistic model, this kind of medium for conscious flow is implicitly based on a *prior* link between mind and matter that presumes mindedness to exist in the universe as a force independent of particles. The flow is then accessed rather than created on a quantum level. In fact, in the panpsychic model, subatomic particles could be an effect rather than a

---

*His book title puts his heart on his sleeve: *The Physics of Consciousness: The Quantum Mind and the Meaning of Life.*

cause of consciousness. This is probably not what Walker and his associates intended, but frankly I don't see any other way that such a Rube Goldberg machine could run. (See the next chapter for "panpsychism.")

At the same time, ferroelectric fields, reverberating networks, and even excitations in quantum tunnels seem too lugubrious for something as gossamer as mind. As gossamer as quarks are, mind still has them beat. Thoughts have *no* curvature, charge, or chirality. Remember too that quantum tunneling is a metaphor and has nothing to do with any actual relationship between computers and consciousness. In addition, as an ostensibly selected function in nature, quantum tunneling does not satisfy the cybernetic version of an original Darwinian objection: why was subjective awareness deemed necessary for computation? If tunneling is a straightforward issue of the class of functions computable by various devices (where a function is a mapping from a set of numbers to a set of numbers), why does a particular expansion of the class of computable functions have to be accompanied by the computation being conscious? In fact, current theories of quantum computing seem to demonstrate that the two are not linked. Quantum computers *per se* would be no more or less conscious than conventional computers; they would be able only to solve *different* "hard search" problems.[34]

●

Higgs's boson has been dubbed "the God particle" because it ostensibly generates the Higgs field through which other particles pass and whereby all mass in the universe is created. Bosons exist at the mercurial borderline of the atomic realm and an abstract mathematical realm; by definition they are not ordinary objects, even small ones (though, as I write [December 9, 2011], researchers at CERN's Large Hadron Collider outside Geneva, Switzerland, claim to have found their telltale tracks). Bosons obey odd statistical rules such that any number of them can occupy identical places and share quantum states. Carriers of force rather than matter, they have zero or integral spin: i.e., no equivalent to angular momentum in their quantum space. Basically massless themselves, they transform nonlinearly in the context of superconductivity. They are also scalar; that is, have magnitude but no direction and are not changed by coordinate system rotations or by

Lorentz transformations (by conversion into each other through different observational frames of reference or space-time relativistic exchanges). In that sense the scalar properties of bosons might be independent of even their own independence—remember, these are quantum effects rather than objects or even energies. What such independence might amount to is anyone's guess. Since we are talking about the stream of mindedness and its possible sources in nature, sci-fi speculation is par for the course.

During the 1960s physicist Hiroomi Umezawa modeled quanta of long-range coherent waves within and between brain cells, placing elusive memory storage and retrieval in the realm of bosons. His theory was later elaborated into a quantum brain dynamics wherein water molecules (comprising seventy percent of the brain) constitute a quantum cortical field. The quanta in this field (dubbed corticons by him) interact with biomolecules generated in its component neurons and propagated along their synaptic network, compacting in a state that allows long-range correlation among the dipoles.

Instead of each of the particles possessing "a different possibility wave, corresponding to the different physical conditions to which it is exposed…, in certain circumstances, many quantum particles may find themselves moving in concert described by the same possibility wave. Such systems are called *Bose-Einstein condensates,* after Indian physicist Nath Bose and Albert Einstein, who independently predicted which kinds of particles (the so-called bosons) would be susceptible to the formation of such collectively occupied quantum states. If enough particles occupy the same condensate, they can form a kind of giant quantum system with peculiar properties that are observable on the macroscopic scale."[35] Among known examples of this behavior are lasers (photons occupying the same optical state), superconductors (linked electrons taking on identical quantum possibilities), and liquid helium (numerous helium atoms generating a quantum-synchronized superfluid that is friction-free).[36] And these are real things, not models of real things.

In the transfer of this rough paradigm to the brain, biomolecules line up along the actin filaments of the cytoskeleton and generate dipolar oscillations in the form of quantum coherent waves. Consciousness and sense of self are then engendered by interactions between energy quanta of the

cortical field (corticons) and biomolecular waves originating from the neuronal network, particularly the dendrites, their quantum states producing
two complementary representations, one of self and the other of the external world. Consciousness is the recognition by each of the existence of the
other (as per theories of cerebral monitors at a macro level).

I still think this is a lick and a promise, but then don't forget those mu-
nificent quantum-processing algae volunteering to back up all this weird-
ness, well, sort of: "Crucial components for the photosynthetic process are
antenna proteins, which absorb light and transmit the resultant excitation
energy between molecules to a reaction centre.... The water-dwelling algae
are in effect highly miniaturized quantum computers.... They have mas-
tered the process of photosynthesis so well that they can convert sunlight
into electrical energy with near-perfect efficiency.... They do so by having their light-harvesting proteins 'wired' together through a phenomenon
known as quantum coherence, enabling them to transfer energy from one
protein to another with lightning-fast speed and so reduce energy loss
along the energy conversion pathway."[37]

Yes, but are these primordia of internal luminosity or just light?

●

In the 1980s mathematical physicist Roger Penrose, dismissing a computer
model for consciousness in the brain as inadequate, proposed a nonalgorithmic computing process—a function *not* reducible to algorithms, hence
non-computable. "Noncomputable" means that "the performance of the system could not be produced by any algorithmic procedure; more to the point,
it could not be *approximated* by an algorithmic procedure";[38] for instance,
river eddies and planetary rings, complex systems in continuous states, are
only "weakly nonalgorithmic." Consciousness has to exceed that bar.

Penrose cites the very reasoning used to create mathematical systems
as one example of a strong nonalgorithmic constraint. He then aligns this
abstract function with the subatomic noncomputable function based on
the random choice of position that follows the collapse of a quantum wave
into a particle under hypothetical conditions of quantum gravity. He later
refined this waveform into a second, autonomous kind of collapse where
multiple quanta, each with its own tag of spacetime curvature, become

260

unstable when separated by a Planck length of more than 10^{-35} meters (see below) and therefore collapse on their own accord without interaction with the environment. Penrose called this collapse "objective reduction" and, through its mechanism, linked the brain directly to spacetime geometry[39] though he could not assign its effect to any neuronal anatomy—no surprise there.

After reading Penrose's book, research physician Stuart Hameroff decided to fill this gap with the possible computing functions of microtubules and other subneuronal components of the cytoskeletons of brain cells.[40] The main supporting structures of the cytoskeleton, microtubules seem to "have properties which make certain quantum-mechanical phenomena (e.g., super-radiance) possible" and they already "play a key role in neuronal functioning."[41] Together, Hameroff and Penrose ultimately argued that tiny elements in neuronal microtubules are small enough to be essentially quantum, hence also potential sites for conscious entry.

The degree of quantum uncertainty or "unreality" of any object is defined as its "Planck's constant." Named after physicist Max Planck who identified it in 1899, this physical constant is actually a measurement of proportionality relating the energy of a photon to its frequency (or a measurement of the proportionality between the momentum and quantum wavelength of any particle); it reflects the size of energy quanta in quantum space-time. While this number is high for an atom as befits its size and wavelength—in fact, as large as the atom itself—it is very low for any classical object (effectively nil), which leads the ordinary world to stay ordinary despite quantum underpinnings.

Planck's constant is so low in everyday life that quantum effects are far too small to be noticed. They arise only, at least as far as we can tell, in experimental circumstances involving atoms and particles. "For ordinary events, the leeway afforded by Planck's constant of action is insignificant. For all practical purposes, most ordinary events have only one possible outcome. The smallness of Planck's constant explains why Newtonian physics worked so well for such a long time."[42] Until the search for origins got into microscopic properties of matter, reality was indeed very real. However, once it got small enough, it also became unreal or at least something other than real. But then I wonder what physicists expected to find under matter: a gateway into another universe, a microcosmic generator of materiality, a primal creationary field, nothing at all? Either stuff comes apart into something else or it spills all the beans, and we know it doesn't spill the beans because, then, what might they be?

The trouble, once again, with grafting consciousness by quantum wave collapses is that nothing operating within the strata of neurons and brain shows properties at the scale of atomic objects—with high enough values of Planck uncertainty. In order to generate consciousness-like effects Penrose and Hameroff had to get down and dirty at the atomic level. The smaller the particle, the larger the extent of the particle's quantum realm— the higher the uncertainty component, the more thoughtlike. Enter microtubules, the ultimate quantum-neural transistor.

Microtubules have pore diameters of 14 nanometers and are composed of tubulin protein dimer subunits, each with hydrophobic pockets 8 nm. apart, perhaps containing delocalized pi electrons. Smaller nonpolar regions incorporate indole rings, themselves rich in pi electrons separated by an even more minute 2 nm.—close enough, Hameroff believed, to become quantum-entangled: to jump the Newtonian barrier into quantum space. Becoming locked in phase and using bosons to form a Bose-Einstein condensate, their microtubule outputs might extend across the synaptic gap junctions between neurons, transducing quantum activity into macroscopic portions of the brain. The city then transduces itself into the acorn. I know, it is a little too facilely presto-chango, but there is no alternative because there are no intermediate steps anyway. To pose intermediate steps you would (again) have to grok what consciousness *is*.

As waves collapse into particles and actuality collapses probability, the aforementioned non-computable influences flood into the brain from the fundamental geometry of space-time, leading to gamma-wave synchronization (which is also a correlate of consciousness in conventional neuroscience models). Hameroff called his enhanced model "orchestrated objective reduction," and he and Penrose subsequently collaborated on developing it as an intracellular model of consciousness.[43] But it is more an "abstract uncertainty" model than a "consciousness" one.

Other theoreticians followed along similar speculative lines. In 2005 physicist Gustav Benroider described "a connection between nerve functions and consciousness. He hypothesized that quantum coherence could last for a long enough duration in ion channels of the nervous system with adjacent lipids and proteins to allow quantum entanglement. These ion channels control the flow of electrical charges across nerve membranes, facilitating the transmission of neural impulses."[44]

Henry Stapp identified "the migration path of calcium ions from channel to vesicle as the crucial locus of conscious intervention into otherwise classical synaptic activity."[45] Although ordinary synapses are at the classical level, neurotransmitter vesicle releases across synaptic gaps may be affected by tiny electrically activated tunnels—calcium ions channels—putting synapses and their responses to electrical signals just across the quantum-realm threshold as entry points for the quantum basis of consciousness in the human brain. These biological gates would supersede the requirements of tinier, colder, or more superconductive systems, providing instead "warm, wet switches" for quantum effects, in a large, disorderly environment.[46] Each ion instead of selecting one path in the classical sense explores "all possible paths open to it without actually committing itself.... [I]f it is mind that makes the measurement..., then it does not have to exert a force on these ions but merely to make a choice among several simultaneously presented alternatives."[47] Semantic content is still missing, as it lies inside the cognitive operations of the brain itself. Exactly what magical nugget could the ions provide to make the brain conscious, and what too is the prior standing state of the brain's *potential* consciousness?

Even apart from the major dilemma of representation and meaning across a quantum-seepage range, leverage from nano to micro states is a lost cause. You are never going to get atomic effects into neural effects without transferring the uncertainty states of atoms into contemplative, psychosomatic, and behavioral uncertainty states with the same literal content-range—an absurdly tall order. Can a particle wave-collapse really be the basis of a poetic metonymy or an imaginary number? After all, without a real theory of consciousness, anything can pose for a reel or two as a suitable option.

If we specify leverage itself rather than whether anything conscious is being leveraged, there are countless candidates: "In Penrose's model, gravity, despite its intrinsic weakness, is able to influence the quantum realm because, unlike all of its stronger cousins, which act within spacetime, the gravitational force acts directly on the spacetime structure itself, a feature that gives this tiny force immense leverage, perhaps enough to force quantum waves to behave in a nonlinear manner and produce an objective collapse from many states of possibility to one state of actuality."[48] As an uncertainty state becomes an actual state, the neuronal-cerebral field suddenly somehow personifies and privatizes itself: the synaptic flow meets the gravitational space-time flow in a mutual uncertainty wave: pop goes the weasel, and "Joe" appears, ready to ride.

The search for leverage from the quantum to the real is what led Erwin Schrödinger to propose his famous cat in the box—an animal that is either fed or poisoned according to the triggering of a quantum switch, hence is alive and dead at the same time. The uncertainty state of the cat's life, which has quantum subtlety, translates its effect from a realm where Planck's constant is high enough (the switch) to a realm where it is miniscule: the unfortunate feline.[49]

A mind that could control a quantum gate with the output of a probability wave splitter could produce quantum "effects in the material world entirely out of proportion to the tiny range of motions allowed by the Heisenberg uncertainty principle. In the case of Schrödinger's cat, the razor-sharp decision of such a mind amounts to a matter of life or death."[50] In the case of microtubules, it is a matter of being or nothingness.

The trouble (by the way) with Penrose's quantum collapse under quantum gravity is that it would pretty much affect any microorganism. "Every living thing smaller than a flea would enjoy a Schrödinger cat-like existence, living no actual life but only lots of merely possible lives."[51] Dead-alive amoebas and rotifers should be wreaking ontological havoc all over the place. Having done away with all life in a droplet of pond water (and then some), Penrose admitted "that [his] Planck mass [was] embarrassingly large, and he [began to work] on ways to calculate quantum gravity effects that lead to smaller crucial collapse masses."[52]

●

My own inclination at this point would be to give up on quantum-mechanical relationships in neural tissue and go for the greater backdrop:

a conscious universe *is* a quantum universe. Breaking the synaptic threshold down to smaller and smaller triggering components has minimal explanatory value regarding leverage from quantum effects to awareness states, but it suggests something more fundamental—that there is no separate, isolated quantum threshold because everything arises in a universe in which quanta knit creation together. In this game we don't need discrete quantum special effects because *the universe itself is a quantum effect.*

In chaotic situations (those involving nonlinear dynamical systems), huge numbers of natural events are simultaneously interacting where "the magnitude of quantum uncertainty is greater than the 'thread size' of the chaotic filigree."[53] In the cosmos at large, a big chaos field, the continuous "rolling of all the dice" may produce cumulative layers (like consciousness states) obeying quantum rules—famously (and even biblically), "order out of chaos": inexplicably persistent geometries, strange attractors, emergent properties. Chaos theory is, in fact, where consciousness as an emergent property meets consciousness as a quantum effect.

The basis of consciousness may lie among the whole of nature in a meta-gravitational field, its terms of specification so large as to be intangible and, for being intangible, immeasurable. In an arena this vast with a specificity so entangled and nonlocal, "random" and "nonrandom" have no meaning. But it is not, as Einstein feared, God playing dice with the universe; it is a matter of making reality stable. Physicist Heinz Pagels speaks to the matter:

"If you want to build a robust universe, one that will never go wrong, then you don't want to build it like a clock, for the smallest amount of grit could cause it to go awry. However, if things at base are utterly random, nothing can make them more disordered. Complete randomness at the heart of things is the most stable situation imaginable—a divinely clever way to build a sturdy universe."[54]

Once again, the deepest entropy is also the ontological basis of the most intelligent information, as the Algorithm converges with the Archetype.

István Dienes, a more reckless contemporary model-maker, posits a full hiphop "Consciousness Holomatrix":

"A thought for the brain is like a neutrino for the universe.... It is everywhere and nowhere.... We can in principle localise an electron in the brain but we cannot, even in principle, localise a thought...." How then to explain its placement? Well, through quantum activity in the brain, a nonlocal knot is generated—"a defect on the field line which characterizes it as a whole ... a spatially extended object...."[55] This then creates its own internal mechanism of collapse and unfolding:

"Nature obeys mathematical laws, but while for the physical brain these laws are primarily geometrical, both in commutative and noncommutative spaces, for the cognitive brain the underlying mathematical theory is essentially and fundamentally topological.... *[C]onsciousness is a topological effect. The brain decides geometrically; the mind decides topologically....* A (topo)logical excitation emerges as a fundamental quantum of consciousness, forming coherent waves that run through the brain matter.... In this way topology is not a matter of choice but ... fundamental.... A (topo) logical process propagating along a closed information loop (knot) manifests itself as the thought process...."[56] Mathematical properties either generate or literally become mental properties.

Additionally, topological energy is not circumscribed by "finite speed of propagations of interactions. Topological properties are 'tachyonic' and could propagate instantly...."[57] Manifesting everywhere, they join any place to any other place. In such a regime nature could be viewed as integrative, emergent, self-organizing, non-isotropic (anisotropic), and quantum coherent to its core. You wouldn't need seepage, leverage, tunneling, or wave collapses. They're Gothic elaborations and superfluous surrealisms compared to the sheer pristine existentiality of the universe's deep-sunk, in-place crisp consciousness-arising system. It is not only concomitant with reality, concomitant with meaning, concomitant with mathematics and geometry; it is concomitant with itself, every representation of itself, and every continuum that gets wrapped around itself in an attempt to extricate it for examination. Compared to this, other quantum models become mere gizmos or toys.

Dienes takes it even further: "[In] a two-dimensional strip universe with both ends extended to an absolutely remote area ... if someone at infinity twisted and glued the ends of the strip the entire universe would instantly change from orientable to nonorientable…,"[58] reversing the chirality (mirror-image superimposability) of anything passing through it. The topo-brain would then involve Moebius-strip-like transpositions in virtuality or hyperspace: "[T]o describe consciousness one doesn't really need spacetime or, more radically, does not have spacetime anymore, but just a tensor [a multidimensional vector-array] product of two-dimensional topologies, much as with string theory … one does not have a classical spacetime but only the corresponding two-dimensional theory describing the propagation of strings. Worldlines are replaced by worldsheets, the interaction vertices in the Feynman diagrams [momentum-conserving eigenstates of particles colliding with anti-particles] are smoothed out, and spacetime exists only to the extent that it can be extracted from that two-dimensional field that encodes information [the Holomatrix]."[59]

Whether this is quantum space or Escher space or Hilbert space is almost beside the point: it is more as though the geometries that arise from consciousness become the *ex post facto* affidavit of consciousness. The particular advantage of Dienes's theory is that it approaches the way that gravity, space-time, topology, and consciousness actually seem to collaborate with one another. It leaves epiphenomenalism in the dust behind a transdimensional Ferrari. We're not just talking about mere excrescences anymore; we're deep within the gears of reality-generating systems. Yes, they transcend our capacity to describe them. Of course they do. You can't ask a tortoise to explain its shell. Yet their language, by comparison to the language of mechanism (and quantum mechanism), tilts toward what a tortoise might say if he could speak: his shell, then the entire universe.

●

Needless to say, neuroscientific researchers do not take any of this modeling seriously for a bevy of reasons: the lack of observable mechanisms whereby microtubule effects could be translated into neuronal effects and conveyed (signified) then from one neuron to another, e.g., "encode … information derived from sensory structures, process it, and then modify the firing of neurons in such a way as to support the consciousness of the stimulus, and perhaps a purposeful response as well";[60] the lack of mechanisms whereby quantum events could be transmitted even from one tubular structure to another;[61] the lack of mechanisms to deter macro-transmitter molecules from inhibiting the spread of quantum coherence;[62] the impossibility of biological tissue surviving the high temperatures, energy, and degree of hydrolysis required for quantum interactions (protoplasm and quantum entanglement can't share a bedroom at their respective scales); and the lack of experimental evidence that quantum coherence even involves super-radiance in microtubules.[63]

While we're at it, what's the difference finally between a Hameroff microtubule and a Cartesian pineal gland? Both are hypothetical and neither has a suitable mechanism for interiorizing the material world. "How is the microtubule supposed to communicate its stuff to the synapses to have the 'Penrose effect'? What precisely is supposed to be the effect on the neuronal membrane and how is it to be achieved? … The release of neurotransmitter vesicles, for example, does not have any characteristic association with microtubules, so far as is known."[64]

Plus, how can microtubules both conduct and be isolated from the neurotransmitters and neuromodulators that they use to carry sensory signals, yet have permeability and impermeability to the same ions as are necessary to fulfill both quantum and neural conditions?

When the quantum realm is coupled irreversibly with the macroframe, we get mainly mixed metaphors—plus why would quantum states not decohere long before they even reached a spatial or temporal scale applicable to neural processing and thought? There are no conventionally testable empirical predictions and no credible arguments that propose ways in which quantum-micro coupling or interaction might produce consciousness.

New research does suggest tentative pathways around *only* one of the objections; e.g., the stringent temperature and pressure regimes. The scientists who found that "light-absorbing molecules in some photosynthetic proteins capture and transfer energy according to quantum-mechanical probability laws instead of classic laws at temperatures up to 180 degrees Kelvin" concluded, "Where our study breaks ground is that we observe the same quantum coherence at normal room temperature … therefore it is occurring in living algae."[65]

But how does it get out of the grasses into the brain and turn sunlight into cinema there?

Penrose, Hameroff, Umezawa, Stapp, Harris, and Dienes weren't able to push their models any further, though ostensibly (unlike me) they knew what they were trying to push. Yet the riddle of quantum-to-micro/macro flow continues to intrigue physicists and philosophers, so tinkering goes on and the models get more sci-fi and topologically weird. Logically or mathematically possible worlds spawn metaphysical realms and artifacts resembling the chimeras of Argentine fabulist Jorge Luis Borges and the floating cities of Polish painter Jacek Yerka. Then things get even weirder.

Quantum systems are not only indeterminate and probabilistic but nonlocal, instantaneous, and entangled without "anything" actually entangling them. This aspect of the quantum probability wave is unprecedented in physics, where any interaction requires direct contact between systems or their coordination within a local field (gravitational or electromagnetic).

First, two particles interact briefly and in a conventional manner; then they travel outside of each other's range. Yet in quantum theory they remain a twinned single object; they have quantum inseparability. Even when they are no longer interacting, their waves remain intermingled.[66]

Once entangled, two atoms replicate one another to a "t": their possibility waves are fused such that an action affecting one of them affects the other in the same way. Again, this is without any contact or mediation: they are literally "quantum-entangled/phase-entangled possibility waves." Like magical fireflies in an untold dance, they change places instantaneously with each other.

However far they disperse, the connection between the atoms is just as strong. A causal force does not have to propagate or project itself to maintain the entanglement; it is innately nonlocal; thus a billion light years is no less intimate than a hundredth of a millimeter. The link is not merely "faster" than light; it is already always there. Its passage is not even invisible; it just is—psychic birthmarks mirroring each other's patterns billions of light years and galaxies apart. Nor can miles of lead shielding prevent such a connection; no mere imposition can. It is unaffected and unmitigated by shielding because it does not actually *cross* space to get there.

Unlike anything else in physics, quantum inseparability transfers from one entity to another like voodoo or a love charm where the effect of burning a doll or creating an amulet on its target is spontaneous:

"[D]espite physicists' traditional rejection of nonlocal interactions, despite the fact that all known forces are incontestably local, despite Einstein's prohibition against superluminal connections, and despite the fact that no experiment has ever shown a single case of unmediated faster-than-light communication ... the quantum world accomplishes its tasks via real nonlocal communication."[67]

You have to ask, what did shamans 40,000 years ago intuit directly from the cosmos and translate into totemic space?[68]

Unlike Hameroff microtubules or Dienes worldsheets, quantum entanglement is not a proposition or model; it is a signed, sealed, and delivered proof, any refutation of which is *incorrect.* The terms of the experiments backing it (John Steward Bell, 1964; John Clausner/Stuart Freedman, 1970—check them out) involve the equivalent of two Schrödinger dead-and-alive cats "traveling at the speed of light in opposite directions."[69] Locked inside their own individual quantum events, the entangled effects do not translate even to a level of observable patterns *caused by* quantum events: "The quantum connection is inaccessible to humans, a private line open to nature alone.... [N]ature can use these superluminal connections for her own purposes, to knit the universe more closely together than was possible with local Newtonian force fields, but humans cannot decode these superluminal communications, which seem to us to be encrypted in an inscrutable random code to which only nature holds the key."[70] This is beyond *both* mathematics and consciousness, but it points to where consciousness might be consorting with reality without a mathematical chaperone.

Another way of interpreting this indecipherability is that "the appearances apparently do not change at all when the distant context is manipulated—the detectable patterns (but not the individual events that make them up) at each end remain exactly the same...." The superluminal messages can be decoded by changing the context at either point but, once they are, they cannot be decoded at the other point "because one random sequence looks exactly like any other...."[71]

It is as though nothing happened or at least, if anything happened, it can't be observed or contextualized as anything, like a symphony of background noise or listening to the silence on a phone line when the other person refuses to speak. For that reason, "superluminal connections and other phase-entangled systems do not violate Einstein's prohibition against faster-than-light travel or signaling because they never show up in our world, the world of appearance: they are 'merely real.' We know these connections are really there beneath the surface [Bell] ... but we are equally assured ... that we will never ever see [them] directly."[72]

●

When I talk about creating rather than controlling reality in Volumes Two and Three of this book, quantum entanglement fits and acquits like a glove.* Again, it doesn't look like a force or an effect at all. It looks like voodoo, like psychic transmission. We can't send messages between solar systems and galaxies, but *nature* can. We can't travel from Earth to a distant galaxy, but *nature* can. We can't use our minds to influence matter in the dining room, let alone in another solar system, but nature can use the equivalent of quantum mind (entanglement) to touch anything anywhere. There is no separation, in fact—given the entangled history of the universe—between any object or particle and any other. No wonder everything works, everything is conscious (or prone to be turned into consciousness), and consciousness works too. Even the physics of the universe can only create a universe; it cannot control it. So science can't either—and description is a form of control. (Meanwhile we and our conscious state of being may be the universe's best guess as to what it itself is: pure creation *sans* control.)

This is also why the quantum-mechanical basis for synchronicity suggests (scandalously) "that separate minds can link via connections that defy ordinary mechanical explanation."[73] Or as Von Neumann proposed, "our human experience is part of a larger experience enjoyed by similar beings—our other selves—in similar universes quite near by (near by in Hilbert space, that is)."[74]

We are operating in infinite dimensional function spaces as a result of the universe's implicit and fundamental quantum entanglement, which is

---

*I know, O. J.'s *didn't* fit; that was Johnny Cochran's whole point, but at this level, you can be innocent and guilty at the same time and by the same evidence.

so deep and so immediate that we don't notice it at all. We are conscious because quantum entanglement already encompasses us moment to moment at the basis of our minded existence, and forms of its atomic translation that we observe in nature are a mere sputter, a limpid leaf on a vast cosmic tree.

As long as I've gotten out of the neural-quantum box (and the infinite regress of referential monitors that preceded it), I would suggest that "large" and "small" may be totally relative insofar as the universe, originating in a very small (subatomic) space, then expanding to create the largest known space, could be conceived in its entirety as a quantum event in a pocket of something else, perhaps something equally "quantum." Ordinary Newtonian space would then represent the outside of that expansion, whereas consciousness as well as the interpolation of quantum effects into matter would represent its inside. Like the speed of light these quantum effects would be independent of reference frames, distances, elapsed times, and orderings of events. Mind becomes a spontaneous scalar fluctuation of the pocket's quantum vacuum—*cogito ergo sum*—and the same electron that is part of the soil and then a sesame seed ends up distributing poetry upon passing into the brain. Such a multiverse (or Consciousness Holomatrix) is generated through its own complex self-computing lattice and objective reduction to space-time topo-lines. It is the living definition of a robust universe.

●

The world is the summation of all that is accessible via a metaphysical index. Our "mind" is a particular way that a receiver-transmitter works at each instant, and as each kind of thing—a consequence of our special mode of trying to know about things at all. Indexation qua mind also has innate epistemic qualities, as it coalesces by manifesting itself (its own trajectories) through further indices, counterfactual logic sets, and conditional logics, always sending a sign-signal that this, and this, and then this is a part of the structure of the plethora, of which not all parts can be ever "known" at once.

Space-time indices allow minds to function as they do. The "story" that we live is the running time of the indices, composing physicality, mental awareness, narratives, etc. Just by existing we make our macrocosmic being "fit" at fractal scales with a cosmic replica of mind, while cosmic scales then lock-and-groove at sub-nano layers. Reality loops into and out of itself as it "knows" itself, creating its sense of self-identity according to rules, motifs, and dynamics of operating reality here-now.

In fact, here-now means nothing, as it is just an indexical statement of my "perspective," the maximal sum of seen-thought-intuited-felt-dreamed being.[75] Consciousness is not so much *itself* quantum-based; it is more that quantum mechanics is the *index* that reflects the conscious basis of *all* matter. Once again, particles and their quantum relationships come closer to arising from consciousness than consciousness does to arising as an epiphenomenon of matter in a classical sense. In fact, both could be seen as simultaneously arising states epiphenomenal to a deeper "real," with the quantum layer representing its transitional zone indexing mind and matter to each other.

At precisely the point at which mind hits its own absolute boundaries within the index, speculative science churns up endless paradoxical propositions because really that's all that could be there: dark matter, event horizons, nano-stringlike forms; a fictive early universe in which the dimension of time is mathematically distorted into one of space so that all four dimensions function like space; another whole universe in "a superposition of many different states, in which the constants of nature like quark masses take different values…" and in which "even a cat can be in a superposition of states, in some of which it is alive and in others dead"; eleven space-time dimensions with "different sets of particles or strings or membranes in spacetimes of various different dimensionalities"; quantum particles entangled such that, if they were dice, each pair would give the same roll light years apart in time or space; or a "universe [that] does not have just a single history but every possible history."[76]

In that vein, Stanford neuroscientist Karl Pribram proposed that "all sensory perceptions are recorded over the entire cerebral cortex of the brain in the form of interference patterns that encode and store the impressions in a holographic manner."[77] His colleague David Bohm explained the functioning of the brain as a hologram in accordance with the waveforms of quantum mathematical functions (a so-called holonomic brain).

Contradictions between quantum and relativity theories indicate a deeper truth, an implicate order from which the universe arose as an explicate order. Reality becomes a "fluctuating interface between two complementary versions of itself,"[78] with the phenomena of classical physics being its explicate order. Mind and matter are *both* projections of implicate into explicate order. Matter in explicate space explains nothing about itself *or* consciousness, but its expressions and anomalies are ontologically equivalent to those of consciousness and point back to the implicate order.[79]

An archetypal web of information "extends over the whole universe and over the whole past, with implications of the whole future."[80] Laws regarding matter arise "between the enfolded structures that interweave and interpenetrate each other, throughout the whole of space, rather than between the abstracted and separated forms that are manifest to the senses (and to our instruments.)"[81]

●

I think that all these lines of modeling are individually important, not because they are "correct" but because they offload consciousness onto systems deep enough to at least bear its immaterial "weight." Each establishes a parallel between two ranges, even if it is not the real ranges or a true parallel. Whitehead laid this one on the line too: "Nature is a structure of evolving processes. The reality is the process."[82]

Discovering whether any such model is valid is all but impossible, even given the technological breakthroughs in digital memory, nanotechnology, crystal structure, and genome mapping. Yet applicability is irrelevant to the larger issue because—and this may come as a surprise if you have not been tracking this feature—the quantum "object" or continuum of objects and emergent field effects including the quantum brain does not open meaningfully *different* ontological territory from the materialists' behavioral and mechanical version of consciousness. It doesn't rescue qualia or banish zombies; it merely runs them as quantum effects and possibility waves at a nano-level.

The stream of consciousness remains an epiphenomenon, albeit a folded-in-upon-itself quantum one—it is now a hallucination of quantized inputs from both internal and external sources set in continuous cognitive streams. It still cannot survive its own illusory basis or provide actual meaning, though it fulfills Whitehead's prerequisite of novelty while increasing the subtlety factor of creation. Until a quantum version of mind enters into the universe through the actual dispersion of thought particles and their superpositions by the Big Bang, it doesn't exist and then, until creatures evolve neural networks on worlds, it is itself *only* a superposed system, like Schrödinger's cat, both dead and alive—in limbo, half-real. The universe has to become *conscious* to become *real*.

Chalmers concludes similarly that, after all the fuss, "quantum theories of consciousness suffer from the same difficulties as neural or computational theories. Quantum phenomena have some remarkable functional properties, such as nondeterminism and nonlocality. It is natural to speculate that these properties may play some role in the explanation of cognitive functions such as random choice and the integration of information, and this hypothesis cannot be ruled out *a priori*. But when it comes to the explanation of experience, quantum processes are in the same boat as any other. The question of why these properties give rise to experience is unanswered."[83] Exactly! When quantum hide-and-go-seek is over, no one has really found consciousness or the acorn into which a whole city can be inverted. Psychiatrist Ronald Milestone raises concurrent philosophical and psycholinguistic issues:

> The paradoxical view that the neural theory cannot explain consciousness while the brain generates it has superficial similarities to quantum theory: energy can be both a wave and a particle. The exact momentum (mass X velocity) of a subatomic particle cannot only be known to within a certain measure (Planck's constant), etc. At the subatomic level, theory that applies at the ordinary level of experience does not apply. New rules must be developed to describe the behavior of entities at this ultra small size. A similar observation is made about the study of consciousness: observations made at the research neurophysiology level do not seem to be able to correlate with the complex details of subjective experience. This had led some theoreticians to propose that the creation

of consciousness in the brain is some type of "quantum effect." This statement is not incorrect, but rather meaningless. Quantum theory is a specific theoretical language for predicting the behavior of observable subatomic particles using a complex mathematical system involving probability theory. *Subjective states, or the verbalization of subjective states, is NOT an output of any quantum theory or calculation!* So there is no way that quantum theory could apply to the "hard problem." Perhaps someone could develop a theory in which "wave equation" calculations on submolecular events in the brain resulted in predictable subjective state reports. No one has, and I would argue that no one will for reasons [of] incompatible language systems. Current discussions of quantum effects are based on poor knowledge of the brain and the attribution of special properties to microtubules. A mystery + a mystery does not add up to a solution![84]

If you're going to look for "mind" in atoms or particles, you have to look for something other than microtubules, tunnels, uncertainty states, and quantum-entangled properties; you have to look for the basis of semantically bound interiorly arising luminosity, for subjective thoughts with semantic content—and you're not going to find that. There is a gap between observed quantum uncertainty, which does not exist as an ordinary thing, and a consciousness-based uncertainty principle, which is a cognitive event, quantal in a different way, plus there is even another gap: between the observation of an electron and the ontological basis of an observation or thought. There is no justification for assigning the two to the same rubric.

Quantum phenomena do not have consciousness in the usual sense because they are not conscious *of* something. It's as important not to reify them gratuitously as it is not to reify consciousness. The "inexplicability" of both mental and quantum phenomena certainly brings the two together in one sense but not necessarily in a sense in which one can subsume or connect each to the other by the same rubric.[85] No matter how small you make the mind-body gap, it still is a gap. The problem is not quantum properties *per se* and their applicability to thought; the problem is the gap between the mind and *any* physical properties of material systems. Nick Herbert gets right to the point:

> What "reality" means in this case is any conceivable mecha-
> nism behind the quantum facts that can reproduce those facts.
> Quantum theory gives us a way to calculate those facts but
> offers no models or mechanisms as to how those facts are ac-
> complished. I say in *Elemental Mind* that we have no models
> of mind, not even bad ones, yet I describe several including
> Culbertson's and Walker's—and as you know Hameroff/Pen-
> rose and Henry Stapp also have provided models. But none of
> these models work. They are all shots in the dark and make no
> contact with the phenomena of mentation.
>
> In the same way Wilhelm Reich's orgone accumulator was
> also a shot in the dark. Like these tentative models of mind,
> Reich's machine was a parody, a prop, a foreboding of some
> future machine for amplifying/modifying subtle energies.
> Quantum tantra is a thrust in the same direction but it's so
> far also a futile gesture along with all the other physics-based
> attempts to connect with our favorite topic. I would not take
> these models very seriously—they are mere Stone-Age lunges—
> longings for a science that does not yet exist.[86]

So where else do the hints of a shadow of that science roam today? Try checking out the sources behind my Cosmic Eternity System in Volume Three. Not that they're it. They're parodies and forebodings too but of a different sort, a kind of extension of the Dienes Holomatrix into alternate dimensions and other planes of consciousness. One way or another we are headed toward life and death: where we come from, where we are going, and how we deal with our systemic mortality. Rick Grush and Patricia Smith Churchland, staunchly materialistic philosophers, pose this matter elegantly while going for its jugular:

> Despite the rather breathtaking flimsiness of the conscious-
> ness-quantum connection, the idea has enjoyed a surprisingly
> warm reception, at least outside neuroscience. One cannot

help groping about for some explanation for this rather odd fact. Is it not even *more* reductionist than explaining consciousness in terms of the properties of networks of neurons? Emotionally, it seems, the two reductionist strategies arouse quite different feelings. After some interviewing, in an admittedly haphazard fashion, we found the following story gathering credence.

Some people who, intellectually, are materialists nevertheless have strong dualist hankerings—especially hankerings about life after death. They have a negative "gut" reaction to the idea that neurons—cells that you can see under a microscope and probe with electrodes, brains that you can hold in one hand and that rapidly rot without oxygen supply—are the source of subjectivity and the "me-ness of me." The crucial feature of neurons that makes them capable of processing and storing information is just ions passing back and forth across neuronal membranes through protein channels. That seems, stacked against the "me-ness of me," to be disappointingly humdrum—even if there are lots of ions and lots of neurons and lots of really complicated protein channels.*

Quantum physics, on the other hand, seems more resonant with those residual dualist hankerings, perhaps by holding out the possibility that scientific realism and objectivity melt away in that domain, or even that thoughts and feelings are, in the end, the fundamental properties of the universe. Explanation of something as special as what makes me *me* should really involve, the feeling is, something more "deep" and mysterious and "other worldly" than mere neurons. Perhaps what is comforting about quantum physics is that it can be invoked to "explain" a mysterious phenomenon without removing much of the mystery, quantum-physical explanations being highly mysterious themselves....

---

*More deference to ping-pong balls in case you hadn't noticed.

> [But] why should it be less scary … degrading … reductionist or counter-intuitive that "me-ness" emerges from the collapse of a wave function than from neuronal activity?[87]

Quantum stuff does seem more ontologically promising than hard drives made of meat, but I don't think that they are in essential opposition. You need quantum uncertainty and entanglement, and you also need that folded-over meat object arising molecularly from an egg with a billion-year history of embryogenically induced layers. They are complementary realities or one reality at two different transparencies. No way to leave either of them out of "it is what it is."

But to hope to surf a quantum wave to personal immortality is a sci-fi trope at best—about on the level of trying to survive transit through a black hole to bubble yourself into another universe. There are better and less exorbitant ways to manifest a quantum-entangled universe inside a classical mechanical one inside a shamanic one. There are paths that are also safely beyond trenchant materialisms. A shaman doesn't need quantum surfboards or black holes to get from embodied "here" to transphysical "elsewhere." He goes by *naguals* and other empowered thoughtforms.

Grush and Churchland clearly don't buy thoughtforms and feelings as "fundamental properties of the universe," but I am not even sure they speak their own entire truth (their "tongues" are "in cheek" at more than one level of unconscious irony, a point that I will come back to in the last refrain of this book).

As our cosmic witness has become a full-time censor, we are cynical about *everything* now: everything in the universe and the meaning of the universe. Counter their cutesy argument, counter the iron-clad position they forge, the quantum brain *is* a powerful metaphor, an alternative universe grounded in a patina of science, a worthy stand-in for the wonders of personal identity in the belly of a beast that requires a physical atlas—a superpositional, every-which-way atlas if need be—for anything to have standing or legitimacy, even to itself.

What I'd say to Grush and Churchland is that the quantum realm puts a ripple into the universe, which allows our existential faith to make a bashful appearance and post alongside our existential despair. Consciousness actualizes quantum possibilities, so we don't need quantum reactions in brain tissue *à la* Penrose and Hameroff to explain it in quantum terms: ordinary reality is extricated from its probabilistic state by sheer observation and immersion—and not just ours. I think that reductionists believe that too—how could they not?—but in a way that they express only by its antithesis. Grief and joy exist absolutely in the universe, whether they are "real" or not; I would bet the house on the fact that Grush and Churchland know that too. Then there are assorted musicians, kachinas, hackers, break-dancers, poetry-slammers, shortstops, *capoeiristas,* and celebrity chefs for whom it's no problem.

The stuff of song, chant, prayer, abstract mathematics, mantra, and hiphop contains intrinsic disjunctions that explode into realities like uncertainty states and wave functions. The quantum world corresponds to the archetypal world of synchronicities, and "the archetypal world of formless possibilities corresponds to the shamanic underworld, the Tao, heaven, the dreamtime, all the adjacent realms of all schools of thought that address, however incompletely or metaphorically, the characteristics of that universe of universes."[88]

But that does not mean that they all represent *the same uncertainty states and waves.* My own provisional preference is that the quantum particle, the quantum brain, and the quantum metaphor function as spin-offs of the quantum intelligence of the universe, rather than that a quantum mind maps onto a quantum brain that generates, like a 24/7 slot machine, a quantum waveform of consciousness.

This is not an easy world to inhabit under the best of circumstances. Even a brilliant modeler of the quantum brain (Bohm) suffered from such extreme depression that he willingly underwent electroconvulsive therapy late in life, a seeming contradiction of his own paradigm, as he was desperate enough to think he had no other option. That's how deep the cynicism and doubt of modernity go. The master of the holonomic brain submitted to the concretization of his dark moods as if they were erasable banks of neurons. But, at some point in this madness, don't we all?

On the other side of the scale, consciousness, whether our own or some alien's or a universal superconsciousness, creates *both* the superposed, quantum-entangled realm of atoms and particles *and* the classic realm of macroscopic and microscopically small objects, entangling the quantum sphere inside—far, far inside itself: *"far away by the rules of seafaring.../far by the rule of its parts/by the law of the proportion of its parts...."*[89] At the same time, it is pulling physical reality out of the yarns of quanta into a classical state by consciousness alone, its oscillation at an innate frequency in the universe at large: *"over the World/over the City/over Man."*[90]

We don't question the Sun or our own existence. Yet something roots us into the Great Rolling Panorama beyond both the mesmer of bosons and the synapsing of neurons. Finally "the whole notion of quantum consciousness is based on the hope that elementary quantum events are not really random but represent the coded carrier medium for some mind ... the external signs of some hidden inner experience—your experience...."[91]

Meanwhile, tell me that all that music—Bach's organs, the didgeridoos, the horns, Peruvian and Romanian panpipes, Pachebel's "Canon," Bob Marley and the Wailers, "Lady Radnor's Suite," Buddy Holly and the Crickets, Steve Roach's "Structures From Silence"—isn't coming from somewhere.

CHAPTER NINE

The Ontology and Cosmology of Consciousness

So no, consciousness is not obvious. We do not know what it is, how it arises, and where, if anywhere, in the physical world it resides or what boundaries, if any, it respects. We do not begin to understand its relationship to matter, a domain inside of which it flat-out doesn't belong (however much gerrymandering is done to shoehorn it in). We don't know what happens to our own "beingness" when we die, which is a big deal, as big a deal as deals get. Most humans want to know where they go and whether they go anywhere or are just swiped. Where will we be in ten thousand years? Where were we ten thousand years ago?

For now an eerie, improbable glow illuminates the darkness, begetting existence. But "just what sort of thing is this 'light' supposed to be? What possibly could be its source…?"[1] "It has seemed to many that with consciousness…, a wholly new thing enters the universe."[2]

Is there an underlying energy or agency that "finds" mind even as mind finds it? Or did an originless epistemology awaken an equally startled universe to itself?

Does consciousness emerge deep enough in the cosmos that seemingly opposite explanations—transpersonal archetypes and molecular dynamical systems—coincide?

●

Moment to moment we continue to take introspective awareness for granted, at face value, as an ordinary, humdrum thing, which it is not. It only seems so because there is no way either to look at it objectively or get out of it, no bead on it at all. That eerie glow is always there, and nothing else is there. "[E]ven where consciousness is not mysterious it is mysterious why it is not mysterious."[3] How could something so intimate and seamless also be so intangible and ineffable? How could a radical, renegade property bust into the physical universe? How could something that is nothing be everything?

"No property we ascribe to the brain on the basis of how it strikes us perceptually … seems capable of rendering perspicuous how it is that damp gray tissue can be the crucible from which consciousness arises fully formed. That is why the feeling is so strong in us that there has to be something *magical* about the brain-mind relation…."[4]

But does that mean that it *has to be magical* or (more to the point) that its magical aspect *has to be delusional too?*

To arrive at a provisional ontology of consciousness, we either have to extrapolate from latent properties of cellular networks or, failing that, look outside them for an innate sentient quality. If that quality should arise extrinsic to entropy, gravity, and thermodynamics, then consciousness has opened a dual ontology of the universe. And not from someplace exotic either—right at the neighborhood store and on the local ballfield.

You continue to assert that consciousness is an *emergent* property of life—of neurons. But where do emergent properties come from if not from *intrinsic* properties that were already present in some form in the first place? How can something absolutely novel emerge in something, by all accounts, utterly barren of it? If consciousness were not *in* objects to begin with, how could those objects install or brew ontological states by mere enhancement, augmentation, or rearrangement of their molecules and ping-pong balls? "How could possession of a meta-state confer subjectivity or feeling on a lower-level state that did not otherwise possess it? Why would being an intentional object or referent of a meta-state confer consciousness on a first-order state? A rock does not become conscious when someone has a belief about it. Why should a first-order psychical state become conscious simply by having a belief about it?"[5]

Blake's Tiger calls out to us from the eighteenth century: *"In what distant deeps or skies/Burnt the fire of thine eyes…./And when thy heart began to beat/What dread hand? & what dread feet…?/What the hammer? What the chain?/In what furnace was thy brain?"*[6] We have not advanced one iota on this riddle; in fact, we have slid backwards into the mechanism of hammer and chain.

Isn't our occasion bizarre and preposterous, mate? Matriculation and chrysalis inside an egg—an egg, for crissakes!—inside (another) sista's body, donning a cloak of mincemeat through which a mind, *your* mind, glows as a property of the very frock that you have borrowed to become yourself—and then you look out from its taut rags, rags all the same, at the whole of creation in an *Imago Mundi*—as *someone?*

You didn't arrive honorifically by chariot or UFO. A bloke and a babe copulated and nine or so months later a miniature replica of them appeared with full capacity for independent existence and awareness. How do you explain—knock on wood and pinch yourself!—this behest of cosmic view and physical emanation?

I think that any arrangement whereby an intelligent identity gets into any world, as long as it involves entering time from outside, is strange and mysterious—especially when each entity hits town without any notion of where it came from, what it is, or what is to become of it. Layers of tissue invaginating around a central axis and then folding over each other while kindling organs of absorption and suction, a neural ladder, and ganglia are as likely (or unlikely) as anything else. The universe's method of writing itself on its own body makes a rugged kind of, literal sense. Where else might you come from? To get here—or anywhere—awake and alert? How else might anything gain purchase on a cosmic view? There are no portals or doorways in infinity, no breaks in the chain of meaning except by rupture of birth or death.

You must understand a thing about us that is all over the place but rarely spoken of: consciousness is dangling somewhere that is nowhere—it doesn't come from anywhere else and doesn't lead anywhere else—and *we* are that dangling thing.

If consciousness is not an integral and absolute panel in the universe with an intrinsic meaning and discrete cause, then (as noted many times throughout this folio) it is an artificially fabricated epiphenomenon that invents its pseudo-meaning on the fly by falling for its own chemical hallucination. Either way, we construct our own meanings, but in the second instance they aren't meanings; they are straw houses in another wind.

●

The term "panpsychism" has been used historically to denote an innate sphere of consciousness, a fundamental property of creation commensurate or senior to substance "on a par with such irreducible phenomena as gravity, light, mass, and electrical charge,"[7] as primal as heat, and "occupy[ing] as fundamental a place ... as matter, energy, space, time and numbers."[8] According to panpsychism, mind and matter are arranged in nature in precisely the reverse order to that which materialism has assigned them: mind is not the cart but the horse—"the fundamental substance of the world"[9] inculcated into nature prior to cells or DNA, before even atoms or quanta. Mind does not in fact exist in or arise out of the physical world, the physical world exists in and arises out of mind.[10] The starry firmament itself began as mind—not mind in our egoic sense but mind as particles with interior light or the innate potential for self-illumination.

"Particle" is the wrong word because it is a "matter" noun. Let's say that "intelligencoids" gave rise to quarks and strings, electrons and molecules, and these forms remain inside matter, seeping through, for instance, in the visions of the *ayahuasquero* but also in the glow and pungency of daily life. "[E]verything is made of mind, [and] matter consists of fleeting vibratory patterns in some vast field of consciousness."[11] In the Star Card of the tarot, an androgynous angel spills vibrations out of a jug, separate rivulets that trickle across rocky terrain—an ibis that is not an ibis looks on.

Under panpsychistic declaration, mental properties cannot ever be reduced to physical properties and there are no emergent phenomena either (there don't have to be): "the basic physical constituents of the universe have mental properties, whether or not they are parts of living organisms."[12] In dualism nonemergence is, in effect, "emergence," as the dual nature of "matter" is its concomitant status as mind. The molecules of the most inert, lumpish doorknob or piece of sod or random particle are minded too, each imbued by its primordial essence. Every speck of dust is an incipient Buddha. That's been said on more than one occasion.

Then why aren't common rocks or for that matter disk operating systems (DOS), which are meticulously arranged in analytic and memory grids, *more* intelligent and subjectively conscious? Some idealists aver that they are, in a latent or dormant fashion. Others hold that their intelligence is implicit but too crude and diffuse to manifest: "Although all matter is sentient to some degree, most of this awareness is of very low quality and is not functionally coupled into matter's behavior in any important way."[13] In the brain, however, nature has concentrated its sentience in high-carat ore: mind may be a near product of the cortex, but rich congeries of intelligencoids (corticons) are the prior components of neural tissue. Matter is *always* intrinsically minded.

Another way of looking at "intelligencoids" is not so much as a psychic flow through nature (like electromagnetism or gravity) but as a whole other reality—a counterweight, a balancing panel and mirroring to the cosmos, which interpenetrates matter and brings intelligence to bear at certain vibrations. Though it follows thermodynamic and gravitational laws while still endowing matter with mental properties, it cannot be detected by matter-based devices that measure only material oscillations.

Animistic cosmologies originated and prevailed mostly in pre-literate societies before a formal distinction was made between mind and matter. Animists do not think of the two modes of nature as separate systems at any level: everything is endowed with both, everything is alive, and everything has a spirit or soul. The animist "is willing to grant the gift of inner life not only to humans and so-called higher animals but to every arrangement of matter in the known universe ... trees, rocks, stars, and spiral nebulas."[14] It would not occur to the Stone Age hunter to consider any morsel of matter void of vital essence or to make a distinction between "alive" and "dead" for anything in the universe.

Consciousness in its collectivity may be no less than a replica universe, an equally expanding holographic monogram of the *actual* Big Bang. Ordinary consciousness is then neither matter's by-product nor its source but its "other" side.

●

Not long ago I met Christian de Quincey, a philosopher (by trade) who gave the title *Consciousness from Zombies to Angels* to a recent book. At a dinner gathering he recalled how he had gotten hooked on the topic in his native Ireland. He was seven years old when he pulled down a family encyclopedia and opened by chance to a page on dinosaurs. After staring at the pictures and considering their implications, he asked his parents key questions. Their answers (and those subsequently of his teachers) brought him to the realization that humans evolved from balder or more blunderous creatures, all the way back to tiny bacteria—an apparently endorsed and agreed-upon account that struck him as more than a bit odd.

Christian wasn't so much curious about palaeontological odds and ends as he was about wanting to get at the special thing he had inside him that made him real to himself.

First he went eyeball to eyeball with the household's cherished pet. Did the dog have the "thing"? 'Yes,' he concluded. Then he stared closely at the family fish and, after a consideration, decided that they had "it" too. His next candidate, a worm in the garden, after a longer *tête-à-tête,* offered an inconclusive shrug.

This childhood quest guided his life thereafter. Christian first studied biology but quickly recognized that the answer wasn't there. A creature's self-knowing was not of the slightest interest to scientists, only its chemistry and behavior. Next he tried psychology, but its modus was behavioral, concerned only with the *mechanism* of consciousness, not consciousness itself.

Initially he was delighted by the Western mode of philosophical inquiry, "like a fine meal," he told us, "a banquet, at which I ate for a long time—but left always hungry."

He moved on to the East, and there he discovered what he had been looking for: mind in spades. But it was not entirely satisfying because he also wanted to understand the trajectory of consciousness, a curiosity which began for him with giant lizards—how cells at the status of bacteria could have matriculated into capable people. This was *not* the stuff of the Buddha.

And there was still the matter of whether that garden worm had "it"!

Christian ultimately concluded that it did, but so did the soil; his verdict was that consciousness and matter are coeval: two faces of the same coin. Everything has intrinsic consciousness, but everything is also material.

Physicist Nick Herbert calls this mode of dualism "interactionalism": "The world of mind needs matter as a relatively stable medium in which

287

to express itself, and the material world needs mind to make its existence 'meaningful.'"[15] The Big Bang (or its mirage) inaugurated what looks to us from inside its popped corn like an implosion but is actually a transitional phase of "interpenetrating mental and physical worlds."[16]

Functionalists dismiss all such metaphysics, without even taking them seriously, in "no biggie" fashion: 'So we're conscious and proud of it, so what!' Maybe consciousness appears to arise *sui generis* but that's only because we can't get at the mirage's mechanism, which has to be the same as all other mechanisms under heaven and earth: animal, vegetable, mineral. We don't have to write a recipe for consciousness for it to be organic chemistry down to the last grinds. Nature has always had one answer, and it's not metaphysical. There is much else in nature that doesn't declare its parentage or cavalierly tip its hand. Who are we to tell nature how to run its own show? We're the grunts.

We may be trying to extract psychological properties from an incomplete knowledge of physical objects, or trying to get normative concepts from mere descriptors. We may be projecting how we think the brain should operate onto what we even think it is. If we knew what the brain really was, we would identify its role in consciousness differently.

Philosopher Alvin Goldman further derogates the idea of the "possession of a meta-state conferring subjectivity on a lower-level state that did not otherwise possess it":[17] a refutation of both panpsychism and the "beingness buzz" that consciousness gives its bearers. He is okay with pure meta-state emergence (properties *ex nihilo*) but not with meta-state nonemergence (properties from archetypal intelligence). Then he parses an elusive but critical point: "It would be completely unexpected if all the causal antecedents of conscious mental processes were themselves conscious. In other words, conscious mental processes emerge out of the neural processes that give rise to them. It would be absurd to expect these emergent conscious neural processes to precede the neural processes they arise from...."[18] Which is of course what panpsychism advocates do. For Goldman, emergence is literally the creation *ex nihilo* (or at least from entropy and chaos) of utterly new qualities and states. Neurobiologist Owen Flanagan chimes in:

"The suspicious phenomenalist must explain how such consciousness can be realized in myriad brain structures it interacts with for subsequent information processing. The brain hardly seems like a system that doesn't listen to itself."[19]

He is saying, I believe, that just because the brain itself is not exuding consciousness-like dialogue bubbles or shooting out cartoon tentacles and does not possess elements that look like the operations of thought-waves does not mean that it is not the sole origin and cause of consciousness or that consciousness has been introduced to the biological platform from a nonmaterial source—it does not mean that the "mind" phenomenon doesn't have solely concrete antecedents. It means only that we do not presently recognize them or their mechanism. Flanagan is asking how a state of information could *otherwise* get realized in the chambers of an indubitably physical brain without an explicit physical cause.

For functionalists consciousness exists in the universe not because of its intrinsic nature but because of the random mutational activity of mundane "vanilla" molecules operating inside the biosphere's machinery. Biology arises from atoms fabricated in stars, organized in molecular fields on planets—catalytic displacements from equilibrium generating never-before-in-the-universe autonomous agents, temporarily irreversible phenomena. Things don't just drop into the universe from outside and then thermodynamic forces and atoms take over the heavy lifting—not unless God is in the house.

It is a fallacy of misplaced concreteness even to hunt for a material basis for consciousness in order to prove its rootedness in matter. The *prima facie* material basis of everything in the universe should be sufficient evidence. Consciousness, magical as it seems, doesn't even need a metaphysical sponsor. Wood is not fire, gasoline is not horsepower, and electricity is not television, and none of the former, in each instance, resembles the latter, materialists are fond of pointing out, but they nonetheless directly convert into it.

There had to be some sort of universe, so it might as well be an emergent one. The operation of the universe doesn't have to be explicable or even pertinent; it just has to be *something*.

Subjective consciousness cannot be predicted in advance. The universe is an unimaginable universe. Think about that. But it goes even further.

Astonishingly, relations between objects and properties transcend their material qualities such that "any causal system whose input states, output states and other internal states bear the right relation to one another has the psychological states in question, regardless of whether its physical make-up consists in neural networks, computer chips or beer cans and bits of string."[20] Those damnable ping-pong balls again! Juggle them long enough and they stir into alive holograms. Emergence is emergence is emergence and then some and does not need to justify its occurrence or explain its route and repertoire. After all, you don't need to confer properties of sorcery on molecules to make hills and dales and caves full of quartz crystals. As noted earlier, phenomenal representation with its "high degree of semantic transparency ... may have been 'recruited' for [its own very] purpose."[21]

But by whom? The Darwinian supervisory panel? Quantum relations among subatomic particles? Chaos theory? Entropy's antientropic side?

And what if the brain is not the source of the voice to which it is listening (or not)—is not where the "buck" stops? What if the relationship between mindedness and the cerebral cortex is more complicated (and more than just more complicated) than neural webs, hierarchies of monitors, innate semantic transparencies, and feedback loops?

When you come down to it, consciousness is not a single voice anyway but multiple tracks that talk to one another such that each has the capacity and free will to craft credible personae and logical arguments, and to experience viewpoints and ambivalences of differing perspectives. What sort of internal representation is going on inside consciousness as part of its very fabric that makes it referentially variant (and variable) and not only creates a blanket of referentiality but discriminates separate within singular views? Do we just throw it all at the behest of those trillion trillion quantum switches, or is that a lame materialist copout? How do we get private integrities out of low-grade panoplies of switches? How do we cross the *actual* gap between agency (which brains pretty much handle) and meaning, which requires an epistemological conversion?

Maybe "mind" is not really "inside" the neurological stream but only facilitated or accessed by it. This would allow for the evolution or emergence of cognitive, linguistic processes, yet simultaneously make those processes the "receptor" channels for currents already circulating in the universe (noosphere-like) that then get localized. So comprehensibility-intelligibility is generated, constructed (socioculturally) and encoded-decoded-communicated as part of a greater telemetric extended structure.[22] The brain would then be a solid-state semi-conductor that is not the source of its own signal. This would satisfy both dualism and panpsychism.

Consciousness could be a field effect independent of networks of ganglia or even of matter. The brain might provide an antenna into a larger "consciousness" field. Our access to mindedness via ganglia could be in the form of sulci folding into an inverted tuning portal, a station for patching into a Cosmic Internet through which information is already flowing—not languaged information, not even anything cognitive, but a nonlocal current containing the rudiments of mentation and meaning.

As noted in Chapter Three, the brain binds a pattern of synchronous oscillations within the 40 to 70 Hertz range. No one knows why that frequency is selected and not others, whether it is random or specified. Reality is unlikely to be a panpsychic podcast at a mandatory wavelength.[23] But if it were, what would be the broadcast's source?

Perhaps what neuroscientists will one day arrive at is a nonlinear matrix within the brain—a combination of lobes and sulci—that contains or consolidates a tuning capacity. Or the tuner might be a series of cellular and subcellular relationships impenetrable to three-dimensional detection. Entry to its field effect could arise from an arrangement of cytoplasmic confederacies. Or each cell (or organelle/microtubule) in the brain may itself be a tiny tuner—together they pick up the broadcast (or enough to index each species distinctively to it according to its body-mind field).

Such hardware could still have evolved in Darwinian fashion—whorls of sulci and cerebral structures "naturally selected" but instead of as matrices for phenomenal intelligence *per se,* for adaptation to meta-phenomenal nonlocal signals, subtending a preexisting consciousness field as it were from beneath or within, like some sort of four- or five-dimensional turnip enveloping a specialized tuner. The flow of incipient conscious energy through the brain is what helps to mold the brain into a receiver of that energy—Darwin and Lamarck in collaboration.

Late nineteenth-century chemist and physicist "William Crookes believed that the synapses between nerve cells behaved in a manner analogous to a radio coherer ... picking up mental broadcasts from some etheric Elsewhere.... His model of brain-as-mental-radio not only suggests a mechanism for ordinary consciousness but could also be used to explain the phenomenon of disincarnate entities who speak their minds through human channels [see below]: under the right circumstances the human brain's biological coherers might be capable of picking up more than one 'station.'"[24] Nick Herbert adds this surprise to the quantum-brain discussion:

"Most quantum models of consciousness are similar to Crookes's coherer proposal in that they consider the synapses to be a sensitive receiver of mental messages that originate outside the brain. The main difference between the coherer model of mind and quantum consciousness models is that the quantum psychologists assert that mind is somehow resident in Heisenberg's quantum potentia rather than in the electromagnetic ether."[25]

Even without such a model, there is no reason to assume that the brain did not evolve in response to the *prior* existence of consciousness, personal as well as general, and that its relation to consciousness is not as genatrix or cause but as attunement to biological networks (in general) and the phase states of individual animals (in particular).

●

It comes down to this too: for all their biotechnological skills and staunch beliefs about the molecular basis and neural location of consciousness, scientists cannot reproduce its system (whether they *should* be able to or not is a separate matter). They cannot manufacture introspective mind out of molecules or fibrils, even highly refined and customized ones—and they have had no lack of trying. Only a fertilized egg can do it, can create that proper relation of input and output states from scratch, can put billions of virginal filaments, strands, synapses of pure-carat, *bona fide* avoirdupois ballast in the proper referential order within a spiraling, thickening, self-generating mass-construct, and then light it up from inside and give it its own "self." Only an egg (or the clone of an already conscious organism). And an egg is about as dumb as Humpty gets, cerebrally anyway. It doesn't have a neuron, let alone a brain.

Mind is so vast, abstruse, and nonlinear that we can't locate it; there is no obvious command control unit. Yet nature just goes out and mass-produces meta-machines that are fundamentally and spectacularly incommensurate with the most advanced mechanical devices—though, in principle, both operate transparently as well as thermodynamically under the same regime. Each is molecular, though they have very different raw-material recipes and industrial protocols. Each has quantifiable internal states, semi-closed circuitry, and an ascertainable inventory of moving parts accessible to microscopy. Each is subject to naked inspection and enforcement of codes. No phantom switches, intangible generators and currents, hidden sources, or forever-invisible threads and networks should be needed to run either factory.

Yet a living machine has properties of an entirely different entelechy. A biont is naturally alive and awake while a zombie engine is inanimate and void—indifferent. No one knows why or how cell matrices self-form and then commandeer squadrons of molecules, metabolize them, and get that charming purple glow inside themselves. A scientific theory of consciousness should be able at least to give us a rough blueprint for making it as well as a tentative map of where its major centers are located and how they are networked to one another.[26]

Suppose, as a thought experiment, you were to formulate a perfect, mess-free pet, a robotic dog complete with artificial fur, plastic fangs, and "a computer program that produces complex interactive behavior indistinguishable from the way that a real dog would behave."[27] No need for dog food or neutering. But then your fastidious customer asks you to improve on it and give it actual feelings and make it really love her—where would you look to alter its circuitry? What would be your first conceptual step toward executing her requested renovation?

Scientists have enough trouble generating AI simulacra out of silicon circuits and rare metals; what could they cook up with an egg's mere potpourri of ingredients? How would they make even a single ocular panel? Neurologist Charles Sherrington speaks to the dilemma:

> The eye's parts are familiar even apart from technical knowl-
> edge and have evident fitness for their special uses. The

likeness to an optical camera is plain beyond seeking. If a craftsman sought to construct an optical camera, let us say for photography, he would turn for his materials to wood and metal and glass. He would not expect to have to provide the actual motor power adjusting the focal length or the size of the aperture admitting light. He would leave the motor power out. If told to relinquish wood and metal and glass and to use instead some albumen, salt and water, he certainly would not proceed even to begin. Yet this is what that little pin's-head bud of multiplying cells, the starting embryo, proceeds to do. And in a number of weeks it will have all ready. I call it a bud, but it is a system separate from that of its parent, although feeding itself on juices from its mother. And the eye it is going to make will be made out of those juices. Its whole self is at its setting out not one ten-thousandth part the size of the eye-ball it sets about to produce. Instead it will make two eyeballs built and finished to the one standard so that the mind can read their two pictures as one. The magic in those juices goes by the chemical names protein, sugar, fat, salts, water. Of them 80% is water.[28]

And that's about it in a nutshell.

●

What does the egg "know" that science doesn't? How did it get so clever and efficient by its own algorithms? What shifts does it choreograph beneath its mask of common molecules, ordinary enzymatic signals, and codable DNA networks that cause matter to change its very nature and become endowed with an inner life? Where is its script written both below and above discernment? What is missing from our astute tabulation and how does it get itself into the hash? What substance, quality, or relation does the zygote conjure and mete that biotechnologists and professional skeptics and magicians do not spot even as they track its every step with hawk eyes and magnetic-resonance imaging and laser devices? If we could see or measure every nano-move and sleight of molecule or subcellular organelle in a fertilized, self-organizing egg, could we retrieve the working blueprint? Or does the egg do something that can't ever be stalked by bio-technicians, that they won't ever see, something of an order forever hidden from even the most scrupulous linear detection? And how would it hide while still having an actionable grip on the physical realm?

Thus far, at the start of the second decade of the twenty-first century, scientists have no idea what it is about cells and embryogenic designs that concocts mindedness. They cannot predicate it or draw it into their own golems by chemicals or electrodes. They cannot start with a supply of rare and raw materials and make even a perspicacious worm.

Say you are an exceptional electrical engineer who endorses panpsy-chism and wants to fabricate a robot with a tuning portal into the "con-sciousness" podcast, guess what? Same problem, different form: "how to build a maximally attractive dwelling for eventual habitation by external sentient entities. If you wanted to build a conscious robot—like Hal 9000, for instance—what sorts of parts would you order? And who would sup-ply them: an electronics shop, a biology tank, a physics lab, or some new specialty shop stocking wares at present inconceivable?"[29]

If you don't know what consciousness is, installing it in the mother-board of your machine is just as difficult as fabricating it from scratch. Scientists can complain all they want that matter isn't panpsychic, but then where in bloody Orkney did mind come from? They can say "mat-ter is inert but consciousness *emerges* from it anyway" till the cows come home, but then, remember, emergence means just that they don't have a clue. "[T]he belief that matter is 'dead' has the same experimental status as the opposite animistic belief that matter is 'alive.' Both beliefs rest on an equal logical footing, although the animist can in his favor point to at least one material system that is 'alive' while the materialist cannot point to any kind of matter that he knows with certainty is 'dead.' The real status of 'inanimate' objects awaits its resolution in a deeper kind of science than we currently possess."[30]

We cannot count or wire our way to mind and meaning any more than we can count or buy our way to prosperity, happiness, justice, and enlightenment. Some things just *are*—white on rice. They have to do with what creation itself is, where it is sourced—Creation with the capital C, not just this slit of time-space-awareness in which we happen peremptorily to dwell.

Functionalist scientists inherit, though they hardly trouble over its contamination of every proposition they put forth, the supreme luxury of already existing themselves, of getting to sit back in their lawn chairs and divans (or stand at their podiums) and expostulate for free—thermodynamically free. They can backdate explanations for phenomena that have already occurred and otherwise characterize a reality that unequivocally exists as if it were happenstance, desultory, and as likely *not* to exist. They can get away with just about anything they want without sawing off the limb whereon they roost because, regardless of their allegations, they continue to bask there.

They can be heedless and brazen and vulgar in a way that they wouldn't hazard if their theories zapped them into the states they imply. They might not be so dare-devil if the set-up were more up-for-grabs, if they could turn themselves into zombies by designating themselves as such. But it isn't that kind of universe. They already are here and are going to stay, no matter what. They are having fun and getting rewarded, and no mere theory is going to put an end to that. They exist and know they exist, so zippo is at stake. They can give consciousness any minimalist, random, emergent, or epiphenomenal basis they want because, no matter how reductive or bleak, it will not remove them from the game. They themselves are not at risk. They can taunt the lion because it is behind bars and drugged to the hilt. They can walk the mean streets because cops and military dudes are patrolling them. They can pretend to be both hip and brave.

Yet if matter hadn't represented itself to itself, there would be no Plato, no Darwin, no you—no soldiers on the battlefield, no crabs, no great apes. There would be no discussion—no pretense or prerogative, no lien on existence, no falling trees or colliding asteroids; there would be no "no." Consciousness alone "ontologically generates, not just the capacity to ascribe significance, but phenomenologically significant states."[31]

Nature in its vast, random idiocy has manufactured objects and systems far more complex than anything Nobel-level microbiologists can replicate— and this is itself a tautological conceit since we in our shallow ego identities and transient minds are only a passing effect of that same entropic, source-less concrescence: *thought thinks itself.* Look around. There are no other phenomenologically significant states, certainly no independent ones, either scooting about this planet or among the stars.

Though we pretend to have access, at least conceptually, to a superior and more efficient intelligence than otters or geese, we are stuck in their same subcategory with their same bastard origin and, if that is insulting or otherwise unsatisfactory, we are left having to claim an internal quantum leap such that either we transcend our own ingredients and etiology, or the evolutionary pathway culminating presently with us (throw in our machines too) has supernal guidance and is smarter and more discrete than its components. In other words, nature is cheating; antientropy rules!

Scientists and philosophers make this assertion implicitly, though they certainly don't admit to it—and would go to hell and back to avoid even the appearance of an admission—in fact, this whole affair is vaguely embarrassing to most of them: Either they (and we) are dolts and imposters or we are upstart chimps issuing metaphysical edicts that violate our own self-definitions and laws. Neither admission is survivable, but the former is less incriminating than the latter, so that is what is maintained as the profession's working disclosure: we have fake and illusory minds—illusions of minds. We fantasize our own existence, which is a chemicoelectric effect, an epiphenomenal artifact generated by trillionfold assemblies of molecules melding into networks of neurons synapsing into screening-room ganglia. *We are not real.*

Yet scientists don't otherwise experience or perform a hallucinated reality; they live garden-variety metaphysical lives. They toast each other, they bowl, they buy cars, they have pressing ambitions and desires. By duplicity and perfidy to their own anthem, they engender the paradox that cleaves to the heart of our civilization—and make no mistake about it, this is no minor anomaly; it speaks to how we treat the planet, each other, and our own lives. Cover-up has become knee-jerk, automatic:

"Ask people about their own freedom of action, and they will claim some type of free-will mentalism. Ask them about taking a physical aspirin for a mental headache, and they will smoothly switch to dualism."[32] Ask a scientist or a science major what his mind is and he will give a dry organic-chemistry explanadum, then he will drop it as if it never mattered in time to honor his reservation at the Ritz.

Though there isn't any middle ground here, most citizens of modernity occupy the ersatz middle. Those who believe that consciousness is merely biochemical activity still behave as if they were superlative individuals with private experiences that they righteously own (and that are worth owning, in fact more than all the diamonds in Africa and tea in China), while even dualists concede that the brain is the sole proximal source of mind. All parties accept that their consciousness depends on components that are not conscious and will never be conscious, but scientists go on acting as if that were not the case, and dualists conveniently forget to render unto Caesar and pay their taxes to molecular reality: "Each of these positions involves a cluster of beliefs, most of which are unconscious most of the time."[33]

Though nearly every scientist rotely professes confidence in a thermodynamic accident followed by the creation of life through the happenstance emergent properties of interacting molecules and amino acids, none of them behave as though they are mere jumbles of atoms or chemical concatenations. *They may claim to be zombies, in fact passionately, but they don't act like zombies. They don't address themselves or each other as zombies. They don't conduct the internal dialogues of zombies.* If a functionalist's girlfriend leaves him, he doesn't dismiss it as zombie heartbreak. If he crashes his car and suffers third-degree burns, he doesn't experience zombie pain and zombie anguish. If mugged on the street, he doesn't issue a zombie fortune cookie to commemorate the event. "Meaning"—that despised and discarded runt—tides him from moment to moment (and is his for the price of admission).

"Zombie" roles are affectations and pretenses to allow scientists to maintain academic standing and corporate salary scales without confessing, even to themselves, the fatal crevasse between their ideology and their phenomenology. Yes, I know, to assign a solely physical source to consciousness should not lessen our experience of it or negate our sorrows and passions. But that quickly becomes a disingenuous stance.

Hallucinating scientists act like kings of their hallucinated kingdom, official spokesmen for the gods. When they love, they are as ardent and romantically urgent as if they (and their lovers) were real. When they have children, they treat them as genuine beings and demand real behavior from them, even superb accomplishments. They are incorrigible hypocrites: proclaiming one thing, living another. They are fecklessly proposing a world without real consciousness while behaving "*as if* it were conscious" and "simultaneously denying that is what they are up to."[34]

Cutting off our own heads is not without consequences. This schizophrenia is a cause and a symptom of modernity's isolation and dementia. What do you think the radical Islamic jihad is about—I mean, at its base in the dusty villages and refugee camps of the Middle East, where political and religious imperatives meet? They don't want us to think that we can get away with blasphemy. They don't want us to flaunt divine law and go unscathed.

But science is the new Catholic Church, the new Crusades, the postmodern Papacy. In place of the Father, the Son, and the Holy Ghost, it provides its trinity of Matter, Genes, and Immaculate Emergence. The Inquisition that it runs is not as obviously brutal as that of the old church but much more efficient. No soldiers with torches and spears—it doesn't torture or slaughter its victims: no screws, rack, or stake. In fact, like the witch in the gingerbread house in the forest visited by Hansel and Gretel, it pampers the heathen with pleasures and delicacies, simulations though they be, keeping them addicted and loyal, fattening them for the kill. Then it leaves them hapless and bereft in luxury high-rises or homeless on the streets, plus in every intermediate phase of suburban blues. It is efficient because, after drawing folks indiscriminately with one type of honey, it leaves them to conduct their own demise.

For materialist functionalists it is clear what is at stake: "[I]f something non-physical sometimes makes a difference to the motions of physical particles, then physics as we know it is wrong. Not just silent, not just incomplete—wrong. Either the particles are caused to change their motion without the benefit of any force, or else there is some extra force that works very differently from the usual four. To believe in the phenomenal aspect of the world, but deny that it is *epi*phenomenal, is to bet against the truth of physics. Given the success of physics hitherto, and even with due allowance for the foundational ailments of quantum mechanics, such betting is rash!"[35]

Orthodoxy comes back to a matter of particles and what causes them to move and change. If mind can't bump particles, it exists only epiphe-nomenally—it doesn't exist. A colleague of the above speaker concurs, while dismissing the last vestiges of panpsychism: "Attributing specks of proto-consciousness to the constituents of matter is not supernatural in the way postulating immaterial substances or divine intervention is; it is merely extravagant."[36] Adios, ciao, ta-ta....

I like and respect science in its ideal form as an empirical, self-correcting system. Yet a modern imposter is leading us in exactly the opposite direc-tion—and in more ways and on more levels than one. We all know that consciousness can't change particles in the manner that heat or gravity do, but that is to ask consciousness to play in the wrong arena and then pro-claim that it doesn't cut it. It's like saying that pigs can't fly or that you can't score a touchdown in baseball, but even that's not quite right.

Consciousness is not responsible for changing particles. Yet it changes them nonetheless.

I can cite many instances of something nonphysical making a difference in the motion of physical particles, most of them not very extravagant. Through human intention, consciousness has reshaped the Earth, erect-ing megalopolises connected by causeways, turning molten stones into rolling and flying appurtenances, replicating neural hardware in digital devices fashioned of the alchemies of other tiny stones. How else to ex-plain markets, railroads, factories, newspapers, radios, tuxedos, crisscross-ing concourses filled with wires and lights, washing and drying machines in homes, cameras, iPads, and iPods? How else except as immaterial con-sciousness acting directly on matter?

We land vehicles on Mars. We spy on the metabolism of the Sun. We listen to the drum roll of the Big Bang.

I dig the objection of science to my argument: the causal chain behind all this stuff goes unimpeded from heat to information to life to behavior and arrives at the same constructs and connivances without panpsychism, without primordial consciousness or gods; but I find that a stretch, an-other cover story that doesn't come close to matching the scale or scope of what it is trying to cover.

Consider a single automobile through cosmic-outsider eyes. Imagine what it took to scrape that operating buggy off a bare Pleistocene landscape: the taming of the circle into the gear, the mining and forging and mass production of metals, the design and standardized molding of an internal combustion engine, the interlocking of electrical grids, the sequences of chemistry lodged in a battery, the technology of electron distribution, the conversion of jungle sap into tires, the slithering of silicon circuits inside a dashboard, the forging and setting of glass, the intricate contraptioning of the defogger and windshield wipers, the seat-adjusting levers, the connection between the steering column and the wheels, etc.—then to put it all together so that it not only converts fossil fuels, moves, turns, achieves NASCAR speeds, and maintains its processes over decades but can be replicated pretty much identically by the hundreds of millions—let alone to extract and refine its fuels and construct and maintain loops of relatively smooth pavements on which to operate these objects. Yet every single vehicle was pulled out of the interglacial Earth and is traveling in ordinary fashion along imaginal highways laid atop Stone Age lava beds and floodplains, rotating its gears up and down mountain passes.

This is the extraordinary level to which mind has impacted not only molecular substance but reality—and these are matters that involve the tilt of the entire universe, the placement of a sentient panel inside it. An explanation of the cosmos must encompass an explanation of consciousness because consciousness is part of the cosmos. It's as straight-arrow and square-shooting as that.

Plus who says consciousness is only epiphenomenal? I say that it moves particles just as effectively as gravity—not as immediately, not as directly or in as rugged a fashion but in different ways and resulting in the sorts of intricate, nonrandom designs that gravity and thermodynamics can't dream of calibrating on their own. It works *with* gravity and heat to do exquisite, astounding things that untaught gravity and heat cannot.

Then there are "extravagant" examples. Denver psychiatrist Jule Eisenbud, who specialized in paranormal referrals during the 1960s, took on a client, Ted Serios, a Chicago elevator operator, who seemingly had the bizarre ability to project images onto film from his mind. On certain occasions when his photograph was taken, what appeared on the developed film was not a portrait but either an extraneous image (like a double-decker bus or the Eiffel Tower) or a double exposure of Serios with an image coming out of his head or chest—in videos the image literally bloomed out of his body. This dude could also apparently redirect images viewed in other venues, including places across oceans, imprinting them onto film without even seeing them.*

Eisenbud brought Serios to the University of Colorado in Denver, where he tested his psychic photography and remote viewing. An engineer and a physicist were among the scientists witnessing the carefully designed experiments. First, Serios was stripped down and checked for concealed devices. Then he was placed in a Faraday cage restricting any mode of known information transfer, e.g., to prevent illicit signals from reaching him. The purchase of film was randomized to preclude conspiratorial hijinks, unlikely as they were. Yet to the astonishment of his audience Serios succeeded in transmitting a series of both personal and remote images onto Polaroids.

Afterwards the researchers all went to lunch together (without their talented subject). No one chose to discuss what had just happened. They made small talk: local restaurants, faculty gossip, college basketball. Finally Eisenbud asked their opinions. They agreed that the guy was remarkable; then they went back to chitchat.

Eisenbud concluded that the event was so profound and shocking that they had no terms or context for it; they had no way to discuss it; they were in denial.[37]

This is true of countless instances of telepathy, telekinesis, psychokinesis,$^\Omega$ future sight (precognition), and remote viewing, cases witnessed by both mainstream researchers and parapsychologists. Some of these have intruded

---

*This separate long-distance ability known as "remote viewing" has been studied extensively in other circumstances by numerous researchers.

$^\Omega$Like telekinesis, psychokinesis (or PK) is the movement of matter by mind: "usually gross physical changes (of state, of position) in objects for which normal physical explanations seem to be lacking but which appear to be associated with the explicitly indicated or inferred 'mental' behavior of particular persons."[38] The term is most often applied to autonomous break-ins like poltergeists.

in monitored experimental situations. Mentation has directly changed matter—perhaps not in the way that rebar and ball bearings under momentum do: more (no big surprise here) like waves and particles in quantum-mechanical settings. And I do not mean the brand of jejune metaphor that New Agers routinely sponsor—*these* guys do it on the spot, *actually!* Familiar accounts from the Earth's general population (going back to antiquity) include haunted houses, accurate premonitions, poltergeists, "clocks suddenly stopping or dishes inexplicably shattering or crashing at the time of someone's death, … the unexplained movement or levitation of small objects or furniture."[39] Nick Herbert summarizes—and, remember, he is a physicist:

> "Spontaneous cases of psychokinesis of the poltergeist variety have been investigated for decades. In these situations, a single person, usually an adolescent, is the focus of the occurrence of loud noises, moving or flying about of heavy objects, and unexplained disturbances of electrical equipment. The Rosenheim poltergeist, for instance, which took place in a Bavarian law office in 1967, was centered around Annemarie, a 19-year-old employee of the firm. The effects included flickering of light bulbs, which also unscrewed themselves from their sockets and exploded, as well as movement of heavy filing cabinets and strange percussive noises. The Rosenheim phenomenon was studied by investigators from the electric company, two physicists, and a professional psychic investigator. Some of the phenomena were recorded on videotape, but no normal explanation was ever discovered."[40]

This account is reminiscent of a nineteenth-century host of purported PK phenomena investigated by William Crookes in the company of psychic Daniel Douglas Home:

> These included the levitation of tables, chairs (with people in them), and other objects, and of Home himself, as well as an ostensible teleportation and the handling of red hot coals by Home with no injury to his own skin and with only negligible

damage to cloth materials in which they were sometimes held. (A handkerchief solicited from witnesses for this purpose was subjected to chemical analysis by Crookes and found not to be impregnated by a substance capable of providing such an effect.)

[T]he undisputed star of these séances, which were witnessed by at least two well-known scientists besides Crookes, was an accordion which on a number of occasions was seen floating about the room in good light ... and, under conditions rendering contact or connection with the keys impossible (according to Crookes), playing chords and tunes like "Home Sweet Home," "Auld Lang Syne," "The Last Rose of Summer," and "Ye Banks and Braes." During one such rendering, which lasted about 10 minutes, "The fingering of notes was finer than anything I could imagine," wrote Crookes, and during it "we heard a man's rich voice accompanying it in one corner of the room, and a bird whistling and chirping."

If this seems as though it might have been a hard act to follow, consider that on one occasion a heavy dining room table around which the witnesses were sitting became so excited, according to Crookes, that it began to vibrate and at length "actually jumped up and down keeping accurate time with the music."[41]

It is difficult to overlook the comedic effects of these performances, sitting ducks for mockery and parody (could you imagine the Amazing Skeptical Randi in attendance?). But it is exactly the slapstick that rings *most true,* not only because, given the attendance of earnestly observing scientists, the pranks are downright hilarious. They demonstrate a lively sense of humor (e.g., a personality) in whatever is teleporting and improvising behind the veil. A bit over the top for mere fakery, they reveal mischief and stunt—a much more fantastic gambit than the skeptics realize or permit themselves to consider. Even the separation of the living and the dead operates sometimes as if it were a huge strategic masquerade—but at what a scale!

I know, the whole thing could have been an elaborate hoax: Doug Bower and Dave Chorley stamping advanced mathematical concepts in crop circles of yellow rapeseed with their boards and rope is written all over it. Every crop circle and every séance worldwide: a fraud. Next!

Nick Herbert reports on more routine experimental procedures involving a Schmidt machine, "a box with an/off switch and a circle of a dozen lights.... To test for the presence of psychokinesis, a human subject watches the lights and tries to 'will' the light to rotate in a selected direction. Most people are unsuccessful at this task, but certain subjects at certain times have been able to move the lights in a particular direction in such a manner that the results exceed what would have been expected by chance by odds of several thousand to one."[42]

A related series of studies was conducted by a medical doctor in Dallas, Texas, Larry Dossey, his goal being to test the role of prayer in healing. Groups were given the names of random patients and asked to use any method they wanted to support their recovery. Those in the prayed-for group were five times less likely to require antibiotics and three times less likely to develop fluid in their lungs. None of the prayed-for needed endotracheal intubation (not true for the control group), and fewer died. Distance had no effect on the phenomenon; the data held up equivalently for patients in Florida and California. It was as if the prayers and the people were quantum-entangled through mere mentation of their names. Dossey concluded: "If the technique being studied had been a new drug or a surgical procedure instead of prayer, it would almost certainly be heralded as some sort of 'breakthrough.'"[43]

Telepathic and psychokinetic effects will never show up on a reliable enough basis to survive double-blind experiments. Telekinetic carpenters will never save themselves physical exertion by mentally hammering nails into walls. Even renowned spoon-benders don't consistently bend spoons (and can't begin to melt scrap metal into a useful utensil). But it's not a problem; it's the way the universe is constructed or reality's Chinese boxes are arranged and set inside their own insides. They are defined by "their sporadic occurrence and short duration [which] make them difficult to study scientifically."[44] Yet intermittency doesn't automatically mean they don't occur or aren't "real."

Disqualification by fickleness or fitfulness would be an overreach, don't you think? I don't see how just anyone can proclaim that inconsistent events are not permitted under the rules. What rules? Whose rules? After all, the universe sets the official criteria for what can and can't happen, and events don't otherwise have to obey any particular rules. They don't even have to "happen" in an ordinary sense. Remember the overly popular line "it is what it is"; well, that includes psi phenomena too.

Yes, paraphysical effects are apparently observer-sensitive, repelled by skeptical belief systems, which cast the mental or "evil eye" equivalent of electromagnetic interference fields. Consider this: telekinesis is meaning-based, not object-based. Meaning follows laws of consciousness not of matter, even when it alters matter. Tied to consciousness *inside* consciousness, it is stubborn, shy, and will not perform—*cannot* perform—with the Amazing Randi and his buddies gandering: a quantum "quantum effect." Imagine gravity behaving that way. But telekinesis isn't gravity (though gravity may be a corollary or sidekick to telekinesis in the universe at large).

As well, many anthropologists have observed routine long-distance telepathic communication between members of Australian Aboriginal and African bands. Nonlocal transference of specific sensory stimuli like electrophysiological responses have been shown to occur in interhemispheric correlations of brain EEG readings of meditating subjects, even when each meditator was placed in a Faraday cage. According to one lavish model, quantum-entangled nonlocal interactions collapse the wave function of a stimulated brain into a correlated nonstimulated brain, eliciting a spontaneous identical state.[45] If you don't like that, come up with something better! I guess you can claim the experiment was biased and the outcome was sloppy observation (or chicanery). But you can't claim that every single time, or maybe (if you are card-carrying professional skeptic) you can. But then you're doing religion under the guise of science.

Plenty of other Fortean, paraphysical, and inexplicable events occur hourly on Earth; they are just not tied to consensus reality in any regular way. They may be real—*real* real—but their reality is not extractable and repeatable; it doesn't go on the Big Scoreboard.

In a certain sense it is amazing that scientists (and science as an institution) can sidestep all episodes for which it has no explanation. Take for example young children speaking languages they were never taught. What's the fallback position for that one?

For years at the Common Ground Fair in Unity, Maine, I have watched border collie trainers control their dogs as they herd harried sheep, goats, and ducks into enclosures.* I always assumed that covert signals linked human and canine, something in the genre of unnoticeable body movements like winks, shifts of expression, high-pitched whistles, or concealed clickers. Yet when the trainer rhetorically asked onlookers to explain how his commands were transferred intelligibly to the compliant pups, no one nailed it, at least to his satisfaction.

Then he answered himself by telling us that the key was *intention.* If he fully and unambiguously intended to communicate an action to a dog, the dog picked it up and carried out his instruction. Other aspects of training or signaling were secondary to this silent communication by will. Without intention, there is no delivery.

Yes, intention *could* mean covert signals that the dogs construed via their own highly developed sensory capacities—subtle but not metaphysical—but it could also mean telepathy.

Aerospace engineer Tienko Ting transmits *chi* potential, not only its flow but the capacity to feel and move it, both face to face and *in absentia* across continents. He literally bestows his own ability to raise chi on others without contact or proximity or even knowledge of who and where they are. Using this same method, he apparently conducts successful long-distance healings.

Ting is not alone in this mode of endowment, nor is chi the only impalpable energy subject to instantaneous long-distance transmission. Reiki masters and psychic healers propose to send cures across the planet telepathically and telekinetically at will. The difference is that, as a trained engineer, Ting considers chi to be a regular, legitimate energy like the physical ones with which he is professionally familiar and deploys in the course of his more conventional aerospace career. He does not redefine or quarantine chi in an entirely different, metaphysical category; it is energy, pure and simple, real energy operating in its own sphere by its own rules. It

---

*I imagined the ducks viewing the collies' Frisbee tricks and their corralling of the sheep as they whisper to one another, "What show-offs! But we're next."

never even enters the phenomenal-versus-epiphenomenal debate because it is operating in a different arena from where the argument is taking place. He believes that he discovered chi specifically because he is an engineer; he thinks like an engineer and has an engineering viewpoint of energy and energy transmission. His childhood exposure to *t'ai chi chuan* and *chi gung*, by contrast, led him nowhere (chi-wise), or more accurately to engineering school, where he then encountered chi on a *physical* basis.

Ting denies that there is anything mystical about chi; it is simply a nonlocal instantaneous form of energy within the spectrum of energies operating in the universe. He classifies his method of chi-movement as a form of wireless communication, creating discrete conditions for delivering energy in the way that radios and satellites create their own conditions to deliver information over long distances without contact—processes that were themselves impossible and inconceivable not so many centuries ago. He adds that chi is somewhat like gravity or electrical current in that it activates energy and creates force though is not itself the energy or force. It operates more like an imbalance of air initiating wind than wind itself. "[T]here is no tangible substance flowing, but there is evidence of energy flow."[46]

Then we have Mr. Automobile Savant: in a single glance at a parking lot he can identify the make, model, and year of every car, including ones just off the assembly line in Europe, though he cannot feed himself, dress himself, or read. His idiot part aside, where does he get the information from? Where would *anyone* get it?[47]

Other savants routinely perform spontaneous prime numbers to eight places but can't give change for a quarter. Many with "high-function autism" perform well above the level of human capacity while they are also psychologically dysfunctional.

Jule Eisenbud cited Croisset, a renowned Dutch clairvoyant who could ostensibly pinpoint events before they happened. In experiments conducted over twenty years Croisset "put down statements about people who, two or three weeks later, on a given specified date, *will* occupy a space,"[48] and these were accurate to an uncanny degree, "well beyond chance expectations."[49]

A hall would be hired, tickets dispensed for its chairs randomly, and then people who received them would enter and take the seats specified on them. Croisset had already written their personal descriptions *before* the tickets were dispensed, noting such characteristics as their height, weight, gender, profession, scar locations (as quirky as on a big toe), surgeries, hobbies (as specific as an indirect allusion to bagpipe music).

These future readings were clearly noncircumstantial anomalous events of a mode that is impervious to scientific experiment or analysis. Eisenbud declared simply, "You see, this type of thing, this ability to simply step over time, as if there were no barrier at all is of such metaphysical, philosophical significance as to render everything else patent pending.... Everything else is almost trivial until you begin to understand what the hell goes on to enable a person to do this. All the work of physics stops at this point and, in fact, because of this, physics in its present form is through."[50]

Nick Herbert noted similarly: "Psychokinesis has important implications not only for consciousness studies but also for physics. If the mind can indeed exert a force on distant matter, then current physics is demonstrably incomplete since it recognizes no mind-based forces whatsoever."[51]

Remote-viewing superstar Russell Targ, a physicist at Stanford Research Institute, is credible enough that during the Cold War he was hired by the Defense Department to conduct remote viewing of military installations in Red China and behind the Iron Curtain and to train uniformed personnel to conduct the same surveillance. He told me succinctly in 2009: "My interest in ESP has been more physics-oriented than from the standpoint of human potential. My feeling is that it tells us something very important about the nature of the world we live in, rather than something special about ourselves. My passion these remaining days is not necessarily to understand precognitive dreams, but rather to understand causality as it is contradicted by precognition. If I can know something days before it is randomly chosen, then there is something wrong with our linear view of causality—which is a pretty big thing to have wrong."[52]

Indeed! But again, I think it is because this is consciousness's universe.

●

Well, if consciousness can transcend brains and override causality, that would explain how automobiles not yet even in car magazines could get into the mind of an "idiot" who needs help with breakfast. He would simply dial into the transpersonal consciousness field, a talent that might be stillborn in most humans but available to infantile savants because they are too naïve to know that it is impossible or too transparent and autistic to get in its way.

The notion of folded-over cerebral cortices and sulci as antennae on a tuner may be simplistic, but that is why I proposed some sort of ganglionic turnip brain. Inside the neural brain we may have another, hyper-brain, an epiphenomenal organ tuned to hyperreality. It could wrap itself around levels of its own holographic hardware again and again, replicating and re-inforcing its own internal matrices many times over as it emerges within its own structure. In this model a vast field of pure information *creates* brains (and DNA and crystals and stars). Less "brainy" animals are simply tuned to lower channels of the same intelligence field.

Perhaps the consciousness field *graves and aligns the brain* at the same time that the brain develops protoplasmic, synaptic antennae and display platforms and, with its own fractal expansion and vegetative coiling, literally designs and generates the frequency of reality it is sucking into its chambers. The cerebral organ (again) localizes, centralizes, and individuates conscious-ness; but it doesn't manufacture it. Not out of molecules, cells, ping-pong balls, beer cans, synapses, or neurons. It creates the reality through which it is itself emergent, as if tentacles of a cosmic intelligence-tuber were discharg-ing information through a sort of Tesla coil: "[A] quantum brain drinks up waves from an ever-present background ocean of pure possibility, the ocean out of which comes everything, mind and matter alike."[53]

Eisenbud said of Croisset, "I don't think [he] foretells the future; I think he creates it."[54]

With such an audacious model for consciousness, we understand how an egg could be hooking up to a paraphysical template while operating an exclusively material program. The wiring (or more properly nonwiring) of consciousness is not in the egg *per se* but in the conditions and bil-lions-of-years-in-development matrix that underlie the egg. It is consti-tuted and incorporated by superposed threads underlying DNA and other "instructions."

A "train" running on tracks like this on this, or just the other, side of the space-time continuum and going thousands of—well, not miles per hour—while gestating forms instantaneously with reference to a different locus of speed, makes things like clairvoyance, telekinesis, precognition, and remote viewing almost ho-hum.

We have to treat the egg as outside space-time, not in its raw ingredients or genetic protocols, not in its morphogenetic activities, not in its relation to shear force and gravity, not in its coordination of subcellular clocks, but in the raw depth and scope of its self-organization, and not even *its* self-organization but the self-organization of the universe that spawned it.

●

The entity known as Seth first appeared on this planet on December 2, 1963, in the form of coherent messages from a male personality received by poet Jane Roberts, initially through a Ouija board and then as a voice in her head. The "spirit" eventually identified itself by name. Over the next twenty years Seth went on to deliver several volumes' worth of text, most of them with Roberts in trance, the sessions either tape-recorded or copied in shorthand by her husband.

Seth/Jane provided a convincing explanation of life and death along with a source map for creation—and with seemingly no obstacles to its transmission across a legendarily impassable barrier that had confounded Home, Crookes, and Frederick Myers (the author of *Human Personality and Its Survival of Bodily Death*), as well as numerous other nineteenth-century clairvoyants who sought to understand consciousness through séances with the dead. Seth skipped all intervening steps—fragmentary dictations, brief and punctuated channelings, cross-correspondences in trance states—to take it to the house. He dictated an entire encyclopedia while elucidating the unseen universe. Before him, the best that could be remitted from "the other side" was half a rusty nail—some interrupted sentences that sounded as though they were coming from an astronaut on a Neptunian moon during a solar storm or a prisoner under duress in a granite dungeon, using an apparatus with a failing battery. "Seth" spoke freely, cheerily, and like a chap who had been gestating for a very long time.

Was he even a single person, or was he a composite energy gestalt, a multi-personhood received in Roberts's mind *as if* one personable old professor chatting sublimely away?

Even though Seth had his own style of expression and knowledge base and disclosed himself as an "energy personality essence no longer focused in physical reality,"[55] Roberts worried that he was merely a subpersonality of herself and came from her own subconscious. Could she have faked his character, even unconsciously, like an actress performing a role? More deviously, could she have put together a mishmash of metaphysical texts to create Seth in the manner that Carlos Castaneda was said to have stitched together multiple shamanic characters from different ethnographies to manufacture Don Juan? Of course. But for now, I am going to focus on what Seth/Jane said and leave the speculative options and their implications for Volume Three. Wherever "Seth" originated, he provided a remarkable hyperspatial cosmology of the universe, an express train shooting past all superpositional quantum-entangled nonlocals.

From the entity's view, our own world of matter resembles shifting castles of fog, psychic material swirling under creative intention: "A true understanding of the way in which an idea becomes physical matter would result in a complete revamping of your so-called modern technology, and in buildings, roads, and other structures that would far outlast those you now have...." Hard to imagine. We're dug in here pretty deep. "You cannot understand the psychic reality that is the true impetus for your physical existence unless you first realize your own psychic reality, and independence from physical laws."[56]

If mind superintends matter, why doesn't the universe look or feel that way? "It is only because you believe that physical existence is the only valid one that it does not occur to you to look for other realities. Such things as telepathy and clairvoyance can give you hints of other kinds of perception, but you are also involved in quite definite experiences both while you are normally waking and while you are asleep.

"The so-called stream of consciousness is simply that—one small stream of thoughts, images, and impressions—that is part of a much deeper river of consciousness that represents your own far greater existence and experience. You spend all your time examining this one small stream, so that you become hypnotized by its flow, and entranced by its motion. Simultaneously these other streams of perception and consciousness go by without your notice, yet they are very much a part of you, and they represent quite valid aspects, events, actions, emotions with which you are also involved in other layers of reality."[57]

Roberts later channeled William James, a more conventional stream-of-consciousness aficionado. Early in his life the tough-minded functionalist proposed astonishingly that, "when finally a brain stops acting altogether, or decays, that special stream of consciousness which it subserved will vanish entirely from the natural world. But the sphere of being that supplied the consciousness would still be intact; and in that more real world with which, even whilst here, it was continuous, the consciousness might, in ways unknown to us, continue still."[58]

James obviously did not keep his metaphysical faith, but he did initially articulate the alternate view that our being, while mediated through the brain, survives its cerebral clockwork after death. Roberts "confirms" the early James, in "James's" own voice, decades later. Speaking of the realm in which he finds himself, the departed "James" comments: "It accepts mental stimuli which imprint it—stamp it—and form it into the habitual grooves of nature's shapes.... [E]xistence takes place initially in mental realms rather than physical ones, it implies states of mind first of all.... I simply use my mind to 'go' where I want and the rest of me follows. My body, real enough to me, can appear or disappear in any given place, however, and each environment is formed by people's belief in it."[59] He speaks later of a "knowing light and atmospheric presence" that are "at every hypothetical point ... wholly here and responsive."[60]

Panpsychism supreme from a master materialist (or his ghost)!

Jane Roberts's channeled universe offers a series of simultaneous domains through which everyone passes in multiple states of being. Incarnation zones are basically interdimensional "schools" on which our more ordinary secular school systems and *madrasas* are modeled. We are here to learn in the deepest sense.

With paternal tenderness Seth tells us to go on playing here—do what we will, splash about, heedlessly and destructively if we choose, killing and maiming and abusing—because we are not (and can never be) more than children with toys. Our actions may cause us to suffer, but we can do no real ultimate harm to the actual universe. There is, however, a price:

"Now, for those of you who are lazy, I can offer no hope: death will not bring you an eternal resting place. You may rest, if this is your wish, for a while. Not only must you use your abilities after death, however, but you must face up to yourself for those that you did not use during your previous existence...."[61]

Let me come back to that one later too, for its dagger lies at the heart of this book:

"You will reincarnate whether or not you believe that you will. It is much easier if your theories fit reality, but if they do not, you will not change the nature of reincarnation one iota."[62]

In the view of turn-of-the-century German mystic scientist Rudolf Steiner, sentient spirits ancestral to us existed in other realms before the Earth and migrated to the physical plane as part of their cosmic evolution.

Your everyday housecat is not once and forever a cat. The conventional bio-cat is only a *Felidae* mutant—after that: a former cat dead, off the ledger for good; a fleeting shape-construct fashioned out of energy, space, and gravitational-molecular constraint.

Mosquito exists—slap! Mosquito never existed. That simple!

But Steiner's cat is *a soul in a cat body.*

It was not the worm that evolved into a frog nor the lemur into an ape; it was an essential being that manifested as a worm at one stage of its development and then evolved *spiritually* to manifest as a fish and a frog at subsequent phases: a progression up the Great Chain of Being. Static by comparison to Darwin's living nature, the Chain of Being converts biological evolution into cosmological involution. After all, the Chain evolves too, but as a procession of archetypes set into worlds, statuses which actual species attain only as they transition through them.

While bodies progress on an ontogenetic level, essential beings matriculate through those bodies to manifest as one or another creature at a particular phase that then evolves into a next different creature at a later phase. These emanations are ordained in exact hierarchically consecutive forms from the beginning of the universe—actual species merely traverse their Chain. Taxonomies on planets may seem to be random arrangements of membranes, but they reflect cosmic eschatologies; they are expressions of a deep function that emerges via atoms and molecules because its sub-waves penetrate everywhere and everything.

In Steiner's anthroposophical parable, the souls of crystals, plants, and invertebrates, which dwelled in tiers of a spiritual domain before they became stones, flowers, and animals, hastened too greedily into matter, thus incarnated too deeply to retain enough etheric plasma or sap needed for introspective consciousness. They sank swiftly and suddenly, so their spirit force was absorbed and held in frozen forms. Minerals and crystals are conscious beings, but ones wedged so deeply into substance that they retain no psychospiritual flexibility. They may "think" they are laughing or commenting wryly, but it comes across as fixed signatures. They are not actually conscious; all of their intelligence is locked in rhomboid and tetrahedronal poses: intelligencoids. Not consciousness in our sense.

As increasingly more of the spiritual vibration is withheld, the substance under it becomes softer and more liquescent, more pliant to metamorphosis. Plants spring from a quantum of restrained spirit, but they are still hard and fixed, without egos or objective consciousness. Steiner proposed, for instance, that more sentient plant forms preceded living plants on Earth: gigantic meditating trees and huge exotic flowers that were mostly invisible because they had not yet entirely passed from an astral into a physical realm—think Tolkien's ents. These were the spiritual forebears of the Earth's botanical kingdom.

Invertebrates like clams, grasshoppers, and spiders were more patient but they too hastened precipitously and landed well inside matter, their intelligence and essence trapped in organs—in claws and stingers, phosphorescence and shells. They are smart critters, but their wisdom is peripheral. They express their consciousness in mechanisms and rituals, not minds. They don't think or plan ahead, for they experience a perfect correspondence between image and object. With no sense of their own presence, they are dreaming as if it were always the day before yesterday.[63]

Spiders arrive spinning webs before they are able to individuate their tapestries as craft. Their entire spirit force is soaked up in the unconscious operation of glands and spinnerets, web-making and pouncing on entrapped prey. Yet it is real to the spider—real energy, real being, realer even than reality itself: it is what *it* must do as well as *everything that must be done.* Crabs scuttle in frenzies of feeding and mating, waving claws and changing colors and bodies … fast asleep. These are souls that took on bodies so abruptly and heedlessly that they left their entire ego potential behind.

Except for the markings that distinguish species from species, nothing personal separates one crab from another, one albatross from another. Their individuality is subsumed in their ancestral type. They are fixed ideas incarnate.[64]

Science has lost this esoteric view in cultivating an equally valid recognition of nature as a randomly raveling design that embodies solely the reality it proposes. No backstage hocus-pocus.

Theosophists, by contrast, were prone to peek behind the wings at the actors without the play: "It is as though something *veiled* were living behind the physiognomy. Something that *craves* to shine through but is withheld by the body's rigidity! This impression becomes positively grotesque and horrible in the case of an insect…. The merciless rigidity and hardness of the casing, out of which the eye, an immobile point, its surfaces walled in, stays there lidless and ever open; that fearful leverwork of the parts of the mouth working mechanically; the hurried jerk and cramped groping of the proboscis; the antennae, always trembling, and yet not looking truly 'alive'—it is as if one saw a ghost, a phantom suspended by invisible threads, that pretends to be alive but in reality is only a moving mechanism."[65]

Something is lost not only to the hornet and grasshopper, sentient beings with psychic and spiritual potential, but to our planet and the universe at large: "If the glow-worm could be transformed into a being who knew that he possessed the secret of making light without warmth," opined Swiss psychologist Carl Jung, "that would be a man with an insight and knowledge greater than we have reached."[66] This is no mere speculative or trivial matter in our present crises of peak oil and climate change.

While hardly a scientific view or presently serviceable piece of alternative technology, it is an intuition of our psychospiritual potential; it points toward how we might extricate forfeited sciences out of unexplored psychic aspects of our body-minds, abilities that are presently locked within us but inaccessible: luminosity without heat, energy *sans* entropy. The glow-worm doesn't know what he possesses and he cannot tell us—and we cannot find its vestige in ourselves.

[Not long after writing this section, I dreamed that the largest asteroid Ceres was found to be maintained at a regular eighty degrees Fahrenheit with a nitrogen-oxygen atmosphere like that of the Earth. A spaceship of scientists and astronauts was secretly dispatched (around 2007) by President W. Bush, clandestine because he wanted the technology, if recovered successfully, to belong to corporate America. Ceres is a long way from here, so the mission was only just arriving (dreamtime 2011).

In my dream the chief scientist reported to President Obama from Ceres that *there was no technology.* "We're the wrong guys to figure this out," said their "tweet." It was followed by a lengthy encrypted message: "There are no computers, no wires, no machines. Well, there is one machine, several huge vats of water in which gigantic rotating brushes are circulating from a hub, but they are not connected to the climate or atmosphere or to anything at all, at least in any manner we can detect. It is summer here, the sky is blue, the air is sweet and fresh, there is ample water, and there are lush gardens but no animals."

The entire Cerean effect was psychically produced and maintained, not as a temporary habitat for traveling aliens but a model left by them at just the right orbital distance for terrestrials to view and emulate, as long as they had the wherewithal to get to the asteroid belt, but only if they could figure out how to look with unbiased eyes, to actually *see* what was happening. Those seemingly irrelevant spiraling brushes were a misdirect as well as a clue.]

●

Rushing into matter with assorted skinks, starlings, and skunks, the flying squirrel hits reality like a meteorite and fossilizes at once into cellular papyrus. Most of its being lands in shape and action. Galloping mammals transition more gracefully. Only the inner skeletons of vertebrates fully mineralize while, at the opposite pole, a sheath of spiritual auras congeals around them, maintaining their connection to higher planes. They have inklings of spirit and even a bit of soul force. Look at a horse or a border collie, even a prowling raccoon. The beginnings, the inklings of philosophy are there.

On Earth, primates suffer the least mineralization and have the most fully retained auras of any zooid. Sharks, crocodiles, chickens, and mice all start as "humans," as coiled knots with a backbone, nerves radiating from a zone of metaphysical becoming physical energy through their primitive streaks at the gastrula phase of development. The more impulsively their spirits raced to migrate (cosmogonically), to incarnate (ontogenetically as their species template in this zone), the less actual consciousness they retained—the more their former wisdom lodged in their organs. They wanted to dwell on this *"isle … full of noises/Sounds and sweet airs that give delight and hurt not,"*[67] but they were fooled by its surface benignity and mirages, and paid a huge spiritual price.

The higher mammals perceive something of the tragedy that has befallen them; they have voices but no language. They try to express the strangeness and isolation they feel, but their words come out as bays, growls, and screeches. Occasionally when their eyes meet ours, they show their desperation, bewilderment, and pure love; their faces reflect the loneliness and the horror of missed opportunity.

Steiner's apes were followed by late Palaeocene and early Pleistocene tribes of hominids. Though they retained enough consciousness to struggle against solidification, these creatures still incarnated an instant too soon (as the Earth now orbited the Sun, and the planet's physical realm was evolving in that metronome). The embryonic gibbon is human but human like an old man, wrinkled and hardened, dead already to the possibility of full ego.

If we yield to a certified palaeontologist at this point, he might say, "Good and well for souls incarnating in bodies; that is beyond reckoning and beyond proof, above my pay grade. But there is an indisputable fossil- and DNA-confirmed continuity from fish to amphibians to reptiles to mammals, from apes to primates to humans. By the archaeological record, by the evidence of the chromosomes, by the demonstrable radiation of forms from other forms, humans *do* descend from apelike ancestors."

Here stealth anthroposophy joins forces with palaeontological heter-ochrony to rescue the appearances, for the metaphysical hand of the mi-totic clock coordinates an esoteric chronology outside time.

To permit human arrival, biological cells have to be deterred from pre-mature skeletalization, while the soul simultaneously has to be dissuaded from headlong incarnation. One act cannot go forward or succeed without the other, though at core they are a singular occurrence. The embryo must remain infantile long enough to allow spirit to retain its spiritual character upon entering matter, the brain to differentiate in sync with its own cog-nitive and emotional intelligence (starting at gorilla scale in a chimplike embryo)—but these too are identical and simultaneous: a panpsychic xe-notransplant. The brain cannot evolve without a spiritual transmigration at its core, but a spirit cannot transmigrate onto this plane without an evolv-ing brain into which to export itself. The mutation and reorganization of genes mirror the transmigration and embodiment of souls; paedomor-phism and neoteny recapitulate spiritual embodiment. The convergence of macrocosm and microcosm marks a watershed phase in the evolution of consciousness: the odyssey of mind into matter. Timewave Zero folds toward its omega point and gives rises to the noosphere.

So in the esoteric tradition man and woman are *not* descended from apes, nor in fact from any other life forms on the Earth; they are indepen-dent shards of spirit hitting the whorl of matter at the precise moment when its locally evolving substance is soft and receptive—paedomorphic enough to receive their psyches. Each species's psychospiritual vibration accepts the physical vibration that is evolving to embody it.

The way in which human bodies are simian is secondary to the way in which a metaphysical subwave is morphing those same body templates into people with egos. Humans represent spiritual energy penetrating mat-ter at its deepest and subtlest declivity (at least so far on this planet), in a form that allows it to preserve the aura of its spiritual aspect under the constraints of a denser dimension. We are the crux of a carrier function bearing novelty into space-time.

Yet the light within is real and ontological. We are not just a bunch of tangled filaments generating hallucinations or quantum switches in microtubules interacting with neurons, kindling mentation; we are Creation's ground luminosity, transmitted through DNA, incarnated in a fissioning blastula and tissue choreographies.

Suspended between worlds, humans recapitulate the biological and spiritual history of the Earth. Our karma and identity give us a soul presence and a psyche, but our raw animal sourcing endows us with a ticket to ride: organs, tissues, and neurotransmitters through which to survive in a jungle of carnal appearances while higher energies and modes of consciousness dance out their karma too.

Again, occult biologists refute our descent from the apes on this basis: the generalized ancestors of men and women never slid that far—baboon- or gorilla-hood—into matter; they were preserved in human form during the heyday of the mollusk and the dinosaur, even during the ascent of the great apes, not here but on another planet. Angels are embodied in apes, but that doesn't make them superior to those apes, because apes are angels too, angels flying too close to the ground, too deep inside matter.

Like Robert Fludd's Divine Man, the human figure and countenance project a spiritual symmetry back through the entirety of the plant and animal kingdoms. We emanate in a way that is both revealed and concealed at every moment so, from a soul standpoint, we look like optical illusions to each other and feel like embodied illusions to ourselves: There's the beast, and then there's the angel, but then there's that beast again.

Intermediate worlds like Lemuria, not constructed of heavy matter and gravitationalized mass in the way that this one is, were different from modern Earth in every way. En route toward incarnation here, we ourselves looked totally different when we dwelled on them. We also experienced different realities in each phase. Just like this one, each zone was graphically and irreconcilably final and real. Fifth Avenues (or their equivalents) ran down each of their concourses.

This account rings true as allegory, even if its worlds were not really planets and we were not men and women on them in the ordinary sense, for it speaks to the hidden apothegm of our situation and allows us to view our scientific projections among guises too vast to tame or elucidate.

Only our strictly genetic material aspect, our bodily substance, discriminating organ motif from organ motif, feature from feature within species, evolved physically on Earth, but our esoteric human identity does not have a biological origin. We are descended biologically and psychologically from the apes *but not spiritually.*

By developing a human form externally, we have recapitulated ancient worlds and nonphysical dimensions that are cached in us metaphysically, deeper than cells and molecules, yet at their basis. At the same time, we have lost who we are by *becoming* it at another level.

These same prior worlds reincarnate in us again and again, as ovum, yolk sac, amniotic ripples in etheric currents; germinating blastula, endoderm raveling through ectoderm, organs emerging out of a coelom between membranes. The gastrulating embryo is an astral moon, a vehicle between moons. Ontogeny recapitulates phylogeny and together they recapitulate cosmogony.

CHAPTER TEN

The Subtexts of Science

I am sure that by now I have lost every scientist and rationalist in the house. But I do not want to lose science itself or rationalism. In fact, I want to capture its precise karmic meaning. Science is unparalleled and indispensable in its territory, and I honor and respect its domain as dutifully as any citizen of modernity. Every day I trust it with my life. These guys put on an incredible show. They grow my food and deliver my water through intricate labyrinths of pipes; they operate functioning cities and intercontinental air-transport systems. They send continuous signals to my television and computer. I am forever in their debt and in awe. Their perseverance, their commitment to fact-checking and truth-identification over epochs merit my admiration and gratitude. Without them we'd all still be fighting the Black Death and looking for an uncontaminated pond, tied to a few-hundred-mile lifetime radius. The Earth is science's planet, for sure.

I just don't think that science gives a complete picture of the universe. Run as a quarantined program, it is brilliant. But as witness to its own contexts, background meanings, and internal paradoxes, it is about as yokel as a high-school dropout. It doesn't understand what it is saying, what it is *really* saying. Yes, it understands what it is saying about mass, electricity, atoms, genomes, neurons, and the like (on which topics it is unsurpassed), but it doesn't understand what it is saying about meaning, value, or context. The state of the planet's ecosystems alone proves that.

Science also enshrouds a mostly unacknowledged subtext. Despite its relative longevity and present hegemony, the Newtonian-Darwinian model is not appreciated for what it actually is: the dissection of the carnal basis of matter, and the degree to which this is also—in fact is primarily, despite itself—a psychic and divine awakening of spirit in substance.

Matter contains a unique knowledge not otherwise available to spirit: the esoteric basis of its own emanation and embodiment. Science is literally an undertaking for testing the objective depth of matter—but, since matter is spirit in another form, it is a vehicle for testing the objective depth of spirit—for plying its fathomless essence in an inextricably concrete medium. Science is pulling us into the extant body of the Divine!

Matter is otherwise empty, a semi-durable dream-state: space and its curvature, energy bonded in shapes. While this is obvious to the guys practicing quantum physics, its extensions go unacknowledged: *There is nothing here,* nothing final or real or lasting. But there is a lot here, in fact everything. So what is going on?

The deeper we travel into matter—and this of course includes experimental research and its applications as a critical part of our submersion—the deeper we also sink into our own true nature, the more we penetrate the intelligence underlying everything, our own inevitability and destiny as well. For we and substance share the same parentage and origin; we are different faces of one coin.

Technology is the aftershock, the tsunami arising from the penetration of the intelligence of matter. Yes, spirit has been imported into substance—long, long ago—but that doesn't mean that spirit *is* substance. Having identified the creationary spark in substance, science has tried to export and extricate it and then appropriate it within its own antiseptic regime. That hasn't accomplished shit.

A year before his premature death in 2000, Terence McKenna posed the basic conundrum from a perspective that most scientists shun:

> The universe goes from simple to complicated. Why? There is no reason for the universe to work like that. Why do you get at first, moments after the Big Bang, an ocean of free electrons at such a state of temperature and energy that no molecular bonds can form? Atomic systems can't even form; the bond strength is overwhelmed by the thermal energy in the system.
>
> Then it cools down and atoms condense, a more complicated thing than electrons by orders of magnitude. [With] further cooling, further nuclear cooking of the most primitive

elements, hydrogen and helium, in gravitationally aggregated masses called stars, heavier elements emerge; [these] were never seen before fusion began to occur in hydrogen masses. And these fusion processes cook out iron, sulphur, carbon. Bingo! Carbon! Molecules—an order of magnitude in their complexity greater than atoms as atoms are to electrons. I'm compressing 13 billion years of emergence here into 30 seconds.

Then out of the molecular soup you get long-chain polymers; out of the long-chain polymers you get molecular transcription systems, pre-biotic stuff; out of that you get non-nucleated DNA; out of that, nucleated DNA; out of that, membranes, organelles, organisms, higher organisms, differentiation of tissue, our dear selves, culture, language, technology....

Why doesn't science take on board, as a major problem in the description of nature, the emergence of complexity? You ask a scientist, and they say, "Well, these are separate domains of nature. How atoms become molecules has nothing to do with how animals become human beings." This is bullshit.... The understanding of the fractal ordering of nature now makes it clear that voting patterns in Orange County, distribution of anemones on the Great Barrier Reef, and the cratering of Europa all follow the same power laws....

If we are going to look for an enormous eruption of emergent phenomena, an enormous unexpected download of novelty, we shouldn't look in a domain of zero space, zero time, zero energy, zero antientropic organization. That's the worst place to look; that's the least likely place where such a singularity would spring out. You should hunt it in domains of immense complexity where you have matter, energy, light, chemistry, language, machines, cultures, intentionality—minds, minds, minds....

We can believe that the universe is under the influence of a strange attractor; we can believe that the universe is pulled toward an ultimate denouement as well as pushed by the unfolding of causal necessity. It's an engine for the generation of complexity, and it preserves complexity, [as] it builds on complexity to ever higher levels. If you entertain this, guess what happens? It's like a light comes on on the human condition....

Who are we in my story? In science's story, we are nobody; we are lucky to be here; we are a cosmic accident; we exist on an ordinary star at the edge of a typical galaxy in an ordinary part of space and time, and essentially our existence is without meaning, or you have to perform one of those existential *pas de deux* where you confirm meaning—one of these postmodern soft-shoes.

But if I'm right that the universe has an appetite for novelty, then we are the apple of its eye. Suddenly cosmic purpose is restored to us. People matter, you are the cutting edge of a 13-billion-year-old process of defining novelty. Your acts matter, your thoughts matter.

Your purpose? To add to the complexity.

Your enemy? Disorder, entropy, stupidity, and tastelessness.

Suddenly you have a morality, you have an ethical arrow, you have contextualization in the processes of nature, you have meaning. You have authenticity, you have hope. You have the cancellation of existentialism and positivism and all that late-twentieth-century crapola....[1]

●

In diametric opposition to McKenna, a cranky Stephen Hawking told a 2011 interviewer: "I regard the brain as a computer that will stop working when its components fail. There is no heaven or afterlife for broken-down computers; that is a fairy story for people afraid of the dark."[2]

I assume he means the dark of nothingness when all goes quiet, which is not dark at all but the absence of either dark or light, rather than the dark of the bardo of death (and the terror of its dazzling brilliance) because he believes in neither. But, again, if this biocomputer is a zombie, where did it get its referent "I," or is it merely a clever simulation? Who is the computer fooling with its fairy tale, and how did an imagination of the universe get into its evolving software? (And what is making that particular model so grumpy and snide?)

Hawking reflects in spades the oddly chauvinistic notion that not only can a human soul not exist but the universe can only create itself from nothing out of nothing for no reason, or we are dipshit rubes. But who made up those cider-house rules? Who made the death of God modernity's most valuable commodity? And, yet again, where did the Void get the tools or the chutzpah to make mind? Hawking has an answer for that one: "Because there is a law such as Gravity, the Universe can and will create itself from nothing. Spontaneous creation is the reason there is something rather than nothing, why the universe exists, why we exist. It is not necessary to invoke God...."[3]

Well, Stephen, to quote my wife when she got angry at me once: "Who are you? You're just some guy I met."

●

I would argue that the sheer ontological issue transcends the biophysical and psychological one, and at a scale that dwarfs quantum switches: something coherent runs through nature as a whole; you can call it function, agency, or consciousness, but it is essentially and fundamentally and existentially different from what nature would be in its absence. If there is life anywhere, there is consciousness there too, at its corresponding frequency, and the gravity of which Hawking speaks so admiringly is just another of its effects. The universe is not just linearly and serially conscious; it is not, or not just, algorithmically and emergently conscious; it is not just panpsychic and/or quantum-entangled. It is conscious as an emanation generating its own proto-gravitational (and quantum gravitational) waves.

When you observe ants in an anthill, you may not be looking at dinner-table consciousness, but not only are you looking at something encompassing memory, knowledge, and identity, you are experiencing the contact reality of another form of mind, not just an incipient form of your own intelligence but *another* form. In sharing cosmic space with it, you are glimpsing the bare lens's edge of the universe's axle—its transgalactic cut. You are observing it writing itself on its own body, not only symbolically or molecularly but actually: fire onto fire. That is what the ant *is:* raw transdimensional mind.

When you disturb old hay such that an eclectic plethora of bugs comes scurrying out of it every which way looking for new shelter, that is as much a burst of reality as a flare on the Sun. Each one is a sally of solar energy and photosynthesis. Yes, and phototropism too.

So what is the universe writing? I would propose that it is writing everything it has to say, which includes us and what we have to say, or we wouldn't be given bodies and ganglia whereby to say it. It also girdles bees and crows and tortoises and what they have to say. It has to or they wouldn't be here, building nests, guarding eggs, defending their zone—they wouldn't be anywhere. And I mean to take this a step further: what ants have to say, in their mute doggerel, what gulls and frogs intone to comrades along shorelines, is exactly and esoterically what we have to say too and what we have been trying to say even yet, and what lies at the heart of all psychospiritual, shamanic, totemic, and magical operations: I AM.

Say this well and you have said everything and can proceed to the next tier. That is what the universe as starry night is expelling with every breath of itself as its vast exhalation intersects each inhalation of our own presence-existence.

●

In visiting giant sea turtles as they paddle, poke, and tumble among the rocks in Turtle Cove, Po'ipu Beach, Kaua'i (July 2010), I see gravity in action. At their sentient root, the terrapins defend their own reality with every proprioception and cold reptile breath. They *cannot be dislodged from it.* Their consciousness is their link to their identity, their integrity, their destiny. *And they know it.* That is, their bodies know it.

Up close they look older, more alien, and less friendly than from across the cove. They are wearing their shaman masks with serious heed. No funny business or truancy here. Their expressions are beyond diffident, but I wouldn't call them either aggravated or serene, though there are qualities of both. Their breath is rough, more like a snake than a bird. What they broadcast is not a human emotion; it is something melding imperiousness, dignity, profundity, ferocity.

I don't mean that they are wise in a human sense, nor do I mean to inflate them in anthropomorphic terms. But they *are* massively intelligent. Their wisdom is the wisdom of their species, their longevity and their acquaintance on a daily basis with the sea and its mysteries.

All sentient realms are intelligent and share a destiny, an antediluvian belief system that cannot be defeated by our modern exegesis, its barges, jetties, and cement. Turtles and other creatures ride hard upon and against human reality.

The difference between what is going on inside a beehive or anthill—where consciousness is its own subject and object both, and neither can be wrenched from the other—and nothing at all is incomprehensibly and immeasurably greater than the distance between a snake's hiss and Einstein's theory of general relativity.

A "mind" awakens to itself as "universe," which is also awakening everywhere locally. And it doesn't ever seem merely wished and dreamed. The more it thinks about itself, the more parts, pieces, moments/instants seem to fit into an objectified electro-chemical, neuromatter "happening" that is up to a point also able to assemble "pieces" into narratives which, though never completely true, seem to be more or less precise, proximate, and emerging through the electro-chemical, neuronal whirlpool—or not.[4]

The hero's voyage is through the vortex of existence in which MIND grasps at its objects of thought, holding to what it knows or tries to believe, thinking itself up as it goes, forgetting yesterday's news in tomorrow's desires.

●

When I ask myself why scientists permit themselves to become peons of nihilism and mere collocations of neurons, why they seek to map their entire sense of being and concede their existence to molecular ciphers, initially I can find no upside other than professional zeal, recreational arrogance, and jock ambition. What is the compensation for being demoted to robots and zombies? The lone upside is that we get to carry out our experiments in a sealed chamber, no interference allowed from elsewhere—we can have pristine results.

But the Big Bang came from outside the chamber, meaning everything came from outside it and continues to come from outside it. Nothing about us is pristine. Even after the Big Bang forged the chamber and its rules, we were still operating inside an equation of mostly unknowns. It is one thing to deduce our position from proximal facts; it is another to argue away as big a fact as consciousness.

But then I realized, if we come from nowhere and exist only as illusory states, then we get to shut down and go away for good at the cessation of this vision. We have a final and permanent escape route via death: the Big Sleep/the Eternal *Nada*. We won't ever have to meet one another again or do any of this or fix what we broke or what got "broke." We can trash and profane the joint, raise creatures in labs and chop them up without fear of karma, run drugs or guns or slaves. Then we just throw ourselves on the pyre of anonymous nebulae, star systems, worlds being born in the godless, soul-less nursery of nature. A maudlin charade of martyrdom. A plaintive, melodramatic tragic hero's swan-song: "Goodbye forever. Ashes to ashes anyway." The perfect excuse, the seamless deliverance.

The "objective" position of science, as you know by now, is that consciousness is a passing effect of neural processes, that personal existence ends with brain death: "Nothing happens. We just become mulch. When our bodies give out, we simply cease to exist. We feel nothing, and we never experience anything ever again. Depressing as fuck, but there it is."[5] Actor Sean Penn was speaking for politically sophisticated humanitarians, defending his relief work in Haiti when he offered his opinion on the matter (October 2011, *Piers Morgan Tonight*): "I don't think there is anything. All will go black and silent."

Black and silent! A pretty vivid end to this vividness. But perfect in its way for all concerned. We don't have to confront the depth and profundity, the immensity of the universe because it will swallow and obliterate us, prior to any grim or gloomy encounter, the way a frog eliminates a mosquito with a quick flick of its tongue, the way Jupiter dissolves a meteoric pebble in its clouds. Until then we are free to skitter like fey bugs on the surface of minded reality. *We have no responsibility to know or liberate anything or anyone, let alone this planet or cosmos. We have no answerability for justice or lasting joy, no imperative for inhabiting the depth of our own mystery. We don't have to undergo the big pain, the big grief, the big betrayal, the big panic—just little ones en route to the masher. We don't have to take our demons to the mat. We disintegrate or evaporate well before that. We return to memoryless nothing. We go back into slumber forever, black and silent, because we were never really anything in the first place.* We get to escape the monstrous enormity of the unconscious abyss and the various gods, demons, and ghosts that guard the archetype and hover about the intimidatingly cavernous entry to ourselves. We don't want to know what lurks therein, and science tells us we don't have to. We can make a clean getaway—uninformed, stupid, innocent forever.

If we're only consoles hatched to monitor other consoles, then who gives a damn? Our existence is incidental and inadvertent, just another console to be discarded with the surplus tin cans lower down in the pecking order, none of them with legitimate pretensions to phenomenology.

The End. They pull the plug. Poof! Into the Great Void. Might as well zone out in couch-potato mode, open another cold one. At least amuse yourself, pass the leisure time between experiments (or social engagements or days at the "shop").

That is modernity's protocol—and it's mandatory. Its acceptance and acknowledgment are a requirement for serious adult activity, even national-security clearance—also for coolness, mature intelligence, career advancement, flirtation, sex appeal. Stay hip, bad, bitchin', gansta, mod, pimptacular. Don't even think of indulging in limpdick metaphysics! Nihilism is the soul of charm, the sword of wit.

But if we are not nothing, if we can't shut ourselves off forever, then a serious existential situation is looming ahead: we have to live! We have to look. Eventually we have to learn the truth about ourselves. And that is trouble. Big trouble indeed! *We have to participate in the cosmic unfolding. We have to travel into the archetype and the unconscious, which is a hell of a lot more terrible for the tremulous to contemplate than the mere gullet of a frog or 10^8 × 1.43 cubic meters of Jovian soup.*

We are in it for the long haul. We are downright responsible, and for everything. We may snooze (or cop out) if we wish, but only for a while. Sooner or later we have to come back, do triage, pitch in, help. That was Seth's message to humanity, whoever or whatever—whenever.

We have to suffer and lose. Our denials will not extricate or conceal us for long. We are called to be stewards and bodhisattvas. Not only that, but being stewards and bodhisattvas is the only option, the only way out.

Science provides the lone statutory exemption to service, which is its secret attraction and claim. It offers a permanent deferral. It lets us off the hook by reducing us to accidents and then mopping up. Why is it not even science these days unless something dies?

While entertaining, feeding, and protecting us from the elements in ways major enough to constitute the bulk of our creature lives, the Church of Western Science is also whispering every second its mantra that we can depart scot-free and blow off everything in the process. Disguising its quitclaim in the most severe judgment of all—a death penalty—it is actually offering the sweetest escape in the universe, with no chance to be roped back in or recaptured: 'Just go to sleep, honey; lovely sleep, sleep forever.' Even Socrates hearkened momentarily to this lullaby:

"Now if you suppose that there is no consciousness, but a sleep like the sleep of him who is undisturbed even by the sight of dreams, death will be an unspeakable gain.... Now if death is like this, I say that to die is gain; for eternity is then only a single night."[6]

That is also what cynics, skeptics, and hard-knocks, tough-guy unsentimentalists, inner-city gangbangers, and Mexican drug-cartel enforcers are pushing (more or less), and unconsciously why they are pushing it: no rules, no ethics, no loyalty, no judgment, no custody, and you only go around once, so grab for all the gusto and booty. Kick all the ass that gets in your way and cut off their heads too.

But that's not what karma says.

There is another, quite different and even more deeply unconscious reason why scientists reduce us to behavioral machineries and mechanical circuits. Their very shtick is an inextricable part of the density we are traveling through, a necessary phase in our awakening. If we squandered our stay here in the catacombs, we would be nowhere, *sans* landmarks or possibility. Matter is delectable, deep, horrific, foul, and essential and, more than that, it is where we abide absolutely as well as conditionally.

Science directs our unbroken and unerring attention there because it can find nothing else, because it has a genius for its deep mechanics and operations and so has surrounded us with a virtual reality and mall world created only of algebra of substance and its termless abstractions. It hasn't made reality real, *but it has* because it has made us mortal, vulnerable, and dense.

But then science has not taken full account of its own diagnosis or ridden it to its natural conclusion. With its machines—telescopes, microscopes, cyclotrons, spectrographs, and the like—the physical quest has come upon a great cosmic revelation—how could it not? The subatomic underpinnings of matter point toward the pure energetic, transphysical basis of not only all things but reality itself. Though science acknowledges pretty much every graviton and dram of this situation, it does not honor its ontological conclusion, meaning the criteria intimated through not just any but an explicit manifestation: a dream within a dream within … within what? I tell you what: an attunement of mindedness to an exquisite frequency in which solid hallucinations *appear* to arise—something solid that isn't actual, to work out things that *are,* whatever they might be (we'll probably be the last to know, at least in that sense, but I'm going to give it my best shot in the next two volumes).

While offering its universal honorable discharge, science paradoxically ensures and enforces that we, as a civilization and as individuals in that civilization, are taking a real journey; that no one is skipping town or otherwise alibiing out. While alive, we are commissioned with creating Reality.

That is why technology is such a blustery, arrogant cop, why it anoints itself judge and jury. It wants its molecules to be the sole currency.

The world is not as ephemeral and evanescent as a dream, but it is made of the same sort of stuff: emptiness and vibration. *It may be a hallucination, but it is not that sort of hallucination.* It is—I'm sure you get this by now—a *real* hallucination. It exists because there is nothing to replace it and no way to shut it off or go somewhere else. Otherwise, we would bar-hop hallucinations, and life would be an endless round of optional albeit lucid dreams. Not good. Not good at all.

Yet these two realities—the real hallucination of a "real" universe and a creative imagination of the same universe—mysteriously come together in such a way that, while we cannot escape the bottomless reality of things, we do change their foundational meanings through our consciousness. The universe is overdetermined from every angle. This means that each possibility is true in its way—panpsychism, natural selection, emergence, zombies, functionalist behaviorism, quantum entanglement, libidinal cathexis, psychokinesis, and the bubblegum aroma of a particular candy-shop to a child—or not so much true as "meaningful." Each possibility overdetermines the meaning of each other, and together they overdetermine creation and pull us irrevocably toward who and what we actually are.

This can't be helped. It is the gravity-well into which each thing is sucked, the only reason why anything is happening at all and whereby all things cohere as meaningful, which is sometimes a horror, sometimes a delight, but always a portal into the same unknown condition that has called us into being at all.

We have no choice, but we never had a choice—this is what we are and what anything is: to think it is to become it, to create it. Only we thought it so long ago that we don't begin to remember what it was that we thought or who we were then. We don't even remember what it turned into when we forgot it, or what that then turned into when we forgot that too. And we are much too far down the road to turn back.

In truth, every scientist knows at heart that he can literally create universes, though not through his calling as a scientist. Every other creature can also create universes. It is arising inside each beaver and crow and one of us at each moment and juncture, expanding in all directions—and it is more than just our experience; it is what our experience is forming and turning us into.

That is why every scientific proclamation and theory, even when etched in numbers and operational signs, is tongue in cheek, though you need a crabwise view to get the irony. Flanagan, Churchland, Hawking, Dawkins *et al.* are comedians, funny "guys," and I mean that in the best sense. I prefer their dark comedy to the cloying sincerity and spiritual bypassing of both psyche and matter by the New Age (and other institutional churches). I like that scientists are not looking for an easy way out, because there isn't one. This hallucination—real or unreal, phenomenal or epiphenomenal—fills every potential crevice and crack with itself.

At the same time, I don't believe that this puzzle can ever be solved by scanning its perimeters—by taking apart shells of matter or probing the fine edges of observable hydrogen. The clues lie in our experience of being in the pickle in which we find ourselves. Not in a metaphorical rendition of the pickle but the pickle. In the candy-shop itself.

All we have to do is *look,* in the way that we look through telescopes and radio-telescopes at a vast, fathomless zone, but inside ourselves at the same universe with the same degree of precision and fidelity. Of course, it is not inside, and it is not *merely* the same—it *is* the same, literally, actually, and eternally, potentiating psychically through spaciousness as well as molecularly and transgalactically around and beyond us. It is very beautiful and very painful because it is very deep. It is scientifically rigorous, but it is also "outrageouser" than a funky monkey.

Scientists (and skeptics and nihilists) have to actively shut down their imaginations, their innate viewing portals, the luminosity from which they themselves are arising, from which their existence is arising, from which their skepticism and nihilism are arising, *not* to see it, *all the time,* not to experience it as everything and as the container for everything.

Every scientist knows that she can literally create universes, as she does it from breath to breath, and then denies it all the way to morning in order to pay homage to her idols.

●

I understand the ascension and hegemony of science. But I do not see why it gets to speak for not only consciousness (an arguable client), but for identity, meaning, and value. These are not its clients; they arise from the universe in a way that is neither tracked nor trackable by physics. The notion that consciousness must reduce entirely to functions and fluctuations of the brain, hence of matter, hence of information generated by entropy and heat, is not some transparent verity or impartial statute. It is propaganda, as surely as the Christian Fundamentalist, Discovery Institute patriarchal horseshit that ends up on Fox News—just a bit more wily and droll.

All deaths under scientific edict are suicides.

If matter is intrinsically and inherently conscious, then the whole universe as well as every theory of the universe turns inside-out and has to be engaged and executed differently—differently epistemologically and differently ontologically. And, again, that is quite another kernel of popcorn from what science intends by the Big Bang.

When revved-up rival groups bring their pit bulls or fighting cocks and set them viciously against each other, the ontological layers aroused flow and fold over each other as syrupy as cake batter being poured. There are no peepholes, no fissures. There are only parties to this affair who are independently aware and orchestrate the acts that will arise and are already arising (the gang-bangers), and there are unaware parties from whom the event flows autonomously out of a different fold in the universe (the pups and fowl). Neither party quite recognizes the other or their movements along absolute tracks that are nonetheless manipulable.

Out of unity comes duality; every two gives rises to a third. But how many additional layers are folding in, what other forces are manipulating the players, and why is the cake being poured in the first place?

You can take those questions out to the galaxies and stars or down into the emerging layers of embryonic and existential structure or to the next panorama littered with barking pit bulls, dead chickens, and testosteroned males, but their ontological status will not change.

Either our feeling of identity and purpose and divine connection is bogus and delusional, or it is the very fabric of an illumination from which we emerge and have lives. If it is rooted somewhere, then that is not only where the flame touches the wick but where the wick touches the thing that it was made of long before any spark came along.

I have written this book to give these two horses an honest run alongside each other: the world of science and the world of spirit. You know my viewpoint, but it's not that I believe it every moment with supreme confidence. I am a skeptic too, not a card-carrying debunker but a doubting Thomas in the way that all denizens of my century are. I have been inoculated with science and nihilism; it is part of the texture and sap of my being as I bob here with you in the backwash, the pool of sanctioned reality. I am calling out to you and the gods and just about everyone else within psychic range: let us bring this home in style!

I am also kicking the can down the road, trying to give you/me/all of us time to work it out. I am providing options for reconsidering our occasion.

My position is twofold: The first is that science is stuck in a trap and has snared civilization in the same trap. The second is that we are in the trap because that is the only place we could presently be on our pilgrimage toward meaning. The ecocatastrophes crescendoing out of our progressive, enlightened technologies are also as they should be, the precise conditioned result of who we are.

Yes, we could ask for a better, happier, safer universe, leaving out all the brutal and tawdry stuff, but it wouldn't be this universe, and it wouldn't be us. So witness the empty, existential truth being wrung by science from a universe that, at base, is nothing masquerading as everything. I will say that again: *is everything masquerading as nothing.*

Deleted Scenes

About three percent of the overall deleted material, these are taken from all three volumes (but mostly this one) and represent an alternate book as heard intermittently from the far end of a crowded restaurant.

to weave the trance and spin the spell

or that the universe itself might not have existed

nonlinear, especially in relation to our incarceration in space and time

The question is: before or after the tree fell in the forest.

What can be shut off isn't real.

You are in queue, always.

closer than gravity, nuclear force, and cell organization, which is an oxymoron because they too are vibrations

That is the esoteric source and *raison* for the world.

In the next phase, immensity and mystery will become the terrain. They will no longer be outside.

back through Greece, Rome, Egypt, Mesopotamia, and the Stone Ages

You get sucked into the intelligence of the pupa wrap.

Ask oil-eating bacteria.

It's Burger King reality.

The universe is such that misdirection and wild goose chases land closer to the target (and source) than rational analysis.

What about lizards and frogs; sticklebacks, clams, snails, fleas, grasshoppers, flukes, jellyfish, amoebas? What about individual cells? Where is the threshold? Or is "being" merely an extension or activation of nothingness?

and simians walked away from prosimians and shrews

of all manner of shape and form and degree but by what proxy and permission?

an unbidden, bidden journey through matter and existence under a mysterious influence and unsourced energy field

the nascent horror-beauty of the universe

If, on the other hand, consciousness can be destroyed—crushed like a computer in landfill—nothing is real, and Stephen Hawking and the neuroscientists got it right. We could live three billion years and it wouldn't matter. It would never change. This is *manifestation*.

The universe is a machine of interdependent origination, a subjectivity that is expanding in all directions.

What is the nature of our attention? Where does *it* originate? Where do you get it from? What holds you to "being" from moment to moment? Why do you not go flying off? What deeper meaning continuously engages your ostensible self?

Eighteenth-century philosopher David Hume wrote: "[W]hen I enter most intimately upon what I call *myself*, I always stumble upon some particular perception or other of heat or cold, light or shade, love or hatred, pain or pleasure.... Were all my perceptions removed by death, and could I neither think, nor feel, nor love, nor hate, after the dissolution of my body, I should be entirely annihilated, nor do I conceive what is further requisite to make me a perfect non-entity." More than two centuries later, this still holds for the scientific world-view: the body is not only the mind but the self and the soul; that is, there is neither consciousness nor soul in reality.

It is more likely that, even within science, a new mental-physical category will arise, something presently unknown, because it is hard to imagine mental phenomena reduced to solely physical coordinates.

And if we are conscious while awake, what are we when asleep? What place do the legions of unconscious contents of our so-called mind have in delineating our identity and being?

We see consciousness change without any change in the brain; yet changes in the brain almost always affect consciousness.

For some people, consciousness means "me," "mind," or telepathy, channeling, souls, spirits, etc., but scientifically oriented folks feel an onus to explain how consciousness is tied to matter in a physical universe.

Paradoxically, no study of consciousness or in fact any mental phenomenon is possible without consciousness first.

Neurons are like computer chips with protoconscious states as they emerged in nature.

The functionalist/behaviorist response takes us to Zombieland.

If we have a clear sense of what the soul or the mind is as an independent, subsistent entity, how do we then relate it to a particular sperm or a particular egg?

Language plays a huge role in the thickening of phenomenal consciousness.

Mental states might also occur in chemically very different beings from us, as long as the proper functional relations occur.

Consciousness may be rooted in the brain, still the capacity of the brain to cause it is fundamentally different from its capacity to move arms and legs.

We are faced with the invasion of an aroused stone.

The moment that person is eliminated, the universe goes about replacing them.

Sorry—no way out.

Life situations put the consciousness in a context.

Something like a Supergalactic Didgeridoo dispatches these notes through the many cosmoses.

Animal bodies are, like human bodies and personalities, conduits and masks for souls.

We cannot fully understand or appreciate meta-consciousness until we grapple with the problem of defining *any* consciousness.

World + Consciousness = Reality. Reality contains consciousness at rock-bottom.

This is almost enough to awaken the sleepy matter of the universe.

If we survive the present trance, as we perhaps will by the skin of our teeth, another awaits us, and then another. We will survive them too, and live through each of them at a level more profound than we can discern (or may presently desire).

absolute humility, total conviction, and the clarity and sincerity of one who doesn't have to sell a bloody thing

Simply shine a bright and discerning enough light, and we are suddenly in a much vaster zone.

Whereas Freud disclosed consciousness and ego in their sufficient biological and energetic states, working off the mere random debris of a Darwinian universe, Jung liberated them to their role as soul-maker and co-creator of the universe.

This physicochemical damnation, a conversion molecularly and psychospiritually into nuclear fire, is a signature act of metaphysics, eschatology, and, yes, science fiction. "Damned" means materiality: suns and galaxies en route back through hydrogen and helium atoms into planetary molecules were once living creatures on long-ago worlds who became unconscious.

I am not saying that zombies can't do this, but they are going to have a hell of a time pulling it off.

Myths and symbols are not only an emergent but an etiological property of mind.

You could be an ostrich or crocodile, could have feathers or scales, but you would be able to create multiform realities and have a meaningful degree of freedom in that range.

We live in prayer and humility or we live in terror.

Even the Big Bang is a fashionable but illusional figment.

The universe is sacred not only when we go about spiritual activities and choreograph ourselves in a sacred way. The world is sacred inside-out and outside-in, in every gesture and call. If it were not sacred, the Creation would not and could not happen.

The unconscious is profound, mutinous, and constantly cascading its contents into poor oblivious consciousness—the conscious realm by comparison is miniscule.

"Carved" is not quite the word, but I want to convey the sense of a village incised inside a walnut.

a honeycombed set of molecularities within a hyperspatial matrix

It was not as though I wanted to go there, not at all. I went because it was more compelling and ineluctable than anything else. And I could think of nothing else to do.

They understand that it is exactly how things should be despite those *"heartaches by the number, troubles by the score."*

They are our bane, but they are our sole opportunity.

We are the lollipop the universe is sucking, its own liqueur of sensation as we are groped and tasted and sprawl deeper into the density of our own stuff.

We are each as vast and multidimensional as a galaxy, linked in a metagalactic network across hyperspace. That's fly.

This is what you have been doing since the beginning, if there is a beginning. I mean, it's not as though you were invented yesterday by some syzygy of sperm and egg and arose Cytherean on the beach of consciousness. All souls are old souls.

Our patterns are impermanent and brief, but they are a sole disturbance of the void.

Even a bare bobbin is more easily proposed than propagated.

Cells are the conduits and vehicles, the chariots and chalices of our metamorphosis.

Joy is an atomic as much as an emotional frequency.

Amid brownish-black lagoons the edges of the islets are streaked with dirty orange. A cobalt blue mountain shading to absolute black at its peak overlooks this bright landscape.

This is always the tarot card *next* to be drawn.

though eerie and forlorn

Beasts and ruffians are becoming love.

Whatever our true origin, it is the destiny that awaits us.

How could there be mind unless mind were inherent in nature? If stone were not implicit in the matter of the Big Bang, the plasma of stars, how did worlds get made, landscapes differentiated, shapes designed? Or was it all thermodynamic happenstance?

Our luminous city inside its acorn!

The *mechanism* of consciousness is never the experience of consciousness.

We are created out of the brains and minds of ancestral creatures, and yet we have immaculate canvases on which to paint with their archaic brush.

Again, the eye cannot look through itself *at* itself.

Nature may be a mistress of caprice and fickleness, but it has adorned its baubles with all the trimmings and then some—and, in so doing, turned the proverbial 'hundred pounds of clay' into a grateful viewer to admire its handiwork.

No operator need apply.

The Moon may distribute tides, but it is not the source of gravity.

How are you going to test for the Holy Ghost?

as they vamp, dowse, howl like wolves in the night

Consciousness is even more indeterminate than the antics of subatomic particles.

Is the universe somehow more textured or vivid to a moose than to a damselfly or stickleback?

Because the eye can detect one quantum of green light (once fully adapted to darkness in experimental situations) it would suggest that sensory perceptions are at the level of quantum effects and thus consciousness might itself be quantum-mechanical.

the infinite promise of fixing everything ultimately that needs fixing. Of course it won't do that, but the promise itself is timeless and radiates its smile across time. It finally comes down to spaciousness and a wink.

It is a verdict rendered by both life and death. It is rendered again, moment to moment.

If the universe is writing the embryo on its body, then we are dictating the universe back on our own bodies.

Ordinary operating reality is esoteric at its core; likewise, the esoteric background expresses itself in a concrete universe.

with its boats, masks, and tombs

You finally have to live—and die—on your own terms. Incarnation is deep—a kaleidoscope of intersecting mazes, sometimes with little reward for weary wanderings. Luckily, though, the complexity that we suspect is real and, though inexplicably beyond simple diagnosis, it is identical to our own complication.

It is because this stuff mostly doesn't happen any longer in churches that the open road, with its bars, cars, bazaars, and *Dancing with the Stars,* is everyone's service, everyone's sacred ground.

They've got all lines of communication blocked, and they own the politicians, the corporations, and media.

Finally breath is more than just lungs and air.

but they aren't, nor will they ever be

It is a single wave, a soundless mandala lotus exploding in the night sky and your DNA-attuned being.

the only thing that finally won't let us sabotage it

to awaken within this life "dream" to something not quite the dream

I more than intuited the truth; I knew it with great and heart-felt conviction. And I held to that conviction despite the battering I took from all sides and the mega-damage I caused. The synchronicities were too persistent, too much like birds who suddenly spoke Latin, like deacons calling me back to a mission every damn time that I was about to abandon it. A paranormal event was trying to get born.

Think about it. You are constructed of atoms and subatomic particles, so you can't know anything foreign to them and everything you know is composed of them at some level. Imagination is literally creation.

I am not saying that any of these is or isn't the universe. I am saying that we need to balance something else against the universe of astrophysics and neuroscience to get anywhere close to a full picture of the manifestation that we are inside of and of which we can't see an actual shape.

Desire is not irrelevant or trivial against the universe's vastness or its collective meaning or meaninglessness. It is the Rosetta stone, the link between ranges of undifferentiated, unassimilated energy. Pursuing desires hones individuality and expresses who you are and are becoming. Desire opens the cosmic cornucopia to your discrete frequency. So whether you get a desire fulfilled or frustrated lies anywhere between minor and irrelevant against the subtle energy contained in the desire itself. Individuality is an exquisite sensation of desire's intricate meanings and elusive properties. Everything else in the universe decomposes, destructs, passes, and is forgotten—the Earth itself, the Sun, the Solar System, even the Milky Way—but individuality, whatever it is, persists on the basis of desire alone—Hesiod called it eros, Freud libido.

Insofar as Creation is a play of pain and pleasure, both are inevitable, so prolonged fulfillment is never in the cards. Yet it is the sheer act of pursuing preferences that makes you distinctively human and binds you to life as yourself.

We forget sometimes that we aren't fun machines or self-improvement courses. We are moving from eternal moment to eternal moment, each packed with information and complete unto itself and the universe—*and, at the same time.*

Grief is the full and only measure of stable joy.

One lives koans, riddles, and signs. One witnesses a universe *inside* inside existence, a universe fructifying and expanding.

The real is always ... real.

for the lifting of karma from our cities and nations, from the broken pledge between the living and the dead

Somalian warlords, overseers of cattle slaughterhouses, Tijuana assassins, producers of kiddie porn, Pyongyang oligarchs, and the like are the ones that will provide the catalyst to get ourselves over the hump into the Jamboree. For now they are imbibing the black radiance while its source radioactivity decays.

That's not only the present intimation here in Hardcore City; it's the activating core of our crisis all the way back to the Stone Age. It's the passion play of our times, fucked up as it has gotten lately. But it was fucked up too for Caesar's troops and the warriors of Charlemagne, for the Crusaders and the roughriders of Genghis Khan—long before Shock and Awe ignited the video-game Xbox. Because it is the same crisis, from the handaxe fratricide of Cain and Peking Man to the nuclear bomb and standard-of-living climate wave. *Everything is at stake.* Don't forget those spirit beings elsewhere and everywhere, watching and rooting for us. Everything is at stake. *Everything in the universe.*

Let us leave these multiple pictures projected against the single dome of the human universe. That is what the Elizabethans, Taoists, and pre-Socratics did re: Lao-Tzu, Robert Fludd: *Utriusque Cosmi, Majoris scilicet et Minoris, metaphysica, physica, atque technica Historia.* Frankly no one is losing sleep over whether we are in an Einsteinian or hyperspatial universe. It may make for lively debates, but we have a prior form uncoiling inside of us. Lao Tzu might have added (and did): "The snow goose need not bathe to make itself white…. To a mind that is still/the whole universe surrenders…. When nothing is done/nothing is left undone…. Every step is on the path. The Tao that can be told/is not the eternal Tao…. As soon as you have made a thought/laugh at it…. At the center of your being/you have the answer;/you know who you are/and you know what you want."

Do you not feel the tangle and turbulence whenever you approach the real in yourself and go deep enough for it to buckle and hurt? And if you peer inside even a little further, to the point where it comes flying back out at you, do you not realize that these attributes are something else entirely, and always have been, from the dawn awakening to the first erotic kiss?

The universe is the pilot of absolute possibility.

Where its neuron stream touches a bridge leading elsewhere, a scotoma forms.

We live in the cosmos's post-traumatic breath.

Endnotes

This is not an academic text. Though I have tried to footnote with care, the housekeeping was overwhelming, and I have lost quite a few page numbers and references along the way and probably mistranscribed some others. Others disappeared in Internet appropriations before I came along. Yet I hope these notes serve their purpose as a general map through my sources.

My capitalization is also not consistent. "Sun," "Moon," "Earth" are capitalized as heavenly bodies but not in their meanings of "sunlight," "satellite," and "soil/mantle."

I have transferred many passages from earlier works of mine (as cited) into *Dark Pool of Light*. I then proceeded to edit and rewrite them *in situ* as if they had been conceived there in the first place. I will refer to these reuses in the Notes as "grafts." "Partially grafted" means that I have mixed old writing with new material.

Note: Foreword contributors used earlier versions of this book to quote from.

Opening Epigraphs

Max Planck, *The Observer*, London, England, January 25, 1931.

Barclay Martin in "The Mystery of Consciousness, Con't.," *The New York Review of Books*, September 29, 2011, Volume LVIII, Number 14, p. 101.

Sir Charles Sherrington, *Man on his Nature* (Cambridge, England: Cambridge University Press, 1963).

Francis Crick, *The Astonishing Hypothesis: The Scientific Search for the Soul* (New York: Simon & Schuster, 1995), p. 3.

Richard Feynman in "The 'Dramatic Picture' of Richard Feynman" by Freeman Dyson, *The New York Review of Books*, July 14, 2011, Volume LVIII, Number 12, p. 40.

William Blake, "The Tyger," *Songs of Experience*, 1794.

Herman Melville, *Moby-Dick; or The Whale*, originally 1851 (Berkeley, California: University of California Press, 1979), p. 348.

Introduction

1. Siegfried Lodwig, personal communication, October 2011.
2. This trope is seminal to my work. Its first unrealized version appeared in *Embryogenesis—From Cosmos to Creature: The Origins of Human Biology* (Berkeley, California: North Atlantic Books, 1986), pp. 335–340. From that template it came into its own in a sequence of sections entitled "Spirit and Matter," "Recapitulation is the miniaturization of astronomic events in tissue," "Conception," "Blood," "Incarnation," and "Birth" in the revised edition of the same book, *Embryogenesis: Species, Gender, and Identity* (Berkeley, California: North Atlantic Books, 2000), pp. 715–723. Then I called it out by name in the Introduction to *Embryos, Galaxies, and Sentient Beings: How the Universe Makes Life* (Berkeley, California: North Atlantic Books, 2003), p. xxix, and interrogated it throughout the text.
3. I am indebted to Ronald Milestone for assistance with the articulation of this distinction.
4. Émile Durkheim, *Elementary Forms of Religious Life,* translated from the French by Joseph Ward Swain (London: George Allen & Unwin Ltd., 1915), p. 225.
5. Edward Dorn, personal communication, circa 1972.
6. Alfred North Whitehead, *Process and Reality: An Essay in Cosmology,* edited by David Ray Griffin and Donald W. Sherburne (New York: Simon and Schuster, 1979), p. 92.
7. ibid., p. 40.
8. ibid., p. 21.

Chapter One. What the Fuck *Is* This?

1. *Hesiod: The Homeric Hymns and Homerica,* translated by H. G. Evelyn-White (Cambridge, Massachusetts: Harvard University Press, Loeb Classical Library, 1914), p. 87.
2. John R. Swanton, *Myths and Tales of the Southeastern Indians* (Creek Stories), *Bureau of American Ethnology Bulletin, No. 88,* Washington, DC: Smithsonian Institute, 1929, www.sacred-texts.com/nam/se/mtsi/mtsi003.htm.
3. Henry R. Voth, *The Traditions of the Hopi* (Chicago: Field Columbian Museum, Publication 96, 1905), p. 1.
4. William James, "The Stream of Consciousness" in Ned Block, Owen Flanagan, and Güven Güzeldere (editors), *The Nature of Consciousness: Philosophical Debates* (Cambridge, Massachusetts: The MIT Press, 1997), p. 82.

Page 8: The section on the surf was grafted from *2010 Kaua'i Trip,* www.richardgrossinger.com/2010/08/2010-kauai-trip.

Pages 8–9: The material on beavers, muskrats, and squirrels is drawn from *Planet Earth*, a 2007 documentary video by David Attenborough.

Page 9: The material on habits of various birds is drawn from *The Life of Birds*, a 1998 documentary video by David Attenborough.

CHAPTER TWO. THE SCIENTIFIC VIEW OF REALITY AND CONSCIOUSNESS

1. David Chalmers, "Facing Up to the Problem of Consciousness," *Journal of Consciousness Studies,* 2: 3, 1995, p. 203.
2. Owen Flanagan, "Conscious Inessentialism and the Epiphenomenalist Suspicion," in Ned Block, Owen Flanagan, and Güven Güzeldere (editors), *The Nature of Consciousness: Philosophical Debates* (Cambridge, Massachusetts: The MIT Press, 1997), p. 357.
3. ibid., p. 372.
4. Robert Kelly, "First in an Alphabet of Sacred Animals" in Richard Grossinger (editor), *Ecology and Consciousness: Traditional Wisdom on the Environment* (Berkeley, California: North Atlantic Books, 1992), p. 61.
5. This fragment of Heraclitus is much quoted, but I can't find a source for its translation. For instance, see George F. Butterick, *A Guide to the Maximus Poems of Charles Olson* (Berkeley: University of California Press, 1980), p. 82.
6. I am indebted to Frederick Ware for this distinction.
7. Julian Jaynes, *The Origin of Consciousness in the Breakdown of the Bicameral Mind* (Boston: Houghton Mifflin Co., 1976), p. 15.
8. Nick Herbert, *Elemental Mind: Human Consciousness and the New Physics* (New York: Dutton, 1993), p. 12.
9. Sydney Shoemaker, "The Inverted Spectrum" in *The Nature of Consciousness,* p. 649.
10. Güven Güzeldere, "The Many Faces of Consciousness: A Field Guide" in *The Nature of Consciousness,* p. 42 (see note 2, above).
11. ibid., p. 25.
12. Richard Rorty, *Consequences of Pragmatism* (Minneapolis: University of Minnesota Press, 1982), p. 183.
13. William James, "Does 'Consciousness' Exist?" (1904) in Güzeldere, "The Many Faces of Consciousness: A Field Guide," in *The Nature of Consciousness,* p. 13 (see note 2, above).
14. John Searle, "Reductionism and the Irreducibility of Consciousness" in *The Nature of Consciousness,* p. 454.
15. Owen Flanagan, "Prospects for a Unified Theory of Consciousness or, What Dreams Are Made Of" in *The Nature of Consciousness,* p. 97.

16. Francis Crick and Christhof Koch, "Toward a Neurobiological Theory of Consciousness" in *The Nature of Consciousness,* p. 287.

17. Daniel C. Dennett, "The Cartesian Theater and 'Filling In' the Stream of Consciousness" in *The Nature of Consciousness,* p. 84.

18. Nick Herbert, *Elemental Mind,* p. 47 (see note 8, above).

19. ibid., p. 94 (slightly rearranged).

20. Güven Güzeldere, "The Many Faces of Consciousness: A Field Guide" in *The Nature of Consciousness,* p. 34.

21. Nick Herbert, *Elemental Mind,* pp. 93–94.

22. ibid., p. 94

23. ibid., p. 98.

24. This paragraph was constructed from a variety of sources: Daniel C. Dennett and Marcel Kinsbourne, "Time and the Observer" in *The Nature of Consciousness,* p. 142; Daniel C. Dennett, "The Cartesian Theater and 'Filling In' the Stream of Consciousness" in *The Nature of Consciousness,* p. 83; Patricia Smith Churchland, "Can Neurobiology Teach Us Anything About Consciousness?" in *The Nature of Consciousness,* pp. 134 and 136; Nick Herbert, *Elemental Mind,* pp. 100–104; and Owen Flanagan, "Prospects for a Unified Theory of Consciousness" in *The Nature of Consciousness,* p. 106.

25. Deepak Chopra, "'Rain Man' and the Connected Life," intentBlog, www.intentblog.com/archives/2005/10/rain_man_and_th.html, October 21, 2005.

26. Joseph E. LeDoux, *Synaptic Self: How Our Brains Become Who We Are* (New York: Penguin Books, 2003), p. 94.

27. Sir Charles Sherrington, *Man on his Nature* (Cambridge, England: Cambridge University Press, 1963), p. 257.

28. ibid.

29. Henry Stapp, www.nonlocal.com/hbar/stappost.html, 1995.

30. Nick Herbert, *Elemental Mind,* p. 15.

31. Sigmund Freud, *An Outline of Psychoanalysis* (New York: Norton Publishers, 1949), pp. 13–14.

32. Daniel C. Dennett and Marcel Kinsbourne, "Time and the Observer" in *The Nature of Consciousness,* pp. 143–144.

33. ibid., p. 143.

34. John Tyndall, *Fragments of Science,* Volume II (New York: Appleton and Co., 1868), p. 86.

35. Nick Herbert, *Elemental Mind,* p. 23.

36. John Searle, "Reductionism and the Irreducibility of Consciousness" in *The Nature of Consciousness,* p. 453.

37. Colin McGinn, "Can We Solve the Mind-Body Problem?" in *The Nature of Consciousness,* p. 529.

38. Colin McGinn, "Consciousness and Content" in *The Nature of Consciousness*, pp. 298–299.

39. ibid., p. 297.

40. Thomas H. Huxley, *Methods and Results* (New York: Appleton Co., 1901), p. 191.

41. Colin McGinn, "Can We Solve the Mind-Body Problem?" in *The Nature of Consciousness*, p. 531.

42. Patricia Smith Churchland, "Can Neurobiology Teach Us Anything About Consciousness?" in *The Nature of Consciousness*, p. 127.

43. Brian Loar, "Phenomenal States" in *The Nature of Consciousness*, pp. 597–598 (slightly rearranged).

44. ibid., pp. 608–609.

45. Owen Flanagan, "The Robust Phenomenology of the Stream of Consciousness" in *The Nature of Consciousness*, p. 93.

46. Colin McGinn, "Can We Solve the Mind-Body Problem?" in *The Nature of Consciousness*, p. 531.

47. ibid., pp. 530–531.

48. Nick Herbert, *Elemental Mind*, pp. 115–116 (slightly rearranged).

49. David Lewis, "What Experience Teaches" in *The Nature of Consciousness*, p. 585.

50. Colin McGinn, "Can We Solve the Mind-Body Problem?" in *The Nature of Consciousness*, p. 537.

51. Thomas Nagel, "What Is It Like to Be a Bat?" in *The Nature of Consciousness*, p. 524.

52. Daniel C. Dennett, "Quining Qualia" in *The Nature of Consciousness*, p. 639.

53. Colin McGinn, "Can We Solve the Mind-Body Problem?" in *The Nature of Consciousness*, p. 531.

54. Noam Chomsky, *Reflections in Language* (New York: Pantheon Books, 1975), quoted in "Can We Solve the Mind-Body Problem?" by Colin McGinn in *The Nature of Consciousness*, p. 540.

55. Robert Van Gulick, "Understanding the Phenomenal Mind, Part I" in *The Nature of Consciousness*, p. 563.

56. Paul M. Churchland, "Knowing Qualia: A Reply to Jackson" in *The Nature of Consciousness*, p. 576.

57. Patricia Smith Churchland, "Can Neurobiology Teach Us Anything About Consciousness?" in *The Nature of Consciousness*, p. 130.

58. Patricia Smith Churchland, *Matter and Consciousness* (Cambridge, Massachusetts: The MIT Press, 1988), p. 285.

59. John Searle quoted in "The Mystery of Consciousness, Con't.," *The New York Review of Books*, September 29, 2011, Volume LVIII, Number 14, p. 101.

60. Patricia Smith Churchland, "Can Neurobiology Teach Us Anything About Consciousness?" in *The Nature of Consciousness*, pp. 130 and 133.

61. David Armstrong, "What Is Consciousness" in *The Nature of Consciousness*, p. 721.

62. John Searle, "Reductionism and the Irreducibility of Consciousness" in *The Nature of Consciousness*, p. 451.

63. The discussion of neurotransmission borrows from Marcia Angell, "The Epidemic of Mental Illness: Why?" in *The New York Review of Books*, June 23, 2011, Volume LVIII, Number 11, p. 20; and Nick Herbert, *Elemental Mind: Human Consciousness and the New Physics*, pp. 57–58

64. Nick Herbert, *Elemental Mind*, pp. 96–97.

65. ibid., p. 58.

66. Francis Crick and Christhof Koch, "Toward a Neurobiological Theory of Consciousness" in *The Nature of Consciousness*, pp. 284–285.

67. Owen Flanagan, "Conscious Inessentialism and the Epiphenomenalist Suspicion" in *The Nature of Consciousness*, p. 358 (slightly rearranged).

68. Owen Flanagan, "Prospects for a Unified Theory of Consciousness or, What Dreams Are Made Of" in *The Nature of Consciousness*, p. 103 (this quote includes text from neuroscientists Rodolfo Llinás and U. Ribary, "Coherent 40-Hz oscillation characterizes dream state in humans," *Proceedings of the National Academy of Sciences of the United States of America, 90*, 1993, pp. 2078–2081).

69. Jerry Fodor in "Meaning in Mind" in *Fodor and his Critics*, edited by B. Loewer and Georges Rey (Oxford, England: Blackwell, 1991), p. 285.

70. Colin McGinn, "Can We Solve the Mind-Body Problem?" in *The Nature of Consciousness*, p. 529.

71. Güven Güzeldere, "The Many Faces of Consciousness: A Field Guide" in *The Nature of Consciousness*, p. 31.

72. Colin McGinn, "Can We Solve the Mind-Body Problem?" in *The Nature of Consciousness*, p. 529.

73. ibid.

74. Nick Herbert, *Elemental Mind*, p. 21.

75. John Searle, "Reductionism and the Irreducibility of Consciousness" in *The Nature of Consciousness*, pp. 456 and 457.

76. Colin McGinn, "Can We Solve the Mind-Body Problem?" in *The Nature of Consciousness*, p. 529.

77. John Searle, "Reductionism and the Irreducibility of Consciousness" in *The Nature of Consciousness*, p. 458.

78. Colin McGinn, "Can We Solve the Mind-Body Problem?" in *The Nature of Consciousness*, p. 534.

79. Güven Güzeldere, "Consciousness—Perception of What Passes in One's Own Mind?" in *The Nature of Consciousness,* p. 804.

80. Tim Shallice, "Modularity and Consciousness" in *The Nature of Consciousness,* p. 272.

81. Michael Tye, "A Representational Theory of Pains and Their Phenomenal Character" in *The Nature of Consciousness,* p. 330.

82. Colin McGinn, "Can We Solve the Mind-Body Problem?" in *The Nature of Consciousness,* p. 541.

83. ibid. p. 534.

84. Siri Hustvedt, *The Shaking Woman, or A History of My Nerves* (New York: Henry Holt and Co., 2010), pp. 116–117.

85. John Tyndall, *Fragments of Science,* Volume II (New York: Appleton and Co., 1868), pp. 86–87.

86. Colin McGinn, "Can We Solve the Mind-Body Problem?" in *The Nature of Consciousness,* p. 541.

Pages 14–18: From "The second law of thermodynamics" to "bats, blowfish, and baboons" was grafted from Richard Grossinger, *Embryos, Galaxies, and Sentient Beings: How the Universe Makes Life* (Berkeley, California: North Atlantic Books, 2003), pp. 10–11, 20–23, 109–110.

Pages 23–24: The description of the reticular function was partially grafted from Richard Grossinger, *Embryogenesis: Species, Gender, and Identity* (Berkeley, California: North Atlantic Books, 2000), p. 451; the fearful reptilian brain stem was partially grafted from Richard Grossinger, *The Bardo of Waking Life* (Berkeley, California: North Atlantic Books, 2008), p. 62.

Page 37: From "For all its garish meat and explicit wiring" to "mystery object" was grafted from Richard Grossinger, *2013: Raising the Earth to the Next Vibration* (Berkeley, California: North Atlantic Books, 2011), p. 559.

Pages 38–39: From "Neurons alone activate" to "flirt with their own algorithms" was grafted from *Embryos, Galaxies, and Sentient Beings: How the Universe Makes Life,* pp. 405–406.

Page 42: From "Though we try to overwhelm the paradox of consciousness...." to "are not 'wires'" was grafted from *Embryos, Galaxies, and Sentient Beings: How the Universe Makes Life* (Berkeley, California: North Atlantic Books, 2003), pp. 405–406.

CHAPTER THREE. CONSCIOUSNESS: EVERYTHING AND NOTHING

1. Julian Jaynes, *The Origin of Consciousness in the Breakdown of the Bicameral Mind* (Boston: Houghton Mifflin Co., 1976), p. 1.

2. Güven Güzeldere, "The Many Faces of Consciousness: A Field Guide" in Ned Block, Owen Flanagan, and Güven Güzeldere (editors), *The Nature of Consciousness: Philosophical Debates* (Cambridge, Massachusetts: The MIT Press, 1997), p. 1.

3. Georges Rey, "A Question about Consciousness" in *The Nature of Consciousness*, p. 462.

4. Daniel Dennett, *Consciousness Explained* (Boston: Little, Brown, and Co., 1991), p. 21.

5. William Seager, *Metaphysics of Consciousness* (London: Routledge, 1991), pp. 223–224.

6. John Searle, *The Rediscovery of the Mind* (Cambridge, Massachusetts: The MIT Press, 1992), p. 93.

7. Jerry Fodor, "The big idea," *Times Literary Supplement*, No. 4657 (July 3, 1992), p. 5.

8. Thomas Nagel, "What Is It Like to Be a Bat?" in *The Nature of Consciousness*, p. 519.

9. George Miller, *Psychology: The Science of Mental Life* (New York: Harper and Row Publishers, 1962), p. 25.

10. Ludwig Wittgenstein, *Tractatus Logico-Philosophicus* (1921), translated by D. F. Pears and B. F. McGuinnes (London: Routledge and Kegan Paul, 1974), p. 57.

11. Güven Güzeldere, "Consciousness—Perception of What Passes in One's Own Mind?" in *The Nature of Consciousness*, p. 804.

12. Colin McGinn, "Can We Solve the Mind-Body Problem?" in *The Nature of Consciousness*, p. 541.

13. Stuart Sutherland, "Consciousness," *Macmillan Dictionary of Psychology*, Second Edition (London: The Macmillan Press, 1995), p. 95.

14. Colin McGinn, "Consciousness and Content" in *The Nature of Consciousness*, p. 304.

15. René Descartes quoted in Güven Güzeldere, "Consciousness—Perception of What Passes in One's Own Mind?" in *The Nature of Consciousness*, p. 802.

16. Colin McGinn, "Consciousness and Content" in *The Nature of Consciousness*, p. 302.

17. ibid.

18. ibid., p. 307.

19. ibid., p. 302.

20. ibid., p. 307.

21. ibid., p. 305.

22. William G. Lycan, "Consciousness as Internal Monitoring" in *The Nature of Consciousness*, pp. 766–767.

23. Brian Loar, "Phenomenal States" in *The Nature of Consciousness,* p. 601.

24. Fred Dretske, "Conscious Experience" in *The Nature of Consciousness,* p. 784.

25. ibid., p. 786.

26. Colin McGinn, "Consciousness and Content" in *The Nature of Consciousness,* p. 300.

27. William G. Lycan, "Consciousness as Internal Monitoring" in *The Nature of Consciousness,* p. 761.

28. ibid., p. 762 (rearranged).

29. Georges Rey, "A Question about Consciousness" in *The Nature of Consciousness,* p. 463.

30. Robert Van Gulick, "Understanding the Phenomenal Mind, Part II" in *The Nature of Consciousness,* p. 439.

31. ibid., pp. 437–438.

32. Gilbert Harman, "The Intrinsic Quality of Experience" in *The Nature of Consciousness,* p. 668.

33. Daniel Dennett, *Consciousness Explained* (Boston: Little, Brown, and Co., 1991), p. 218.

34. Martha J. Farah, "Visual Perception and Visual Awareness after Brain Damage" in *The Nature of Consciousness,* pp. 204, 214, and 225.

35. Daniel C. Dennett, "The Cartesian Theater and 'Filling In' the Stream of Consciousness" in *The Nature of Consciousness,* p. 83.

36. Alvin I. Goldman, "Consciousness, Folk Psychology, and Cognitive Science" in *The Nature of Consciousness,* p. 121.

37. Patricia Smith Churchland, "Can Neurobiology Teach Us Anything About Consciousness?" in *The Nature of Consciousness,* p. 129.

38. Daniel C. Dennett and Marcel Kinsbourne, "Time and the Observer" in *The Nature of Consciousness,* p. 143.

39. ibid., p. 167.

40. ibid., p. 145.

41. William R. Uttal, "Do central nonlinearities exist?" Cambridge, England: *Behavioral and Brain Sciences* 2: 2, 1979, p. 286.

42. Daniel C. Dennett, "The Cartesian Theater and 'Filling In' the Stream of Consciousness" in *The Nature of Consciousness,* p. 84.

43. Daniel C. Dennett and Marcel Kinsbourne, "Time and the Observer" in *The Nature of Consciousness,* p. 144.

44. ibid., p. 149.

45. ibid., p. 163.

46. ibid., p.162

47. Nick Herbert, *Elemental Mind: Human Consciousness and the New Physics* (New York: Dutton, 1993), p. 46.

48. John Searle, *The Rediscovery of the Mind* (Cambridge, Massachusetts: The MIT Press, 1992), p. 95.

49. ibid.

50. Güven Güzeldere, "The Many Faces of Consciousness: A Field Guide" in *The Nature of Consciousness,* p. 10.

51. Robert Van Gulick, "Understanding the Phenomenal Mind, Part II" in *The Nature of Consciousness,* p. 439.

52. Colin McGinn, "Consciousness and Content" in *The Nature of Consciousness,* p. 302.

53. Thomas Nagel, *The View from Nowhere* (Oxford, England: Oxford University Press, 1986), pp. 7–8.

54. *The Philosophical Writings of Descartes,* Vol. 1, translated by John Cottingham, Robert Stoothoff, and Dugald Murdoch (Cambridge, England: Cambridge University Press, 1992), p. 174.

55. John Locke, *An Essay Concerning Human Understanding,* Vol. 1, 1690 (New York: Dover Publications, 1959), p. 138.

56. Robert Van Gulick, "Time for More Alternatives" in *The Nature of Consciousness,* p. 183.

57. Robert Van Gulick, "Understanding the Phenomenal Mind, Part II" in *The Nature of Consciousness,* p. 439.

58. Güven Güzeldere, "The Many Faces of Consciousness: A Field Guide" in *The Nature of Consciousness,* p. 37.

59. *The Secrets of the Inner Mind: Journey Through the Mind and Body* (New York: Time-Life Books, 1993).

60. Daniel C. Dennett and Marcel Kinsbourne, "Time and the Observer" in *The Nature of Consciousness,* p. 141.

61. William James, "The Stream of Consciousness" in *The Nature of Consciousness,* p. 78.

62. William James quoted in Bernard J. Baars, "Contrastive Phenomenology: A Thoroughly Empirical Approach to Consciousness" in *The Nature of Consciousness,* p. 193.

63. Bernard J. Baars, "Contrastive Phenomenology: A Thoroughly Empirical Approach to Consciousness" in *The Nature of Consciousness,* p. 194.

64. Francis Crick and Christhof Koch, "Toward a Neurobiological Theory of Consciousness" in *The Nature of Consciousness,* p. 284.

65. William James quoted in Bernard J. Baars, "Contrastive Phenomenology: A Thoroughly Empirical Approach to Consciousness" in *The Nature of Consciousness,* p. 187.

66. John Searle, "Who is computing with the brain?" *Behavioral and Brain Sciences* 13: 4, 1990, p. 365.

67. David Armstrong, *A Materialist Theory of the Mind* (New York: Humanities Press, 1968), p. 94.

68. Georges Rey, "A Question about Consciousness" in *The Nature of Consciousness*, p. 466.

69. Michael S. Gazzaniga, "Right hemisphere language following bisection: A 20-year perspective" (*American Psychologist* 38, 1983), p. 536.

70. Owen Flanagan, "Conscious Inessentialism and the Epiphenomenalist Suspicion" in *The Nature of Consciousness*, p. 366.

71. John Locke, *An Essay Concerning Human Understanding*, Vol. 1 (New York: Dover Publications, 1959).

72. Owen Flanagan, "Conscious Inessentialism and the Epiphenomenalist Suspicion" in *The Nature of Consciousness*, p. 360.

73. Robert W. Thatcher and E. Roy John, *Foundations of Cognitive Processes* (Hillsdale, New Jersey: Lawrence Erlbaum Associates, 1977), p. 294.

74. Nick Herbert, *Elemental Mind*, p. 80.

75. ibid., p. 7.

76. I am indebted to Ronald Milestone for educating me about Turing machines and consciousness-simulating computers. The closing paragraphs of this chapter partially represent him speaking through me or me through him.

77. Georges Rey, "A Question about Consciousness" in *The Nature of Consciousness*, p. 478.

CHAPTER FOUR. DEGREES OF CONSCIOUSNESS:
PROTOCONSCIOUSNESS, PRECONSCIOUSNESS, AND THE FREUDIAN
UNCONSCIOUS

1. Nick Herbert, *Elemental Mind: Human Consciousness and the New Physics* (New York: Dutton, 1993), p. 24.

2. Jean-Paul Sartre, *Being and Nothingness*, translated by H. Barnes (New York: Philosophical Library, 1956), p. 173.

3. Robert Van Gulick, "Understanding the Phenomenal Mind, Part II" in Ned Block, Owen Flanagan, and Güven Güzeldere (editors), *The Nature of Consciousness: Philosophical Debates* (Cambridge, Massachusetts: The MIT Press, 1997), p. 440.

4. ibid., p. 441.

5. ibid., p. 440.

6. William G. Lycan, "Consciousness as Internal Monitoring" in *The Nature of Consciousness*, p. 766.

7. Ned Block, "On a Confusion about a Function of Consciousness," *Behavioral and Brain Sciences*, 18 (2), 1996: pp. 230–231.

8. ibid.

9. ibid.

10. Owen Flanagan, "Conscious Inessentialism and the Epiphenomenalist Suspicion" in *The Nature of Consciousness*, pp. 368 and 370.

11. ibid., p. 359.

12. Güven Güzeldere, "Consciousness—Perception of What Passes in One's Own Mind?" in *The Nature of Consciousness*, p. 801.

13. Siri Hustvedt, *The Shaking Woman, or A History of My Nerves* (New York: Henry Holt and Co., 2010), p. 114.

14. John Searle, "Breaking the Hold" in *The Nature of Consciousness*, p. 497.

15. Colin McGinn, "Consciousness and Content" in *The Nature of Consciousness*, p. 301.

16. Thomas Nagel, "What Is It Like to Be a Bat?" in *The Nature of Consciousness*, pp. 520–521.

17. ibid., p. 521.

18. David Lewis, "What Experience Teaches" in *The Nature of Consciousness*, p. 579.

19. William G. Lycan, "Consciousness as Internal Monitoring" in *The Nature of Consciousness*, p. 767.

20. Thomas Nagel, "What Is It Like to Be a Bat?" in *The Nature of Consciousness*, pp. 522–523.

21. Peter Tompkins and Christopher Bird, *The Secret Life of Plants* (New York: Harper & Row, 1973).

22. Masaro Emoto, *The Hidden Messages in Water*, translated by David A. Thayne (New York: Atria Books/Simon & Schuster, 2005). Yes, I know, Emoto's advanced degree is bogus and his "research" has been inflated to the max by every New Age wannabe with a healing shtick or new energy to promote, but that doesn't mean that there isn't a profound truth at the heart of it all, perhaps one more profound than anything Emoto is saying.

23. See Richard Grossinger, *2013: Raising the Earth to the Next Vibration* (Berkeley, California: North Atlantic Books, 2011), pp. 565–574.

24. Güven Güzeldere, "The Many Faces of Consciousness: A Field Guide" in *The Nature of Consciousness*, p. 33.

25. ibid., pp. 1–2.

26. John Searle, "Breaking the Hold: Silicon Brains, Conscious Robots, and Other Minds" in *The Nature of Consciousness*, p. 493.

27. Nick Herbert, *Elemental Mind*, p. 19.

28. John Searle, "Breaking the Hold: Silicon Brains, Conscious Robots, and Other Minds" in *The Nature of Consciousness*, p. 493.

29. Nick Herbert, *Elemental Mind*, p. 19.

30. John Searle, "Breaking the Hold: Silicon Brains, Conscious Robots, and Other Minds" in *The Nature of Consciousness,* p. 494.

31. René Descartes, *The Philosophical Writings of Descartes,* Vol. 1, translated by John Cottingham, Robert Stoothoff, and Dugald Murdoch (Cambridge, England: Cambridge University Press, 1992), p. 195.

32. Gottfried W. Leibniz, *Selections* (New York: Charles Scribner's Sons, 1951), pp. 374–378.

33. Siri Hustvedt, *The Shaking Woman, or A History of My Nerves* (New York: Henry Holt and Co., 2010), p. 38.

34. Sigmund Freud, "Mental Qualities" in Clara Thompson, Milton Mazer, and Earl Wittenberg (editors), *An Outline of Psychoanalysis,* revised edition (New York: Random House, 1955), p. 21.

35. Siri Hustvedt, *The Shaking Woman,* p. 18.

36. ibid.

37. Sigmund Freud, "Mental Qualities" in *An Outline of Psychoanalysis,* p. 21 (see note 34, above).

38. Sigmund Freud, "The Theory of Instincts" in *An Outline of Psychoanalysis,* p. 5.

39. ibid.

40. Sigmund Freud, "Mental Qualities" in *An Outline of Psychoanalysis,* p. 19.

41. Sigmund Freud, ibid., pp. 16–17.

42. Güven Güzeldere, "The Many Faces of Consciousness: A Field Guide" in *The Nature of Consciousness,* p. 20.

43. Sigmund Freud, "Mental Qualities" in *An Outline of Psychoanalysis,* p. 7.

44. This film was shown to my psychology seminar at Amherst College in 1964 and, though there are numerous online references to a Freud-Jones exchange, I can find no direct citation either of the comment or the film.

45. Sigmund Freud, *The Interpretation of Dreams,* translated from the German by James Strachey (New York: Basic Books, 1955). The original German edition was 1899.

46. Siri Hustvedt, *The Shaking Woman,* pp. 131–132.

47. Marcia Angell, "The Epidemic of Mental Illness: Why?" in *The New York Review of Books,* June 23, 2011, Volume LVIII, Number 11, pp. 20–22.

48. ibid., p. 20.

49. ibid., pp. 21–22.

50. ibid., p. 21.

51. ibid.

52. Marcia Angell, "The Illusion of Psychiatry," in *The New York Review of Books,* July 14, 2011, Volume LVIII, Number 12, p. 20.

53. Jacques Derrida, *Of Grammatology,* translated by Gayatri Chakravorty Spivak (Baltimore: Johns Hopkins University Press, 1974). I am loosely summarizing

Derrida and then in a sense deconstructing the Deconstructer, though I would not be so immodest as to think that I have accomplished even a scintilla of this.

Pages 84–86: Some of the exegeses on Freud, dreams, and free connection were grafted from "A Primary Reading List," www.richardgrossinger.com/2010/04/a-primary-reading-list/

Pages 86–87: The comparison of psychoanalysis and shamanism was partially grafted from Richard Grossinger, *Planet Medicine: Origins,* Revised Edition (Berkeley, California: North Atlantic Books, 2005), pp. 177–179.

Pages 94–95: The section on Jacques Derrida was grafted from "A Primary Reading List."

CHAPTER FIVE. SYSTEMIC CONSCIOUSNESS: NONCONSCIOUSNESS AND
THE LOSS OF CONSCIOUSNESS

1. Güven Güzeldere, "The Many Faces of Consciousness: A Field Guide" in Ned Block, Owen Flanagan, and Güven Güzeldere (editors), *The Nature of Consciousness: Philosophical Debates* (Cambridge, Massachusetts: The MIT Press, 1997), p. 20.
2. Fred Dretske, "Conscious Experience" in *The Nature of Consciousness,* p. 773.
3. Robert Van Gulick, "Understanding the Phenomenal Mind, Part II" in *The Nature of Consciousness,* p. 438.
4. William G. Lycan, "Consciousness as Internal Monitoring" in *The Nature of Consciousness,* p. 762.
5. David Armstrong, quoted in William G. Lycan, "Consciousness as Internal Monitoring" in *The Nature of Consciousness,* p. 755.
6. John A. Bargh, "Bypassing the Will: Toward Demystifying the Nonconscious Control of Social Behavior" in Ran R. Hassin, James S. Uleman, and John A. Bargh, *The New Unconscious: Social Cognition and Social Neuroscience* (Oxford, England: The Oxford University Press, 2005), p. 54.
7. Güven Güzeldere, "The Many Faces of Consciousness: A Field Guide," in *The Nature of Consciousness,* p. 20.
8. Martha J. Farah, "Visual Perception and Visual Awareness after Brain Damage" in *The Nature of Consciousness,* p. 209.
9. John A. Bargh, "Bypassing the Will" in *The New Unconscious,* p. 52 (see note 6, above).
10. Jerry Fodor, "Meaning in Mind," in *Fodor and his Critics* (Oxford, England: Blackwell, 1991), p. 12.
11. Nick Herbert, *Elemental Mind: Human Consciousness and the New Physics* (New York: Dutton, 1993), p. 60.
12. Jack Glaser and John F. Kihlstrom, "Compensatory Awareness: Unconscious

Volition is Not an Oxymoron," in *The New Unconscious* (see note 6), p. 171 (slightly rearranged).

13. ibid.

14. Owen Flanagan, "Conscious Inessentialism and the Epiphenomenalist Suspicion" in *The Nature of Consciousness*, p. 362.

15. For a discussion of Feldenkrais method, see Richard Grossinger, *Planet Medicine: Modalities*, Revised Edition (Berkeley, California: North Atlantic Books, 2003), pp. 209–228.

16. Bernard J. Baars, "Contrastive Phenomenology: A Thoroughly Empirical Approach to Consciousness" in *The Nature of Consciousness*, p. 190.

17. Benjamin Libet, "Unconscious Cerebral Initiative and the Role of Conscious Will in Voluntary Action," *Behavioral and Brain Sciences*, Volume 8, Issue 4, 1985, p. 536.

18. Owen Flanagan, "The Robust Phenomenology of the Stream of Consciousness" in *The Nature of Consciousness*, p. 90.

19. Tyler Burge, "Two Kinds of Consciousness" in *The Nature of Consciousness*, p. 433.

20. Bernard J. Baars, "Contrastive Phenomenology: A Thoroughly Empirical Approach to Consciousness" in *The Nature of Consciousness*, pp. 190–191.

21. Tyler Burge, "Two Kinds of Consciousness" in *The Nature of Consciousness*, p. 432.

22. Alvin I. Goldman, "Consciousness, Folk Psychology, and Cognitive Science" in *The Nature of Consciousness*, p. 111; and Tyler Burge, "Two Kinds of Consciousness" in *The Nature of Consciousness*, p. 430.

23. Owen Flanagan, "Conscious Inessentialism and the Epiphenomenalist Suspicion" in *The Nature of Consciousness*, p. 370.

24. Owen Flanagan, "The Robust Phenomenology of the Stream of Consciousness" in *The Nature of Consciousness*, p. 90.

25. William James, "The Stream of Consciousness" in *The Nature of Consciousness*, p. 81.

26. ibid., p. 80.

27. ibid., p. 81.

28. William G. Lycan, "Consciousness as Internal Monitoring" in *The Nature of Consciousness*, p. 764.

29. Siri Hustvedt, *The Shaking Woman, or A History of My Nerves* (New York, Henry Holt and Co., 2010), p. 110.

30. ibid., p. 109.

31. Alfred North Whitehead, *Process and Reality* (New York: The Free Press/Macmillan, 1929), p. 161.

32. Daniel C. Dennett and Marcel Kinsbourne, "Time and the Observer" in *The*

Nature of Consciousness, pp. 153–158.

33. ibid., p. 162.

34. Ariel Dorfman quoted in Dennett and Kinsbourne, ibid., p. 141.

35. David Armstrong, "What Is Consciousness" in *The Nature of Consciousness,* p. 721.

36. ibid., p. 722.

37. ibid.

38. Owen Flanagan, "Prospects for a Unified Theory of Consciousness or, What Dreams Are Made Of" in *The Nature of Consciousness,* p. 101.

39. Wilder Penfield, *The Mystery of the Mind: A Critical Study of Consciousness and the Human Brain* (Princeton, New Jersey: Princeton University Press, 1975), p. 39.

40. Ned Block, "On a Confusion about a Function of Consciousness" in *The Nature of Consciousness,* p. 376.

41. David Armstrong, "What Is Consciousness" in *The Nature of Consciousness,* p. 723.

42. Fred Dretske, "Conscious Experience" in *The Nature of Consciousness,* p. 778.

43. Mervyn LeRoy (director), *Random Harvest,* Metro-Goldwyn Mayer, 1942.

44. Larry J. Stettner, group email, 2010.

45. Oliver Sacks, *Awakenings* (New York: Doubleday and Co., Inc., 1974), p. 9.

46. John E. Upledger, *Somatoemotional Release: Deciphering the Language of Life* (Berkeley, California: North Atlantic Books, 2002), pp. 39–44; and John E. Upledger, personal communication, in *Planet Medicine: Origins,* Revised Edition (Berkeley, California: North Atlantic Books, 2005), pp. 318–319.

47. John Searle, "Breaking the Hold" in *The Nature of Consciousness,* p. 501.

48. Bonnie Bainbridge Cohen, *Sensing, Feeling, and Action: The Experiential Anatomy of Body-Mind Centering* (Northampton, Massachusetts: Contact Editions, 1994).

49. Emilie Conrad, *Life on Land: The Story of Continuum* (Berkeley, California: North Atlantic Books, 2007).

50. John E. Upledger, *Lessons Out of School: From Detroit Gangs to New Healing Paradigms* (Berkeley, California: North Atlantic Books, 2006), pp. 270–280.

51. Siri Hustvedt, *The Shaking Woman,* p. 106 (see note 29).

52. ibid.

53. ibid., p. 109.

54. Daniel Keyes, *The Minds of Billy Milligan* (New York: Random House, 1981).

55. See Elizabeth M. Carman and Neil J. Carman, *Cosmic Cradle: Souls Waiting for Birth* (Fairfield, Iowa: Sunstar Publishing, Ltd., 1999), p. 17; David Chamberlain, *The Mind of Your Newborn Baby* (Berkeley, California: North

Atlantic Books, 1998); David Chamberlain, *Windows on the Womb: Your First Nine Months* (unpublished manuscript, tentatively North Atlantic Books, 2013); and the website www.birthpsychology.com.

56. A number of people have said this or something close to it: T. S. Eliot, Woody Allen, Groucho Marx, Spike Milligan, George James Grinnell.

57. Michael Tye, "A Representational Theory of Pains and Their Phenomenal Character" in *The Nature of Consciousness*, p. 339.

58. Edoarado Bisiach, "Understanding Consciousness" in *The Nature of Consciousness*, p. 249.

59. Owen Flanagan, "The Robust Phenomenology of the Stream of Consciousness" in *The Nature of Consciousness*, p. 90. (Flanagan is actually quoting William James while applying his concept to a neurological discovery made decades after his death.)

60. Thomas Nagel, "Brain bisection and the unity of conscious experience," *Synthese* 22, 1971, p. 409.

61. Siri Hustvedt, *The Shaking Woman*, p. 51 (see note 29).

62. Francis Crick and Christhof Koch, "Toward a Neurobiological Theory of Consciousness" in *The Nature of Consciousness*, p. 280.

63. Alvin I. Goldman, "Consciousness, Folk Psychology, and Cognitive Science" in *The Nature of Consciousness*, p. 111.

64. Siri Hustvedt, *The Shaking Woman*, p. 51.

65. ibid., p. 47.

66. Tanya L. Chartrand, William W. Maddox, and Jessica L. Lakin, "Beyond the Perception-Behavior Link: The Ubiquitous Utility and Motivational Moderators of Nonconscious Mimicry" in *The New Unconscious*, p. 356 (see note 6, above).

67. Martha J. Farah, "Visual Perception and Visual Awareness after Brain Damage" in *The Nature of Consciousness*, p. 224.

68. Tim Shallice, "Modularity and Consciousness" in *The Nature of Consciousness*, p. 258.

69. Martha J. Farah, "Visual Perception and Visual Awareness after Brain Damage" in *The Nature of Consciousness*, pp. 208–209.

70. ibid., p. 209.

71. ibid., p. 210.

72. ibid., p. 209.

73. ibid., p. 225.

74. ibid., p. 213.

75. ibid., p. 205.

76. Georges Rey, "A Question about Consciousness" in *The Nature of Consciousness*, p. 465.

77. Ned Block, "On a Confusion about a Function of Consciousness" in *The Nature of Consciousness*, p. 393.

CHAPTER SIX. QUALIA OR ZOMBIES?

1. Daniel C. Dennett, "Quining Qualia" in Ned Block, Owen Flanagan, and Güven Güzeldere (editors), *The Nature of Consciousness: Philosophical Debates* (Cambridge, Massachusetts: The MIT Press, 1997), p. 620.
2. Edward Tolman, *Purposive Behavior in Animals and Men* (orig. published 1932) (New York: Irvington Publishers, Inc., 1967), p. 251.
3. Thomas Nagel, "What Is It Like to Be a Bat?" in *The Nature of Consciousness*, p. 519.
4. ibid.
5. Daniel C. Dennett, "Quining Qualia" in *The Nature of Consciousness*, p. 622.
6. Michael Tye, "A Representational Theory of Pains and Their Phenomenal Character" in *The Nature of Consciousness*, p. 337.
7. Thomas Nagel, "What Is It Like to Be a Bat?" in *The Nature of Consciousness*, p. 524.
8. Güven Güzeldere, "The Many Faces of Consciousness: A Field Guide" in *The Nature of Consciousness*, p. 27.
9. Daniel C. Dennett, "Quining Qualia" in *The Nature of Consciousness*, p. 619.
10. Frank Jackson, "Epiphenomenal Qualia," *Philosophical Quarterly*, XXXII: 127, 1982, p. 135.
11. Güven Güzeldere, "The Many Faces of Consciousness: A Field Guide" in *The Nature of Consciousness*, p. 41.
12. David Pearce, "Review of *Mind, Brain and the Quantum* by Michael Lockwood," Hedweb ("Naturalistic Panpsychism") www.hedweb.com/lockwood.htm.
13. Güven Güzeldere, "The Many Faces of Consciousness: A Field Guide" in *The Nature of Consciousness*, p. 41.
14. Thomas H. Huxley, *Methods and Results* (New York: Appleton Co., 1901), p. 240.
15. ibid., pp. 243–244.
16. Nick Herbert, *Elemental Mind: Human Consciousness and the New Physics* (New York: Dutton, 1993), p. 3 (slightly rearranged).
17. Ned Block, "Inverted Earth" in *The Nature of Consciousness*, p. 680.
18. Gilbert Harman, "The Intrinsic Quality of Experience" in *The Nature of Consciousness*, p. 668.
19. ibid., p. 671.
20. Ned Block, "Inverted Earth" in *The Nature of Consciousness*, p. 677.
21. Charles Dunbar Broad, *The Mind and its Place in Nature* (orig. published 1925) (London: Routledge and Kegan Paul, 1962), p. 71.

22. Gilbert Harman, "The Intrinsic Quality of Experience" in *The Nature of Consciousness*, p. 668.

23. John Searle, "Reductionism and the Irreducibility of Consciousness" in *The Nature of Consciousness*, p. 454.

24. Robert Van Gulick, "Understanding the Phenomenal Mind, Part I" in *The Nature of Consciousness*, p. 562.

25. Ned Block, "Inverted Earth" in *The Nature of Consciousness*, p. 689.

26. David Lewis, "What Experience Teaches" in *The Nature of Consciousness*, p. 584.

27. Brian Loar, "Phenomenal States" in *The Nature of Consciousness*, p. 600.

28. Ned Block, "Inverted Earth" in *The Nature of Consciousness*, p. 690.

29. Frank Jackson, "What Mary Didn't Know" in *The Nature of Consciousness*, p. 567.

30. Brian Loar, "Phenomenal States" in *The Nature of Consciousness*, p. 599.

31. Ned Block, "Inverted Earth" in *The Nature of Consciousness*, p. 689.

32. Daniel C. Dennett, "Quining Qualia" in *The Nature of Consciousness*, p. 623.

33. Thomas Nagel, "What Is It Like to Be a Bat?" in *The Nature of Consciousness*, p. 527.

34. Thomas Nagel, *The View from Nowhere* (Oxford, England: Oxford University Press, 1986), p. 47.

35. David Pearce, "Review of *Mind, Brain and the Quantum* by Michael Lockwood" (see note 12, above).

36. Robert Van Gulick, "Understanding the Phenomenal Mind, Part I" in *The Nature of Consciousness*, p. 564.

37. John Searle, "Breaking the Hold" in *The Nature of Consciousness*, p. 498.

38. Ned Block, "Inverted Earth" in *The Nature of Consciousness*, p. 688.

39. ibid., p. 684.

40. Robert Van Gulick, "Understanding the Phenomenal Mind, Part I" in *The Nature of Consciousness*, p. 565.

41. Siri Hustvedt, *The Shaking Woman, or A History of My Nerves* (New York, Henry Holt and Co., 2010), p. 141.

42. Sydney Shoemaker, "The Inverted Spectrum" in *The Nature of Consciousness*, p. 651.

43. John Locke quoted in Sydney Shoemaker, ibid., p. 643.

44. Ludwig Wittgenstein, "Notes for Lectures on 'Private Experience' and 'Sense Data'," edited by Rush Rhees, *The Philosophical Review*, 77 (1968), pp. 284 and 316.

45. Francis Crick and Christhof Koch, "Toward a Neurobiological Theory of Consciousness" in *The Nature of Consciousness*, p. 279.

46. Sydney Shoemaker, "The Inverted Spectrum" in *The Nature of Consciousness*, p. 651.

47. Ludwig Wittgenstein, *Philosophical Investigations* (Oxford, England: Blackwell Scientific Publications, 1958), p. 100.

48. Alvin I. Goldman, "Consciousness, Folk Psychology, and Cognitive Science" in *The Nature of Consciousness*, p. 112.

49. Fred Dretske, "Conscious Experience" in *The Nature of Consciousness*, p. 785.

50. ibid., p. 785–786.

51. Nick Herbert, *Elemental Mind*, p. 134 (see note 16).

52. ibid.

53. Daniel C. Dennett and Marcel Kinsbourne, "Time and the Observer" in *The Nature of Consciousness*, p. 167.

54. Christopher Peacocke, "Sensation and the Content of Experience: A Distinction" in *The Nature of Consciousness*, p. 341.

55. William G. Lycan, "Consciousness as Internal Monitoring" in *The Nature of Consciousness*, p. 765.

56. ibid.

57. Daniel C. Dennett, "Quining Qualia" in *The Nature of Consciousness*, pp. 619–620.

58. Frank Jackson, "What Mary Didn't Know" in *The Nature of Consciousness*, p. 567.

59. Daniel C. Dennett, "Quining Qualia" in *The Nature of Consciousness*, pp. 626–627.

60. ibid., p. 638.

61. Carlos Castaneda, *Journey to Ixtlan: The Lessons of Don Juan* (New York: Simon and Schuster, 1972), p. 107.

62. David Lewis, "What Experience Teaches" in *The Nature of Consciousness*, p. 589.

63. ibid., p. 591.

64. ibid., p. 594.

65. David Armstrong, "What Is Consciousness" in *The Nature of Consciousness*, p. 726.

66. ibid., pp. 737–728.

67. Edwin G. Boring, "A History of Introspection," *Psychological Bulletin*, 50 (3), 1953, p. 184.

68. Tyler Burge, "Two Kinds of Consciousness" in *The Nature of Consciousness*, p. 427.

69. John Searle, "Breaking the Hold" in *The Nature of Consciousness*, p. 495.

70. Michael Tye, "A Representational Theory of Pains and Their Phenomenal Character" in *The Nature of Consciousness*, p. 332.

71. Daniel C. Dennett, "Quining Qualia" in *The Nature of Consciousness*, p. 638.

72. Thomas Nagel quoted in Güven Güzeldere, "The Many Faces of Consciousness: A Field Guide" in *The Nature of Consciousness*, p. 62.

73. Sydney Shoemaker, "The First-Person Perspective" in *The Nature of Consciousness*, p. 511.

74. Güven Güzeldere, "The Many Faces of Consciousness: A Field Guide" in *The Nature of Consciousness*, p. 39.

75. ibid., p. 28.

76. Astra Taylor (director), *Zizek!*, 2005.

77. John Locke, *An Essay Concerning Human Understanding*, Vol. 1 (1690) (New York: Dover Publications, 1959), p. 520.

78. William James, "The Stream of Consciousness" in *The Nature of Consciousness*, pp. 72 and 74.

79. Sydney Shoemaker, "The First-Person Perspective" in *The Nature of Consciousness*, p. 514.

80. Güven Güzeldere, "The Many Faces of Consciousness: A Field Guide" in *The Nature of Consciousness*, p. 24.

81. ibid., p. 25.

82. Güven Güzeldere, "Consciousness—Perception of What Passes in One's Own Mind?" in *The Nature of Consciousness*, p. 796.

83. Güven Güzeldere, "The Many Faces of Consciousness: A Field Guide" in *The Nature of Consciousness*, p. 40.

84. Alvin I. Goldman, "Consciousness, Folk Psychology, and Cognitive Science" in *The Nature of Consciousness*, p. 112.

85. David Chalmers, "Facing Up to the Problem of Consciousness," *Journal of Consciousness Studies*, 2 (3),1995, p. 201. The thought experiment using zombies was popularized by Chalmers.

86. Güven Güzeldere, "The Many Faces of Consciousness: A Field Guide" in *The Nature of Consciousness*, p. 41.

87. Nick Herbert, *Elemental Mind*, p. 41 (see note 16).

88. Ned Block, "Begging the Question against Phenomenal Consciousness" in *The Nature of Consciousness*, p. 175.

89. Ned Block, "On a Confusion about a Function of Consciousness" in *The Nature of Consciousness*, p. 387.

90. Ned Block quoted in Owen Flanagan, "Conscious Inessentialism and the Epiphenomenalist Suspicion" in *The Nature of Consciousness*, p. 368.

91. Daniel Dennett quoted in Ned Block, "On a Confusion about a Function of Consciousness" in *The Nature of Consciousness*, p. 394.

92. Thomas Nagel, "What Is It Like to Be a Bat?" in *The Nature of Consciousness*, p. 519.

93. Nick Herbert, *Elemental Mind,* p. 41.

94. ibid., p. 24.

95. ibid., p. 42.

96. Georges Rey, "A Question about Consciousness" in *The Nature of Consciousness,* p. 470.

97. David Chalmers quoted in Güven Güzeldere, "The Many Faces of Consciousness: A Field Guide" in *The Nature of Consciousness,* p. 65.

98. Stephen L. White, "Curse of the Qualia" in *The Nature of Consciousness,* p. 704.

99. ibid., pp. 695–696.

100. Georges Rey, "A Question about Consciousness" in *The Nature of Consciousness,* p. 470.

101. ibid., p. 467.

102. ibid., p. 471.

103. William G. Lycan, "Consciousness as Internal Monitoring" in *The Nature of Consciousness,* p. 765.

104. Michael Gazzaniga, *Nature's Mind: The Biological Roots of Thinking, Emotions, Sexuality, Language and Intelligence* (New York: Basic Books, 1992), p. 2.

105. ibid.

106. "I Started a Joke," Barry Gibb, Maurice Ernest Gibb, and Robin Hugh Gibb, the Bee Gees, *Idea,* 1968.

107. Bee Gees, *Bee Gees Anthology* (Milwaukee, Wisconsin: Hal Leonard Publishing Company, 1991), p. 188.

108. Steven Weinberg, "The Universes We Still Don't Know," *The New York Review of Books,* February 10, 2011, Volume LVIII, Number 2, p. 32.

109. Nick Herbert, *Elemental Mind,* pp. 141–142.

110. David Chalmers, *The Conscious Mind* (Oxford, England: Oxford University Press, 1996, p. 95.

111. Nick Herbert, *Elemental Mind,* p. 10.

112. Güven Güzeldere, "The Many Faces of Consciousness: A Field Guide" in *The Nature of Consciousness,* p. 44.

113. John Searle, *The Rediscovery of the Mind* (Cambridge, Massachusetts: The MIT Press, 1992), p. 95.

114. John Searle, "Breaking the Hold" in *The Nature of Consciousness,* p. 494.

115. ibid.

116. Georges Rey, "A Question about Consciousness" in *The Nature of Consciousness,* pp. 470–471.

117. Eugene Wigner quoted in Nick Herbert, *Elemental Mind,* p. 250.

118. Claude Lévi-Strauss, *The Savage Mind,* translation anonymous (Chicago: The University of Chicago Press, 1966), p. 155. *"But the doctor says we'll be all*

right" is a riff off the chorus to Tom Waits's make-it-up-as-you-go song "Had Me A Girl."

119. Frank Loesser, *Guys and Dolls,* Broadway musical, 1950.

120. Claude Lévi-Strauss, *The Savage Mind,* p. 269.

121. Claude Lévi-Strauss, *The Naked Man,* translated by John and Doreen Weightman (New York: Harper & Row Publishers, 1981), p. 687.

122. ibid., p. 694.

123. Frank Loesser, *Guys and Dolls.*

CHAPTER SEVEN. CONSCIOUSNESS AS AN EMERGENT PHENOMENON: THE PSYCHOLINGUISTICS AND PHYLOGENESIS OF MEANING

1. Richard Grossinger, *Embryos, Galaxies, and Sentient Beings: How the Universe Makes Life* (Berkeley, California: North Atlantic Books, 2003), p. 118.

2. *Holy Bible: The New King James Version* (Nashville: Thomas Nelson Publishers, 1979), p. 525.

3. Paul Valéry, quoted in Adam Phillips, *Terrors and Experts* (Cambridge, Massachusetts: Harvard University Press, 1995), p. 3.

4. Curtis McCosco, *Maya Cosmos, Buddha Mind, Quantum World* (Wailua, Hawai'i: unpublished manuscript, 2012).

5. Owen Flanagan, "Conscious Inessentialism and the Epiphenomenalist Suspicion" in Ned Block, Owen Flanagan, and Güven Güzeldere (editors), *The Nature of Consciousness: Philosophical Debates* (Cambridge, Massachusetts: The MIT Press, 1997), p. 360.

6. Patricia Smith Churchland, "Can Neurobiology Teach Us Anything About Consciousness?" in *The Nature of Consciousness,* p. 133.

7. Owen Flanagan, "Conscious Inessentialism and the Epiphenomenalist Suspicion" in *The Nature of Consciousness,* p. 360.

8. ibid.

9. ibid.

10. Terrence W. Deacon, "A role for relaxed selection in the evolution of the language capacity," paper delivered at colloquium entitled "In the Light of Evolution IV: The Human Condition," held December 10–12, 2009, Irvine, California, *Proceedings of the National Academy of Sciences,* Direct Submission, 2009; www.nasonline.org/SACKLER_Human_Condition, p. 5.

11. ibid., p. 4.

12. Terrence W. Deacon, "Relaxed Selection and the Role of Epigenesis in the Evolution of Language" in Mark Blumberg, John Freeman, and Scott Robinson (editors), *Oxford Handbook of Developmental Behavioral Neuroscience* (Oxford, England: Oxford University Press, 2011), p. 735.

13. ibid., p. 738.

14. ibid., p. 736.
15. ibid.
16. ibid., p. 737.
17. ibid., p. 740.
18. ibid., p. 738.
19. ibid., p. 743.
20. Terrence W. Deacon, "A role for relaxed selection in the evolution of the language capacity," p. 4 (see note 10).
21. Terrence W. Deacon, "Relaxed Selection and the Role of Epigenesis in the Evolution of Language," p. 736 (see note 12).
22. ibid., p. 743.
23. ibid., p. 731.
24. Terence McKenna, *Dreaming Awake at the End of Time,* lecture recorded by Sound Photosynthesis, San Francisco, December 13, 1998, quoted in Richard Grossinger, *2013: Raising the Earth to the Next Vibration* (Berkeley, California: North Atlantic Books, 2010), p. 36.
25. Terrence W. Deacon, "Relaxed Selection and the Role of Epigenesis in the Evolution of Language," p. 742.
26. ibid., pp. 742–743.
27. ibid., p. 738.
28. ibid., p. 737.
29. Terrence W. Deacon, "A role for relaxed selection in the evolution of the language capacity," p. 5.
30. Terrence W. Deacon, "Relaxed Selection and the Role of Epigenesis in the Evolution of Language," p. 736.
31. ibid.
32. Nick Herbert, *Elemental Mind: Human Consciousness and the New Physics* (New York: Dutton, 1993), pp. 85–87 (including unsourced quote from Egyptian-born postmodern literary theorist Ihab Hassan).
33. Donald I. Williamson, quoted in Steven Shaviro, *Doom Patrols: A Theoretical Fiction about Postmodernism* (London: Serpent's Tail/High Risk Books, 1997), p. 113.
34. Terrence W. Deacon, "A role for relaxed selection in the evolution of the language capacity," p. 3.
35. ibid.
36. Noam Chomsky, *Aspects of the Theory of Syntax* (Boston: The MIT Press, 1969).
37. Terrence W. Deacon, "Relaxed Selection and the Role of Epigenesis in the Evolution of Language," p. 744.
38. Terrence W. Deacon, "A role for relaxed selection in the evolution of the language capacity," p. 6.

39. Terrence W. Deacon, "Relaxed Selection and the Role of Epigenesis in the Evolution of Language," p. 743.

40. ibid., pp. 745–746.

41. ibid., p. 747.

42. ibid., p. 744.

43. Terrence W. Deacon, "A role for relaxed selection in the evolution of the language capacity," p. 6.

44. Terrence W. Deacon, "Relaxed Selection and the Role of Epigenesis in the Evolution of Language," pp. 747–748.

45. ibid., p. 748.

46. ibid., p. 749.

47. ibid.

48. ibid., p. 748.

49. Donald T. Campbell, "Variation and selective retention in socio-cultural evolution" in Herbert R. Barringer, George Blanksten, and Raymond Mack (editors), *Social Change in Developing Areas: A Reinterpretation of Evolutionary Theory* (Cambridge, Massachusetts: Schenkman, 1965), see pp. 19–49.

50. Terrence W. Deacon, "Relaxed Selection and the Role of Epigenesis in the Evolution of Language," p. 750.

51. Terrence W. Deacon, "A role for relaxed selection in the evolution of the language capacity," p. 3.

52. ibid., p. 6.

53. ibid., p. 3.

54. H. Robert Bagwell, "Integrative Processing," unpublished paper, Portland, Oregon, 1999.

55. ibid.

56. ibid.

57. ibid.

58. Terrence W. Deacon, "Relaxed Selection and the Role of Epigenesis in the Evolution of Language," p. 736.

59. ibid., p. 747.

60. ibid.

61. ibid.

62. Terrence W. Deacon, "Relaxed Selection and the Role of Epigenesis in the Evolution of Language," p. 742.

63. ibid.

64. Patricia Smith Churchland, "Can Neurobiology Teach Us Anything About Consciousness?" in *The Nature of Consciousness*, p. 127.

65. Claude Lévi-Strauss, *Structural Anthropology*, translated by Claire Jacobson and Brooke Grundfest Schoepf (New York: Doubleday Anchor, 1967), p. 178.

66. Erich Neumann, *The Great Mother,* translated by Ralph Mannheim (Princeton: Princeton University Press, 1963), p. 12.

67. Claude Lévi-Strauss, *Totemism,* translated by Rodney Needham (Boston: Beacon Press, 1963), p. 91.

68. Claude Lévi-Strauss, *The Savage Mind,* translation anonymous (Chicago: The University of Chicago Press, 1966), p. 95.

69. Claude Lévi-Strauss, *The Raw and the Cooked,* translated by John and Doreen Weightman (New York: Harper & Row Publishers, 1969), pp. 340–341.

70. Claude Lévi-Strauss, *The Savage Mind,* p. 268.

71. Claude Lévi-Strauss, *Elementary Structures of Kinship,* translated by James Harle Bell, John Richard Von Sturmer, and Rodney Needham (Boston: Beacon Press, 1949). The Alliance Theory did not originate with CLS but was developed by him out of earlier social science and is most identified with his version of it.

72. Terrence W. Deacon, "Heterochrony in Brain Evolution: Cellular versus Morphological Analyses" in Sue Taylor, Jonas Langer, and Michael L. McKinney (editors), *Biology, Brains & Behavior: The Evolution of Human Development* (Santa Fe: School for Advanced Research Press, 2000), p. 45.

73. Terrence W. Deacon, "Relaxed Selection and the Role of Epigenesis in the Evolution of Language," p. 740.

74. ibid.

75. ibid., p. 744.

76. Terrence W. Deacon, "Heterochrony in Brain Evolution," pp. 58–59.

77. Terrence W. Deacon, "Relaxed Selection and the Role of Epigenesis in the Evolution of Language," pp. 737–738.

78. ibid., p. 739.

79. Terrence W. Deacon, "Heterochrony in Brain Evolution," p. 56.

80. ibid., p. 44.

81. ibid., p. 58.

82. Terrence W. Deacon, "Relaxed Selection and the Role of Epigenesis in the Evolution of Language," p. 732.

83. Terrence W. Deacon, "Heterochrony in Brain Evolution," p. 58.

84. Terrence W. Deacon, "Relaxed Selection and the Role of Epigenesis in the Evolution of Language," p. 733.

85. ibid., pp. 733–734.

86. ibid., p. 734.

87. ibid.

88. ibid., p. 735.

89. Terrence W. Deacon, "Heterochrony in Brain Evolution: Cellular versus Morphological Analyses," p. 41.

90. ibid., p. 50.

91. ibid., p. 44.

92. ibid., p. 80.

93. ibid., pp. 52–53.

94. ibid., p. 54.

95. ibid., p. 67.

96. ibid., p. 79.

97. ibid., p. 81.

98. ibid., p. 82.

99. ibid.

100. ibid., p. 83.

101. ibid.

102. ibid.

103. ibid., p. 84.

104. ibid., p. 87.

Pages 179–182: From "Because of the continual miniaturization…" through "…quicker than white on rice" was grafted from Richard Grossinger, *Embryos, Galaxies, and Sentient Beings: How the Universe Makes Life* (Berkeley, California: North Atlantic Books, 2003), pp. 158–160.

Pages 189–190: The description of the cell was partially grafted from *Embryos, Galaxies, and Sentient Beings: How the Universe Makes Life*, pp. 69–70.

Page 190 *et seq.:* From "The order of primates, the human lineage, emerged from a branch…" through the end of Section iv has several passages grafted from "Mind" in *Embryogenesis: Species, Gender, and Identity*, pp. 547–572.

Pages 197–198: The discussion of attractors and strange attractors was grafted from *Embryos, Galaxies, and Sentient Beings: How the Universe Makes Life*, p. 34.

Pages 208–211: From "Simple jellyfish and comb-jellies" through "a replica of the cosmos" was grafted from Richard Grossinger, *Embryogenesis: Species, Gender, and Identity* (Berkeley, California: North Atlantic Books, 2000), pp. 398–420.

Page 214 *et seq.:* Portions of the section entitled "The Birth of the Symbol" regarding the origin of primates, hominoids, hominids, language, the gift, and incest taboo were grafted from the chapter "Mind" in *Embryogenesis: Species, Gender, and Identity*, pp. 547–572. The primary sources for this material were: M. F. Ashley Montagu (editor), *Culture and the Evolution of Man* (New York: Oxford University Press, 1962); J. N. Spuhler (editor), *The Evolution of Man's Capacity for Culture* (Detroit: Wayne State University Press, 1965—particularly Spuhler's "Somatic Paths to Culture"); W. E. Le Gros Clark, *The Antecedents of Man: An Introduction to the Evolution of the Primates* (New York: Harper and Row, 1963);

A. J. Kelso, *Physical Anthropology* (Philadelphia: J. B. Lippincott Co., 1970); Keith L. Moore, *The Developing Human: Clinically Oriented Embryology* (Philadelphia: Saunders Publishing Co., 1977); and Bernard Campbell, *Human Evolution: An Introduction to Man's Adaptations* (Chicago: Aldine Publishing, 1966).

Briefer sections were grafted from other chapters in *Embryogenesis: Species, Gender, and Identity*—"The Origin of the Nervous System," "The Evolution of Intelligence," and "Neurulation and the Human Brain." Additional primary sources for these chapters include: T. H. Bullock and G. A. Horridge, *Structure and Function in the Nervous System of Invertebrates* (San Francisco: W. H. Freeman and Co., 1965); Lloyd S. Woodburne, *The Neural Basis for Behavior* (Columbus, Ohio: Charles F. Merrill Books, 1967), B. I. Balinsky, *An Introduction to Embryology,* Fifth Edition (Philadelphia: Saunders Publishing Co., 1981), and N. J. Berrill and Gerald Karp, *Development* (New York: McGraw-Hill Book Co., 1976).

Sections of material on Claude Lévi-Strauss and *epistemes* (Michel Foucault) were grafted from "A Primary Reading List," www.richardgrossinger .com/2010/04/a-primary-reading-list/.

Page 230 *et seq.*: Parts of Section iv were grafted from "Ontogeny and Phylogeny" in *Embryogenesis: Species, Gender, and Identity,* pp. 325–351, and "The Principles of Biological Design" in *Embryos, Galaxies, and Sentient Beings: How the Universe Makes Life,* pp. 129–138.

CHAPTER EIGHT. THE QUANTUM BRAIN

I am grateful to Nick Herbert for his help in constructing this chapter and to Neil Stillings for serving as a sounding board and articulate critic for some of my notions.

1. Rick Grush and Patricia Smith Churchland, "Gaps in Penrose's Toilings," *Journal of Consciousness Studies,* 2 (1), 1995, p. 10.
2. ibid., p. 27.
3. ibid., p. 10.
4. I am grateful to John Visvader for this perception.
5. Nick Herbert, *Elemental Mind: Human Consciousness and the New Physics* (New York: Dutton, 1993), pp. 213–214.
6. ibid., p. 166.
7. ibid., p. 143.
8. ibid., p. 160.
9. Nick Herbert, personal communication, November 2011.
10. *Quantum Tantra: Investigating New Doorways into Nature* <http://quantum-tantra.blogspot.com/2010/02/algae-master-quantum-science.html>.
11. Jim Melodini, *Sage Visions: Salvia Divinorum, Shamanism, and Quantum Physics* (Cabot, Vermont: unpublished manuscript, 2011).

12. Nick Herbert, *Elemental Mind,* p. 158.

13. ibid., p. 177.

14. ibid.

15. ibid., p. 144.

16. ibid., p. 145.

17. http://dictionary.reference.com/browse/quantum.

18. Nick Herbert, *Elemental Mind,* p. 230.

19. ibid., p. 294.

20. ibid., p. 156.

21. ibid., p. 250.

22. ibid., p. 158.

23. ibid., p. 250.

24. ibid., p. 159.

25. ibid., p. 157.

26. ibid., p. 219.

27. David Chalmers in Galen Strawn, Peter Carruthers, Frank Jackson, and William G. Lycan (editors), *Consciousness and Its Place in Nature* (Exeter, England: Imprint Academic, 2006).

28. Henry Stapp, "Why Classical Mechanics Cannot Accommodate Consciousness but Quantum Mechanics Can," www.nonlocal.com/hbar/qbrain .html#quantumparadigm), no date.

29. ibid.

30. Jim Melodini, *Sage Visions* (see note 11).

31. Nick Herbert, *Elemental Mind,* p. 266.

32. ibid., p. 259.

33. ibid., p. 260.

34. I am grateful to Neil Stillings for this perception.

35. Nick Herbert, *Elemental Mind,* p. 265.

36. ibid.

37. *Quantum Tantra: Investigating New Doorways into Nature* <http://quantumtantra.blogspot.com/2010/02/algae-master-quantum-science.html>.

38. Rick Grush and Patricia Smith Churchland, "Gaps in Penrose's Toilings," pp. 13–14.

39. Roger Penrose, *The Emperor's New Mind* (New York: Doubleday/Vintage Books, 1990).

40. S. R. Hameroff, "Quantum coherence in microtubules," *Journal of Consciousness Studies,* 1 (1), 1994, pp. 91–118.

41. Rick Grush and Patricia Smith Churchland, "Gaps in Penrose's Toilings," p. 11.

42. Nick Herbert, *Elemental Mind,* p. 189.

43. Roger Penrose, *Shadows of the Mind: A Search for the Missing Science of Consciousness* (Oxford, England: Oxford University Press, 1994).

44. Jim Melodini, *Sage Visions.*

45. Nick Herbert, *Elemental Mind*, p. 257.

46. ibid., p. 255.

47. ibid., p. 258.

48. ibid., p. 265.

49. ibid., p. 178.

50. ibid., p. 171.

51. ibid., p. 263.

52. ibid.

53. ibid., p. 188.

54. Henz Pagels quoted in Nick Herbert, *Elemental Mind*, p. 174.

55. István Dienes, "The Quantum Brain and the Topological Consciousness," www.slideshare.net/Dienes/the-quantum-brain-and-the-topological-consciousness-field-presentation, 2011. Much of the information in this article is drawn from August Stern, *The Quantum Brain: Theory and Implications* (Amsterdam: Elsevier Science, 1994), and August Stern, *Quantum Theoretic Machines: What is Thought from the Point of View of Physics* (Amsterdam: Elsevier Science, 2000).

56. ibid. (rearranged)

57. ibid.

58. ibid. (rearranged).

59. ibid. (rearranged)

60. Rick Grush and Patricia Smith Churchland, "Gaps in Penrose's Toilings," p. 25.

61. ibid.

62. ibid.

63. ibid., p. 24.

64. ibid., p. 26.

65. *Quantum Tantra: Investigating New Doorways into Nature,* http://quantumtantra.blogspot.com/2010/02/algae-master-quantum-science.html.

66. Nick Herbert, *Elemental Mind*, p. 231.

67. ibid., p. 182.

68. ibid., p. 180.

69. ibid., p. 181.

70. ibid., pp. 233, 182.

71. ibid., p. 238.

72. ibid., p. 239.

73. ibid., p. 241.

74. ibid., p. 161.

75. From "Such a multiverse..." through "dreamed being" is cocreated with Frederick Ware.

76. "The Universes We Still Don't Know" by Steven Weinberg, *The New York Review of Books,* February 10, 2011, Volume LVIII, Number 2, p. 32. The cat belongs (as noted in the text) to early twentieth-century physicist Erwin Schrödinger, the dice to twenty-first-century quantum-information physicist Michael A. Nielsen, the universe with every possible history to astrophysicist Stephen Hawking.

77. Jim Melodini, *Sage Visions.*

78. ibid.

79. ibid.

80. David Bohm, *Wholeness and the Implicate Order* (Oxford, England: Routledge Ark, 1980).

81. ibid.

82. Alfred North Whitehead, *Science and the Modern World* (Cambridge, England: Cambridge University Press, 1926), p. 90.

83. David Chalmers in Galen Strawn, Peter Carruthers, Frank Jackson, and William G. Lycan (editors), *Consciousness and Its Place in Nature.*

84. Ronald Milestone, "Solving the Consciousness Problem," unpublished paper, Redondo Beach, California, 2012.

85. I am grateful to John Visvader for clarifying this point.

86. Nick Herbert, personal communication, November 2011.

87. Rick Grush and Patricia Smith Churchland, "Gaps in Penrose's Toilings," pp. 27–28.

88. Jim Melodini, *Sage Visions.*

89. Charles Olson, "Enyalion," in Richard Grossinger (editor), *Ecology and Consciousness: Traditional Wisdom on the Environment* (Berkeley, California: North Atlantic Books, 1992), p. 216.

90. ibid.

91. Nick Herbert, *Elemental Mind,* p. 228.

Page 265: From "The basis of consciousness" to 'random' and 'nonrandom' have no meaning" was grafted from Richard Grossinger, *Embryos, Galaxies, and Sentient Beings: How the Universe Makes Life* (Berkeley, California: North Atlantic Books, 2003), p. 165.

CHAPTER NINE. THE ONTOLOGY AND COSMOLOGY OF CONSCIOUSNESS

1. Georges Rey, "A Question about Consciousness" in Ned Block, Owen Flanagan, and Güven Güzeldere (editors), *The Nature of Consciousness: Philosophical Debates* (Cambridge, Massachusetts: The MIT Press, 1997), pp. 472–473.

2. David Armstrong, "What Is Consciousness" in *The Nature of Consciousness,* p. 725.

3. Colin McGinn, "Consciousness and Content" in *The Nature of Consciousness,* p. 305.
4. ibid., p. 297.
5. Alvin I. Goldman, "Consciousness, Folk Psychology, and Cognitive Science" in *The Nature of Consciousness,* pp. 111–112.
6. William Blake, "The Tyger" in *Songs of Experience,* 1794.
7. Nick Herbert, "The Consciousness Wars," *Omni,* October 1993, p. 56.
8. Thomas Nagel, *The View from Nowhere* (Oxford, England: Oxford University Press, 1986), pp. 7–8.
9. Nick Herbert, *Elemental Mind: Human Consciousness and the New Physics* (New York: Dutton, 1993), p. 27.
10. Jan-Markus Schwindt, "Mind as Hardware and Matter as Software," *Journal of Consciousness Studies,* 15 (4), 2008, pp. 22–23.
11. Nick Herbert, *Elemental Mind,* p. 185.
12. Thomas Nagel, "Panpsychism" in Thomas Nagel, *Mortal Questions* (Cambridge, England: Cambridge University Press, 1988), p. 181.
13. Nick Herbert, *Elemental Mind,* p. 120.
14. ibid., p. 24 (rearranged).
15. ibid., p. 186.
16. ibid. p. 187.
17. Alvin I. Goldman, "Consciousness, Folk Psychology, and Cognitive Science" in *The Nature of Consciousness,* p. 113.
18. Owen Flanagan, "Conscious Inessentialism and the Epiphenomenalist Suspicion" in *The Nature of Consciousness,* p. 362.
19. ibid., p. 361.
20. Stephen L. White, "Curse of the Qualia" in *The Nature of Consciousness,* p. 695.
21. Robert Van Gulick, "Understanding the Phenomenal Mind, Part II" in *The Nature of Consciousness,* p. 440.
22. Frederick Ware and I co-created the delivery in this paragraph.
23. Ned Block, "On a Confusion about a Function of Consciousness" in *The Nature of Consciousness,* p. 381.
24. Nick Herbert, *Elemental Mind,* p. 247.
25. ibid., pp. 247–248.
26. Nick Herbert, *Elemental Mind,* p. 33.
27. ibid., p. 120.
28. Sir Charles Sherrington, *Man on his Nature* (Cambridge, England: Cambridge University Press, 1963), p. 105.
29. Nick Herbert, *Elemental Mind,* p. 33.
30. ibid., pp. 13–14 (with minor typo correction).

31. David Pearce, "Review of *Mind, Brain and the Quantum* by Michael Lockwood," Hedweb ("Naturalistic Panpsychism").

32. Bernard J. Baars, "Contrastive Phenomenology: A Thoroughly Empirical Approach to Consciousness" in *The Nature of Consciousness*, p. 187.

33. ibid.

34. Colin McGinn, "Consciousness and Content" in *The Nature of Consciousness*, p. 300.

35. David Lewis, "What Experience Teaches" in *The Nature of Consciousness*, p. 590.

36. Colin McGinn, "Can We Solve the Mind-Body Problem?" in *The Nature of Consciousness*, p. 539.

37. This account is mainly constructed from my personal discussions with Jule Eisenbud during the 1970s; for more on this topic, the interested reader can check out two of his books: *The World of Ted Serios: 'Thoughtographic' Studies of an Extraordinary Mind* (Jefferson, North Carolina: McFarland & Company, Publishers, Inc., 1966) and *Parapsychology and the Unconscious* (Berkeley, California: North Atlantic Books, 1983).

38. Jule Eisenbud, *Paranormal Foreknowledge: Problems and Perplexities* (New York: Human Sciences Press, 1982), p. 123.

39. ibid., p. 124.

40. Nick Herbert, *Elemental Mind*, p. 193.

41. Jule Eisenbud, *Paranormal Foreknowledge*, pp. 125–126. The internal quotes from William Crookes appeared in "Notes of Séances with D. D. Home," *Proceedings of the Society for Psychical Research*, 1889, pp. 116 and 122, respectively.

42. Nick Herbert, *Elemental Mind*, p. 194.

43. Larry Dossey, *Recovering the Soul—A Scientific and Spiritual Search* (New York: Bantam Books, 1989).

44. Nick Herbert, *Elemental Mind*, p. 193.

45. Jack Sarfatti, "Evidence for Quantum Brain Fluctuations," http://w2.eff.org/Net_culture/Consciousness/the_quantum_brain.article.

46. Tienko Ting, *Natural Chi Movement: Accessing the World of the Miraculous* (Berkeley, California: North Atlantic Books, 2010), p. 55. Much of the preceding account was taken from my conversations with Tienko Ting.

47. Deepak Chopra, "'Rain Man' and the Connected Life," intentBlog, www.intentblog.com/archives/2005/10/rain_man_and_th.html, October 21, 2005.

48. Jule Eisenbud, "Interview with Jule Eisenbud," January 8, 1972 in Richard Grossinger (editor), *Ecology and Consciousness: Traditional Wisdom on the Environment* (Berkeley, California: North Atlantic Books, 1992), p. 152.

49. ibid.

50. ibid., p. 154.

51. Nick Herbert, *Elemental Mind*, p. 196.

52. Russell Targ quoted in Richard Grossinger, *2013: Raising the Earth to the Next Vibration* (Berkeley, California: North Atlantic Books, 2010), p. 168.

53. Nick Herbert, *Elemental Mind*, p. 210.

54. Jule Eisenbud, "Interview with Jule Eisenbud," p. 154.

55. Jane Roberts, *The Seth Material* (Englewood Cliffs, New Jersey: Prentice-Hall, Inc., 1970), p. 4.

56. Jane Roberts, *Seth Speaks: The Eternal Validity of the Soul* (Englewood Cliffs, New Jersey: Prentice-Hall, Inc., 1972), p. 81.

57. ibid., p. 108.

58. William James, *Human Immortality* (New York: Dover Publications, 1956), pp. 17–18.

59. Jane Roberts, *The Afterdeath Journal of an American Psychologist: The World View of William James* (Englewood Cliffs, New Jersey: Prentice-Hall, Inc., 1972), p. 110.

60. ibid., p. 175.

61. Jane Roberts, *Seth Speaks*, p. 143.

62. Jane Roberts, *The Seth Material*, p. 146.

63. Hermann Poppelbaum, *Man and Animal: Their Essential Difference* (London: Anthroposophical Publishing Company, 1960).

64. ibid., p. 85.

65. ibid., pp. 149–150.

66. Carl Jung, "Seminar Report on *Nietzsche's Zarathustra*," X, p. 51f (privately mimeographed), quoted in "Senex and Puer: An Aspect of the Historical and Psychological Present" by James Hillman, in *Puer Papers* (Irving, Texas: Spring Publications, 1979), p. 44.

67. William Shakespeare, *The Tempest*, 1611, Act Three, Scene Two, Lines 129–130.

Pages 292–293: From "For all their skills scientists cannot..." to "...and get that charming purple glow inside themselves" was grafted from Richard Grossinger, *Embryos, Galaxies, and Sentient Beings: How the Universe Makes Life* (Berkeley, California: North Atlantic Books, 2003), pp. 406–407.

Pages 294–296: From "What does the egg know..." to "happen peremptorily to dwell" was grafted from *Embryos, Galaxies, and Sentient Beings: How the Universe Makes Life*, pp. 407–408.

Pages 297–299: From "Scientists and philosophers make this claim..." to "treat them as genuine beings" is an argument posed in different language in Richard Grossinger, *Embryogenesis: Species, Gender, and Identity* (Berkeley, California: North Atlantic Books, 2000), p. 701.

Page 300: "I can cite many instances of something nonphysical making a difference…" *et seq.* See Richard Grossinger, *The Bardo of Waking Life* (Berkeley, California: North Atlantic Books, 2008), pp. 48–50, for another version of this idea.

Pages 314–321: From "In the view of turn-of-the-century German mystic scientist Rudolf Steiner…" to "biologically and psychologically from the apes *but not spiritually*" is adapted from *Embryogenesis: Species, Gender, and Identity,* pp. 704–723.

Pages 314–315: "The Great Chain of Being" was grafted from "A Primary Reading List," www.richardgrossinger.com/2010/04/a-primary-reading-list/.

CHAPTER TEN. THE SUBTEXTS OF SCIENCE

1. Terence McKenna, *Dreaming Awake at the End of Time,* lecture recorded by Sound Photosynthesis, San Francisco, December 13, 1998, quoted in Richard Grossinger, *2013: Raising the Earth to the Next Vibration* (Berkeley, California: North Atlantic Books, 2010), pp. 33–35.

2. Stephen Hawking quoted in "Stephen Hawking says afterlife is a fairy story" by Liz Goodwin, *Yahoo News,* May 16, 2011, http://news.yahoo.com/s/ yblog_thelookout/20110516/us_yblog_thelookout /stephen-hawking-says-afterlife-is-a-fairy-story.

3. ibid.

4. I have improvised this paragraph with Frederick Ware.

5. PatronofPorn, http://forum.xnxx.com/archive/index.php/t-66891.html.

6. I took this translation of Socrates's words from Plato's *Phaedo* from an online source: http://public.wsu.edu/~wldciv/world_civ_reader/world_civ_reader_1 /phaedo.html. For the full text, see http://oll.libertyfund.org/?option=com _staticxt&staticfile=show.php%3Ftitle=766&chapter=93700&layout=html &Itemid=27.

Page 223: The giant sea turtles were grafted from *2010 Kaua'i Trip,* www .richardgrossinger.com/2010/08/2010-kauai-trip.

Index

Page numbers with "n" indicate a footnote on the page.

self-organization (continued)
  and nonlinear aspects of natural
    selection, 205, 206
self-similarity, and evolution, 235
  See also fractals
semantic transparency, 51, 66
semeiology, psychoanalysis as, 89
sensory cortex, 35–37
sensory modalities
  atrophy of, with disuse, 209
  brain lobes corresponding to, 211
  and ego (Freud), 81
  information transmitted from,
    35–37, 52
  models of reality reconstructed
    from, 47–48
  senior, in primates, 191
  and utilitarian definition of con-
    sciousness, 58
  See also editing of sensory input
    (selection and specification);
    qualia; smell sense; vision
Serios, Ted, 301–302
serotonin, 35
Seth (Jane Roberts), 311–314, 330
sexual abuse, memory and, 107–108
shamanism
  and archetype, 280
  and language, 217–218, 220
  and the *nagual*, 150, 150n, 279
  psychoanalytic transference as,
    86–87, 89
  See also tribal/indigenous society
shame, 82–83
shear force, 189, 241–242
Sherrington, Charles, 293–294
shock treatment, 116
silicon-based lifeforms, 72–73
simplicity, arising from complexity, 233
simultanagnosia, 124

Sixty Minutes, 109
size, influences on, 239
skeleton, evolution of primate,
    191–193
skepticism, 136, 304, 306, 331, 334,
    335
skin, information transmitted by, 36
skull, evolution of primate, 191–193,
    221
sleepwalking (somnambulism), 110
Smart, Elizabeth, 105
smell sense
  information transmitted by, 36
  unpredictability of experience via,
    133
  vision replacing, as senior sense,
    191
"socialism", cosmic, 203
society, and mimicry of identity, 123
Socrates, 331
Solar System, center of, 184–185
somnambulism (sleepwalking), 110
songbirds, 215–216
space-time
  arboreal adaptation of brain and,
    194
  gravity and, 264
  as indices, 272–273
  self-organization of fertilized egg
    and, 311
  See also chronology; time
specification. *See* editing of sensory
    input
spectra of consciousness
  clones, 74–75
  complexity of animal lifeforms and,
    63–70
  cyborgs and, 73–74
  Earth-based vs. alien lifeforms and,
    71–72

and memory, 106, 114
tree shrews, 190–191
"tree that falls in the forest" metaphor,
    49, 147, 337
tribal/indigenous society
    as adoring vs. interrogating nature, 4
    brain power of, 228
    as emergent function of brain/
        mind, 219
    "everything arose from nothing," 4–5
    and horror of functionalism, 172
    incest taboo, 229
    marriage and kinship, 229–230
    and plant consciousness, 70–71
    and quantum theory, 254
    religion and, xix
    telepathic communication and, 306
    See also myth; shamanism; totemism
turbulence, 184–186
    See also nonlinear dynamical systems
Turing, Alan, 60
Turing machines, 60–61, 61n
Turing test, 60, 71–72, 160–161, 163
2001: A Space Odyssey, 166

U

Umezawa, Hiroomi, 259
uncertainty principle. See Heisenber-
    gian uncertainty
uncertainty states
    body plans and, 205
    embryogenesis and, 233–234
    gravity and, 263–264
    non-existence of, 276
    See also Planck's constant
unconscious, Freud and the
    animal mind and, 78
    antithesis of, 95
    consciousness, conception of, 78,
        80, 81–82

dreams and, 8, 84–86
and free vs. bound energy
    (cathexis), 79, 79n, 83
and the id-ego-superego triunity,
    79–84, 85, 89, 90
preconscious and, 81, 82–83, 84, 93
repression and, 82–84
systemic unconscious distinguished
    from, 96–97
"unawareness" redefined, 93
and unknowability of mind, 93
See also Freud, Sigmund;
    psychoanalysis/psychotherapy
unconscious mind
    age of, in evolution, 75, 96, 97
    as antithesis, Derrida and, 94–95
    as developmental phase of con-
        sciousness, 76, 86
    irrational acts, slips of the tongue,
        etc., 76
    neural data processing in, 36
    pre-Freudian conceptions of, 76–77
    rationalist dismissals of, 75–76, 86
    trauma and sublimation and, 77
    See also unconscious, systemic;
        unconsciousness
unconscious, systemic
    age of, in evolution, 96, 97
    autonomic functions and, 96
    behavior arising from, 98–101, 110
    and brain anatomy, 122–123
    and cells, consciousness of, 121
    dreams and fugue states and,
        109–115
    and editing of reality by mind,
        101–104
    efficiency and, 98
    and fissioning of consciousness,
        117–119
    Freudian unconscious distinguished
        from, 96–97

About the Author

PHOTO: MIKE MILLS

A native of New York City (1944), Richard Grossinger attended Amherst College and the University of Michigan, receiving a BA in English (1966) and a PhD in anthropology (1975). He wrote his doctoral thesis on his fieldwork with fishermen in Eastern Maine, after which he taught for two years at the University of Maine at Portland-Gorham and five years at Goddard College in Vermont.

With his wife, Lindy Hough, he is the co-founder and publisher of North Atlantic Books as well as its forerunner, the journal *Io*. His works include early books of experimental prose; a series of titles on holistic medicine, cosmology, and embryology; two memoirs; and recent books reexploring these themes, related topics, and aspects of contemporary politics and pop culture.

He and Hough live in Berkeley, California, and Manset, Maine. Their children are Robin, a historical geographer at San Francisco Estuary Institute, and Miranda July, a writer, film director, and conceptual artist.

For more information, see www.richardgrossinger.com.